Handbook of Experimental Pharmacology

Volume 123

Editorial Board

G.V.R. Born, London
P. Cuatrecasas, Ann Arbor, MI
D. Ganten, Berlin
H. Herken, Berlin
K.L. Melmon, Stanford, CA

Springer
*Berlin
Heidelberg
New York
Barcelona
Budapest
Hong Kong
London
Milan
Paris
Santa Clara
Singapore
Tokyo*

Glucagon III

Contributors

J.M. Amatruda, D. Bataille, D. Bennis, H.-P. Bode
P.E. Cryer, C.F. Deacon, H.-C. Fehmann, N. Geary
B. Göke, R. Göke, J.J. Holst, V. Iwanij, M.M. Johnson
T. Kornfelt, P.J. Lefèbvre, J.N. Livingston, I. Mollerup
M.A. Nauck, P. Oreskov, C. Ørskov, G. Paolisso, C. Pavoine
F. Pecker, J. Philippe, A.J. Scheen, J. Skucas, G. Slama
B. Thorens, H.V. Tøttrup, C. Widmann, G.P. Zaloga
P. Zirinis

Editor
Pierre J. Lefèbvre

 Springer

Professor PIERRE J. LEFÈBVRE, M.D., Ph.D., FRCP
Centre Hospitalier Universitaire de Liège
Department of Medicine
Division of Diabetes, Nutrition, and Metabolic Disorders
C.H.U. Sart Tilman
B-4000 Liège 1
Belgium

With 69 Figures and 7 Tables

ISBN 3-540-60989-X Springer-Verlag Berlin Heidelberg New York

Library of Congress Cataloging in Publication Data. Main entry under title: Glucagon. (Handbook of experimental pharmacology: v. 123 Bibliography: p. Includes index. I. Glucagon–Adresses, essays, lectures. I. Lefèbvre, Pierre J. II. Series. [DNLM: I. Glucagon. W1 HA5IL vol. 66 pt. 1–2/WK 801G5656] QP905:H3 vol. 123 [QP572.G5] 615′1s [612′.34] 83-583
ISBN 0-387-12068-8 (U.S.: v. 1)
ISBN 0-387-12272-9 (U.S.: v. 2)
ISBN 3-540-60989-X (v. 3)

This work is subject to copyright. All rights are reserved, whether the whole or part of the material is concerned, specifically the rights of translation, reprinting, reuse of illustrations, recitation, broadcasting, reproduction on microfilm or in any other ways, and storage in data banks. Duplication of this publication or parts thereof is permitted only under the provisions of the German Copyright Law of September 9, 1965, in its current version, and permission for use must always be obtained from Springer-Verlag. Violations are liable for prosecution under the German Copyright Law.

© Springer-Verlag Berlin Heidelberg 1996
Printed in Germany

The use of general descriptive names, registered names, trademarks, etc. in this publication does not imply, even in the absence of a specific statement, that such names are exempt from the relevant protective laws and regulations and therefore free for general use.

Product liability: The publisher cannot guarantee the accuracy of any information about dosage and application contained in this book. In every individual case the user must check such information by consulting the relevant literature.

Cover design: Design & Production, Heidelberg

Typesetting: Best-set Typesetter Ltd., Hong Kong

SPIN: 10497518   27/3136/SPS–5 4 3 2 1 0–Printed on acid-free paper

Preface

Glucagon III supplements *Glucagon I* and *II*, also published in the *Handbook of Experimental Pharmacology* (vols. 66 I, II), updating the encyclopedic information compiled on glucagon, the hyperglycemic hormone produced by the A-cells in the islets of Langerhans of the pancreas. Since 1983, the year of publication of *Glucagon I* and *II*, the increase in our knowledge on glucagon has been immense: the glucagon gene and the glucagon receptor gene have been identified, the biosynthesis of glucagon has been elucidated, and our understanding of the mode of action of the hormone has been revised. The role of glucagon in diabetes is now recognized, and the search for glucagon antagonists is active. Furthermore, glucagon is used not only to treat hypoglycemia but also as a useful agent in emergency medicine, an interesting drug to improve imaging of the gut, and a useful tool to evaluate residual insulin secretion. Recent data also indicate that glucagon might be involved in the regulation of appetite. Another member of the glucagon family, glucagon-like peptide 1 or GLP-1, has now been clearly identified. It is recognized as a major "incretin," stimulating the release of insulin in response to food ingestion. GLP-1 appears to have a great future in the management of diabetes while another glucagon-related peptide, oxyntomodulin, seems to play a major role in the control of gastric acid secretion.

These various topics are covered in the 19 chapters of the present volume, all of which have been written by recognized world experts.

Individuals working on glucagon will find *Glucagon III* to be an unrivaled source of information, joining volumes I and II as the standard references on this subject. Hospital physicians are likely to be interested in the chapters on the role of glucagon in diabetes, the modes of glucagon administration, the place of glucagon in emergency medicine and in medical imaging, the glucagon test for evaluating insulin secretion, and glucagonoma and its management.

I wish to thank the authors for their comprehensive contributions and – almost all of them – for having sent their chapters on the agreed deadline. This permits us to present a perfectly updated volume. Thanks go also to my secretary, Mrs. E. Vaessen–Petit, who, once again, helped me in the editorial process, to Mrs. D. Walker who has been a most efficient contact with the publisher, and to Mr. W. Shufflebotham, an excellent copy editor.

Glucagon I and *II* received an enthusiastic welcome when they were published in 1983. I wish the same fate to the third volume of the trilogy some 13 years later.

Liège
July 1996   PIERRE J. LEFÈBVRE

List of Contributors

AMATRUDA, J.M., Bayer Corporation, Pharmaceutical Division, Metabolic Disorders Research, Building B-24, 400 Morgan Lane, West Haven, CT 06516-4175, USA

BATAILLE, D., INSERM U376, CHU Arnaud-de-Villeneuve, 34295 Montpellier Cedex 05, France

BENNIS, D., Service de Diabétologie, Hôtel-Dieu, 1, place du Parvis Notre-Dame, 75181 Paris Cedex 04, France

BODE, H.-P., Clinical Research Unit for Gastrointestinal Endocrinology, Department of Internal Medicine, Philipps-University Marburg, Baldingerstrasse, 35033 Marburg, Germany

CRYER, P.E., Division of Endocrinology, Diabetes and Metabolism, Washington University School of Medicine, Campus Box 8127, 660 S. Euclid Avenue, St. Louis, MO 63110, USA

DEACON, C.F., Institute of Medical Physiology C, The Panum Instituttet, University of Copenhagen, Blegdamsvej 3, 2200 Copenhagen N, Denmark

FEHMANN, H.-C., Clinical Research Unit for Gastrointestinal Endocrinology, Department of Internal Medicine, Philipps-University Marburg, Baldingerstrasse, 35033 Marburg, Germany

GEARY, N., E.W. Bourne Behavioral Research Laboratory, New York Hospital, Cornell Medical Center, 21 Bloomingdale Road, White Plains, NY 10605, USA

GÖKE, B., Clinical Research Unit for Gastrointestinal Endocrinology, Department of Internal Medicine, Philipps-University Marburg, Baldingerstrasse, 35033 Marburg, Germany

GÖKE, R., Clinical Research Unit for Gastrointestinal Endocrinology, Department of Internal Medicine, Philipps-University Marburg, Baldingerstrasse, 35033 Marburg, Germany

HOLST, J.J., Department of Medical Physiology, The Panum Instituttet, University of Copenhagen, Blegdamsvej 3, 2200 Copenhagen N, Denmark

IWANIJ, V., Department of Genetics and Cell Biology, University of Minnesota, College of Biological Sciences, 250 Biological Sciences Center, 1445 Gortner Avenue, St. Paul, MN 55108-1095, USA

JOHNSON, M.M., Department of Medicine, Pulmonary Critical Care, Bowman Gray School of Medicine, Wake Forest University, Winston-Salem, NC 27157-1009, USA

KORNFELT, T., Novo Nordisk, Hagedornsvej 1, 2820 Gentofte, Denmark

LEFÈBVRE, P.J., Centre Hospitalier Universitaire de Liège, Department of Medicine, Division of Diabetes, Nutrition and Metabolic Disorders, C.H.U. Sart Tilman (B35), 4000 Liège 1, Belgium

LIVINGSTON, J.N., Bayer Corporation, Pharmaceutical Division, Metabolic Disorders Research, Building B-24, 400 Morgan Lane, West Haven, CT 06516-4175, USA

MOLLERUP, I., Novo Nordisk, Hagedornsvej 1, 2820 Gentofte, Denmark

NAUCK, M.A., Department of Internal Medicine, Knappschaftskrankenhaus, Ruhr-University, In der Schornau, 44892 Bochum-Langendreer, Germany

ORESKOV, P., Novo Nordisk, Novo Alle, 2880 Bagsvaerd, Denmark

ØRSKOV, C., Institute of Medical Anatomy, The Panum Instituttet, University of Copenhagen, Blegdamsvej 3, 2200 Copenhagen N, Denmark

PAOLISSO, G., Department of Geriatric Medicine and Metabolic Diseases, II University of Naples, 80138, Naples, Italy

PAVOINE, C., Unité de Recherches INSERM U99, Hôpital Henri Mondor, 94010 Créteil, France

PECKER, F., Unité de Recherches INSERM U99, Hôpital Henri Mondor, 94010 Créteil, France

PHILIPPE, J., Department de Medecine, Division d'Endrocrinolgie et Diabétologie, Hôpital Cantonal Universitaire, Unité de Diabetologie Clinique, 1211 Geneva 14, Switzerland

List of Contributors

SCHEEN, A.J., Centre Hospitalier Universitaire de Liège, Department of Medicine, Division of Diabetes, Nutrition and Metabolic Disorders, C.H.U. Sart Tilman (B35), 4000 Liège 1, Belgium

SKUCAS, J., Department of Diagnostic Radiology, University of Rochester Medical Center, 601 Elmwood Avenue, Box 648, Rochester, NY 14642-8648, USA

SLAMA, G., Service de Diabétologie, Hôtel-Dieu, 1, place du Parvis Notre-Dame, 75004 Paris, France

THORENS, B., Institute of Pharmacology and Toxicology, University of Lausanne, 27, Rue du Bugnon, 1005 Lausanne, Switzerland

TØTTRUP, H.V., Novo Nordisk, Hagedornsvej 1, 2820 Gentofte, Denmark

WIDMANN, C., Institute of Pharmacology and Toxicology, University of Lausanne, 27, Rue du Bugnon, 1005 Lausanne, Switzerland

ZALOGA, G.P., Section on Critical Care, Bowman Gray School of Medicine, Wake Forest University, Winston-Salem, NC 27157-1009, USA

ZIRINIS, P., Service de Diabétologie, Hôtel-Dieu, 1, place du Parvis Notre-Dame, 75181 Paris Cedex 04, France

Contents

CHAPTER 1

The Industrial Production of Glucagon
I. MOLLERUP, T. KORNFELT, P. ORESKOV, and H.V. TØTTRUP
With 7 Figures 1

A. Introduction 1
B. Production 1
    I. Expression and Fermentation 1
    II. Recovery and Purification 1
        1. Recovery 1
        2. Purification 2
        3. Aggregation 3
    III. Pharmaceutical Preparations 3
C. Analysis of Recombinant Human Glucagon 3
    I. Introduction 3
    II. Identification 3
    III. Purity 4
        1. Microbiological Impurities 4
        2. Water 5
        3. Chemical Impurities 6
    IV. Assay 9
D. Stability 9
References 10

CHAPTER 2

The Glucagon Gene and Its Expression
J. PHILIPPE. With 8 Figures 11

A. Introduction 11
B. Structure 12
    I. The Glucagon Gene 12
    II. Genes Encoding Peptide Hormones of the Glucagon
        Superfamily 14
C. Expression 17

	I. Tissue-Specific Expression	17
	II. A-Cell-Specific Expression	19
	III. Regulation of Glucagon Gene Expression	23
	1. Regulation by Insulin	23
	2. Regulation by the Second Messenger cAMP	24
D.	Conclusions	26
References		27

CHAPTER 3

Preprogluagon and Its Processing
D. BATAILLE. With 8 Figures 31

A.	Introduction	31
B.	Tissue-Specific Post-translational Processing of Proglucagon	32
	I. Glucagon-Containing Peptides	32
	II. Glucagon-Related Peptides	38
C.	Role of Prohormone Convertases in Proglucagon Processing	39
D.	Secondary, Postsecretory Processing of Proglucagon-Derived Peptides	41
	I. Glucagon (19-29) or Miniglucagon	42
	II. Oxyntomodulin (19-37)	44
E.	Conclusions	44
References		46

CHAPTER 4

The Glucagon Receptor Gene: Organization and Tissue Distribution
V. IWANIJ. With 4 Figures 53

A.	Introduction	53
B.	Cloning of the Glucagon Receptor	53
C.	Organization of the Glucagon Receptor Gene	60
D.	Tissue Distribution of the Glucagon Receptors	61
	I. Adipose Tissue Glucagon Receptors	61
	II. Kidney Glucagon Receptors	62
	III. Pancreatic Islets Glucagon Receptors	62
	IV. Brain Glucagon Receptors	63
	V. Heart Glucagon Receptors	63
	VI. Intestinal Tract Glucagon Receptors	63
E.	Tissue Distribution of Glucagon Receptor Transcripts	63
F.	Regulation of Glucagon Receptor Gene Expression	65
G.	Structure/Function Analysis of the Glucagon Receptor	67
H.	Human Glucagon Receptor	69
I.	Conclusions	70
References		70

CHAPTER 5

Mode of Action of Glucagon Revisited
F. PECKER and C. PAVOINE. With 14 Figures 75

A. Introduction 75
B. Glucagon Actions Mediated Through Glucagon Receptors 76
    I. Glucagon Receptor 76
    II. Glucagon Action in Liver 77
        1. Glucagon Mobilizes Ca^{2+} in Hepatocytes 77
        2. Glucagon Potentiates the Effect of Ca^{2+}-Mobilizing
           Agonists 78
        3. Regulation of the Cyclic-GMP-Inhibited
           Phosphodiesterase (CGI-PDE) by cAMP-Dependent
           Phosphorylation in Liver 79
    III. Glucagon Action in Heart 79
        1. Adenylyl Cyclase Activation Versus
           cAMP-Phosphodiesterase Inhibition in Heart Cells 79
C. Glucagon is Processed by Its Target Cells 83
D. Action of Mini-glucagon [Glucagon (19-29)] in Liver 85
    I. Pharmacological Concentrations of Glucagon Inhibit
       the Liver Plasma Membrane Ca^{2+} Pump 85
    II. Mini-glucagon is the True Effector of the Liver Plasma
       Membrane Ca^{2+} Pump 86
    III. αs- and βγ-Subunits of G Protein Mediate Inhibition
       of the Liver Ca^{2+} Pump by Mini-glucagon 87
E. Mini-glucagon Action in Heart 89
    I. Mini-glucagon is a Component of the Positive Inotropic
       Effect of Glucagon 89
    II. The Sarcolemmal Ca^{2+} Pump, in Heart, Is a Target
       for Mini-glucagon Action 91
    III. Mini-glucagon Produces Accumulation of Ca^{2+}
       into the Sarcoplasmic Reticulum Stores 93
F. Glucagon and Mini-glucagon Act in Concert 94
G. Conclusion and Perspectives 96
References 98

CHAPTER 6

Pulsatility of Glucagon
P. LEFÈBVRE, G. PAOLISSO, and A.J. SCHEEN. With 3 Figures 105

A. Introduction 105
B. Oscillations in Glucagon Plasma Levels 105
    I. Animal Studies 105
    II. Human Studies 106
C. Pulsatile Glucagon Secretion In Vitro 106

> D. Pulsatile Glucagon Delivery In Vitro 107
> E. Pulsatile Glucagon Delivery In Vivo 107
>   I. Pulsatile Glucagon Administration in Normal Man 107
>   II. Combined Pulsatile Administration of Glucagon
>   and Insulin in Normal Man 108
>   III. Pulsatile Administration of Glucagon
>   in Diabetic Patients 109
>   IV. Pulsatile Administration of Glucagon in Dogs 109
> F. Conclusions 112
> References 112

CHAPTER 7

Glucagon and Diabetes
P.J. LEFÈBVRE 115

A. Introduction 115
B. Diabetogenic Effects of Glucagon 115
C. The A Cell in Diabetes 117
D. Circulating Glucagon Levels in Diabetes 117
E. Glucagon Dysfunction in Diabetes 118
  I. Hyperglucagonemia of Diabetes 118
  II. Defective Glucose Counterregulation 120
F. Role of Glucagon Excess in the Metabolic Abnormalities
   of Diabetes 121
  I. Postpancreatectomy Diabetes 121
  II. Insulin-Dependent Diabetes 121
  III. Non-Insulin-Dependent Diabetes 122
G. Therapeutic Implications 123
  I. Insulin-Dependent Diabetes 123
  II. Non-Insulin-Dependent Diabetes 123
H. Conclusions 125
References 125

CHAPTER 8

The Search for Glucagon Antagonists
J.M. AMATRUDA and J.N. LIVINGSTON 133

A. Glucagon as a Drug Target 133
B. Search for a Glucagon Antagonist 135
  I. Other Effects of Glucagon 135
  II. Glucagon-Receptor Knockout Mice 136

III. Targets for a Glucagon Antagonist	137
1. Synthesis, Processing and Secretion of Glucagon	137
2. Inhibition of the Actions of the Glucagon Receptor	139
IV. Humanized Mice	143
References	144

CHAPTER 9

Glucagon and Glucose Counterregulation
P.E. CRYER. With 2 Figures 149

A. Introduction	149
B. Glycemic Action of Glucagon	150
C. Glucagon Secretion	151
I. Regulatory Mechanisms	151
II. Glycemic Thresholds	153
D. Role of Glucagon in Glucose Counterregulation	153
I. Physiology	153
1. Insulin	153
2. Glucagon	154
3. Epinephrine	154
4. Other Counterregulatory Factors	154
II. Pathophysiology	155
E. Conclusions	155
References	156

CHAPTER 10

Modes of Glucagon Administration
G. SLAMA, D. BENNIS, and P. ZIRINIS. With 2 Figures 159

A. Introduction	159
B. Classic Routes of Administration	160
I. Intravenous Route	160
II. Intramuscular Route	160
III. Subcutaneous Route	161
C. New Routes of Glucagon Administration	162
I. Intranasal Route	162
II. Eye Drops	166
III. Rectal Route	167
D. Conclusions	168
References	169

CHAPTER 11
The Place of Glucagon in Emergency Medicine
M.M. JOHNSON and G.P. ZALOGA 171

A. Hypoglycemia 171
B. Cardiovascular Insufficiency 174
C. Vascular Effects 178
D. Renal/Urologic Effects 180
E. Shock 180
F. Respiratory Effects 181
G. Gastrointestinal Effects 182
H. Radiographic Studies 184
I. Adverse Effects 184
References 184

CHAPTER 12
The Place of Glucagon in Medical Imaging
J. SKUCAS 195

A. Introduction 195
B. Upper Gastrointestinal Tract 195
    I. Esophagus 195
    II. Stomach and Duodenum 196
C. Small Bowel 198
    I. Enteroclysis 198
    II. Retrograde Ileography 198
    III. Peroral Pneumocolon 199
D. Large Bowel 199
    I. Barium Enema 199
    II. Intussusception Reduction 200
E. Biliary Tract 201
F. Other Applications 201
    I. Computed Tomography 201
    II. Ultrasonography 201
    III. Angiography 202
    IV. Magnetic Resonance Imaging 202
    V. Hysterosalpingography 202
    VI. Scintigraphy 203
    VII. Urography 204
G. Side Effects and Contraindications 204
References 205

Contents XVII

CHAPTER 13
The Glucagon Test for Evaluation of Insulin Secretion
A. J. SCHEEN 211

A. Introduction 211
B. Methodological Aspects 211
    I. Classical Test 211
    II. Dose-Response Curve 212
    III. Reproducibility 212
    IV. Influence of Prevailing Glucose Level 213
    V. Combined Stimulation 214
        1. Glucagon-Glucose Test 214
        2. Glucagon-Meal Test 214
C. Comparison with Other Stimuli 214
    I. Oral Glucose Tolerance Test 214
    II. Intravenous Glucose Tolerance Test 215
    III. Meal 215
    IV. Other Tests 216
D. Clinical Applications 216
    I. Insulin-Dependent Diabetes Mellitus 216
    II. Non-Insulin-Dependent Diabetes Mellitus 217
    III. Hypoglycemia 217
    IV. Other Diseases 218
E. Conclusions 218
References 219

CHAPTER 14
Glucagon and the Control of Appetite
N. GEARY. With 2 Figures 223

A. Introduction 223
B. Prandial Glucagon Secretion 223
C. Glucagon Administration and Food Intake 225
    I. Animal Studies 225
    II. Human Studies 229
D. Glucagon Antagonism and Food Intake 229
E. Mechanism of Glucagon Satiety 231
    I. Site of Action 231
    II. Transduction 231
    III. Hepatic Vagal Afferents 232
F. Clinical Aspects 233
    I. Pathophysiology of Glucagon Satiety 233
    II. Therapeutic Potential 233
G. Conclusions 234
References 234

CHAPTER 15

Glucagonoma and Its Management
A.J. SCHEEN and P.J. LEFÈBVRE 239

A. Introduction 239
B. Diagnosis and Localization of the Tumor 240
C. Management of the Glucagonoma Syndrome 242
    I. Surgical Treatment 242
    II. Vascular Occlusion 243
    III. Radiation Therapy 243
    IV. Chemotherapy 244
        1. Streptozotocin 244
        2. Dacarbazine 245
        3. Others 246
    V. Biological Therapy 246
    VI. Antisecretory Peptide Therapy 247
    VII. Symptomatic Treatment 248
D. Prognosis 248
E. Conclusions 249
References 250

CHAPTER 16

Structure and Function of the Glucagon-Like Peptide-1 Receptor
B. THORENS and C. WIDMANN. With 7 Figures 255

A. Introduction 255
B. GLP-1 Receptor 256
    I. Structure 256
    II. Tissue Distribution 258
    III. Binding Characteristics 261
        1. GLP-1 and Related Peptides 261
        2. Exendins 261
    IV. Coupling to Intracellular Second Messengers 262
    V. Cross-talk Between the GLP-1
       and Glucose-Signaling Pathways 264
    VI. GLP-1 Versus GIP in the Stimulation
       of Insulin Secretion 266
    VII. Regulation of Receptor Function 267
        1. Regulated Expression 267
        2. Desensitization 267
        3. Internalization 267
    VIII. GLP-1 in Non-Insulin-Dependent Diabetes 268
C. Conclusions 268
References 269

Contents XIX

CHAPTER 17

Physiology and Pathophysiology of GLP-1
B. GÖKE, R. GÖKE, H.-C. FEHMANN, and H.-P. BODE
With 5 Figures 275

A. The Incretin Concept 275
B. Origin, Processing, Secretion and Fate of GLP-1 277
    I. GLP-1 as Post-translational Product of Proglucagon
       Processing in Gut, Postsecretory Fate 277
   II. Secretion of GLP-1 280
C. Tissue Distribution of GLP-1 Receptors
   and Biological Actions 282
      I. General 282
     II. Endocrine Pancreas 285
    III. Lung 287
    IV. Stomach 288
     V. Brain 289
    VI. Adipose Tissue 290
   VII. Skeletal Muscle 291
  VIII. Others 291
        1. Exocrine Pancreas 291
        2. Liver 291
D. Signal Transduction of the GLP-1 Receptor 291
    I. cAMP Pathway 291
   II. Calcium 293
E. Pathophysiological Relevance? 297
References 298

CHAPTER 18

Potential of GLP-1 in Diabetes Management
J.J. HOLST, M.A. NAUCK, C.F. DEACON, and C. ØRSKOV
With 2 Figures 311

A. Introduction 311
B. Actions of GLP-1 on Blood Glucose in Humans 314
C. Gastrointestinal Effects of GLP-1 in Humans 315
D. GLP-1 and Diabetes 316
    I. Secretion 316
   II. Receptors 316
  III. Effects 317
E. GLP-1 Metabolism in Normal and Diabetic Subjects 320
F. Conclusion and Outlook 321
References 322

CHAPTER 19

Oxyntomodulin and Its Related Peptides
D. BATAILLE. With 5 Figures 327

A. Introduction 327
B. Biological Characteristics of Oxyntomodulin 327
    I. Receptors 327
    II. Acid Secretion 329
    III. Biologically Active Moiety of Oxyntomodulin 330
    IV. In Vivo Mode of Action: Interactions
        with Other Peptides 331
    V. In Vitro Mode of Action 332
    VI. Pharmacology: Search for a "Minimal Oxyntomodulin" 332
    VII. Human Physiology and Pathophysiology:
        Oxyntomodulin-Like Immunoreactivity 333
    VIII. Recent Developments 335
C. Conclusions 336
References 337

Subject Index 341

CHAPTER 1
Industrial Production of Glucagon

I. Mollerup, T. Kornfelt, P. Oreskov, and H.V. Tøttrup

A. Introduction

Glucagon can be produced in several ways: by chemical synthesis (review in Merrifield et al. 1983), from pancreas as a by-product from the production of insulin (review in Pingel et al. 1983) or by recombinant DNA technology utilizing either *Escherichia coli* (Ishizaki et al. 1992) or *Saccharomyces cerivisiae* (Moody et al. 1987). The present chapter describes the most recent development in the production of glucagon, namely production by recombinant DNA technology. In addition, we will summarize the improvements in analytical methods for assessing the purity and potency of the glucagon so produced.

B. Production

As the amino acid sequences of human, porcine and bovine glucagon are identical (review in Bromer 1983), human glucagon produced by gene technology results in the same product as that obtained from pancreas of these species.

I. Expression and Fermentation

Expression and secretion of glucagon are obtained by transforming *Saccharomyces cerevisiae*. Large-scale fermentations are performed in suitable fermentors. The propagation procedure includes growth on agar followed by a seed tank step. The main fermentor is then inoculated and grown as a glucose fed-batch fermentation with a simple medium containing yeast extract as the main nitrogen source. When the fermentation is finished, the glucagon can be recovered from the supernatant.

II. Recovery and Purification

1. Recovery

The objective of the recovery procedure is to isolate glucagon from the fermentation liquid and obtain a stable intermediate product for further processing. This intermediate product contains several impurities from the fermentation broth (e.g., yeast proteins, DNA). The recovery procedure is outlined in Fig. 1.

To isolate glucagon, the fermentation broth is centrifuged to remove yeast particles that would otherwise clog the column. The supernatant is then passed through a cation exchange column at acid pH. Under these conditions glucagon is retained on the column. After elution of glucagon, filtration takes place to remove the last traces of yeast particles, and glucagon is crystallized at the isoelectric pH. The crystals are recovered by centrifugation and constitute a stable intermediate product.

2. Purification

The purification procedure, consisting of three chromatography steps, is outlined in Fig. 2. The first step is anion exchange chromatography performed at

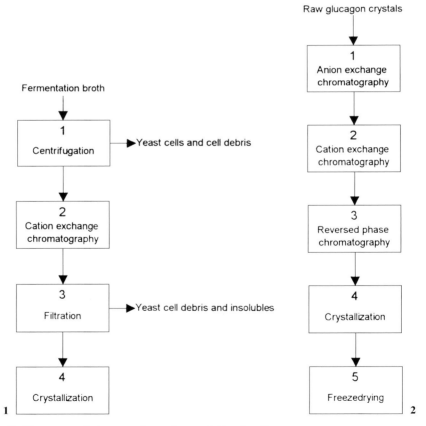

Fig. 1. Recovery of glucagon from fermentation broth

Fig. 2. Purification process of glucagon. Glucagon is isolated by crystallization between each step

a pH above the isoelectric pH; this step removes nucleic acid and yeast proteins. The second step is cation exchange chromatography, removing N-terminally truncated forms of glucagon.

Finally, remaining derivatives of glucagon are removed by reversed-phase chromatography. After each chromatographic step glucagon is recovered by crystallization for intermediate storage. The purification procedure is finalized by isoelectric precipitation before the drug substance is obtained by freeze-drying.

3. Aggregation

Glucagon easily forms noncovalent aggregates, fibrils (review in BLUNDELL 1983). High glucagon concentrations and elevated temperatures are known to induce fibrillation; however, the fibrils may be dissolved and glucagon recovered by alkaline treatment at pH > 10 (BROMER 1983). It has therefore become customary to include defibrillation steps during purification.

III. Pharmaceutical Preparations

Glucagon produced by fermentation is identical to glucagon produced from pancreas (bovine or porcine). Therefore the pharmaceutical preparations are essentially the same regardless of production method. The pharmaceutical preparations were reviewed by PINGEL et al. (1983).

C. Analysis of Recombinant Human Glucagon

I. Introduction

The quality requirements of the health authorities with respect to protein drug substances produced by recombinant DNA technology such as glucagon are constantly increasing. The identity and content of the protein itself as well as of the impurities must be properly documented. Of particular concern is the content of related structures of the native protein/polypeptide-like covalent aggregates, deamidation products and methionine sulfoxidation products. Application of modern analytical methods is necessary for acquisition of information concerning these compounds in the drug substance as well as the pharmaceutical formulations thereof.

II. Identification

To verify the identity of recombinant glucagon and at the same time to document the stability of the plasmid construction used for expression of glucagon, analysis of the so-called primary or covalent structure of the protein from every batch produced is a well-recognized method of controlling this parameter. Amino acid analysis and peptide mapping are two of the methods used most frequently for this purpose. By amino acid analysis it is checked that the glucagon sample contains the correct number of all the amino acids and only these. However, the analysis does not give information concerning the amino acid sequence. This information in provided by peptide mapping of glucagon.

Peptide mapping is an excellent method of checking that the correct amino acids are linked together in the correct sequence. A peptide map of a protein/polypeptide is generated by fragmentation by enzymatic digestion with a suitable endopeptidase followed by reversed-phase high-performance liquid chromatography (RP-HPLC) of the digest.

The resulting chromatogram – called the peptide map – is compared with a reference peptide map. The peak patterns of the two chromatograms should be essentially identical. The individual peaks of the reference map must be fully characterized by amino acid sequence analysis and mass spectrometry. Figure 3 shows a peptide map of a chymotryptic digest of recombinant glucagon. Chymotrypsin is a rather unspecific endoproteinase, which preferentially cleaves at the carboxyl end of aromatic amino acids (phenylalanine, tyrosine, tryptophan), but also after several aliphatic amino acids (e.g., leucine, methionine). The activity of the enzyme is optimal in the pH range 7–9. Due to low solubility of glucagon in this pH interval, the digestion is performed at pH 10.5. In this way both the activity and the specificity of the enzyme are changed so that the digestion results in only five major peptide fragments which together account for the entire amino acid sequence of glucagon (Fig. 3).

The cleavage after Arg-17 is due to the well-known contamination of chymotrypsin with trypsin, which cleaves specifically after the basic amino acids arginine and lysine. Besides the five predominant peaks, which together account for the entire amino acid sequence of glucagon, the peptide map also shows a few additional peaks, which represent secondary chymotryptic cleavage sites.

Coelution by RP-HPLC of the glucagon sample and a glucagon reference substance is also used as supplementary documentation for the identity of glucagon. Finally the bioassay may also serve to document the biological identity of recombinant glucagon.

III. Purity

The impurities present in recombinant glucagon can be divided into chemical and microbiological items.

1. Microbiological Impurities

Recombinant glucagon and the freeze-dried drug product are routinely checked for germ and endotoxin content. Germs are tested according to the standard procedures described in the official pharmacopeia (Pharmacopeia Europea, United States Pharmacopeia). Whereas there is an absolute require-

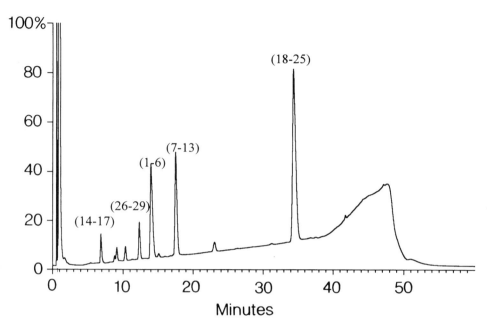

Fig. 3. Peptide mapping of recombinant human glucagon by RP-HPLC. Column, C18, 50 × 4 mm; column temperature, ambient; buffer A, 0.05% v/v trifluoroacetic acid; buffer B, 0.05% v/v trifluoroacetic acid, 60% v/v ethanol; elution, linear gradient from 0% to 47% B-buffer over 35 min, followed by a linear gradient to 100% B-buffer over the following 10 min; flow, 1 ml/min; detection, 215 nm; load, 20 µg chymotryptic digested glucagon. *Numbers* above the individual peaks indicate the corresponding peptide fragment. ↓ indicates the actual chymotryptic cleavage sites; ↕ indicates possible cleavage sites

ment for sterility of the pharmaceutical preparation, a low number of germs in the glucagon drug product is acceptable. Less than 100 germs/g is normally acceptable. Endotoxins are tested for both by measuring the temperature elevation in rabbits which have received an injection of the sample and by the limulus amebocyte lysate (LAL) test. Whereas the biological test in rabbits responds not only to endotoxins but to pyrogens in general, the LAL test is only sensitive to bacterial endotoxins.

2. Water

Water is by far the quantitatively most dominating non-glucagon-related component in recombinant glucagon. Whereas other impurities individually are present in less than 1%, the content of water is usually about 10 times higher, but should not exceed 10%. It has been found that Karl Fischer titration is the most appropriate method for measuring the water content.

3. Chemical Impurities

The chemical impurities of recombinant glucagon consist of the glucagon-related impurities and the foreign impurities.

a) Foreign Impurities

The content of inorganic impurities (salt) in recombinant glucagon is determined as the sulfated ash by gravimetry. The only protein impurity which is not related to glucagon itself is host cell proteins. As these are present only in minute amounts, they are measured by enzyme-linked immunosorbent assay (ELISA). This technique has the sensitivity and specificity required for measuring this kind of impurity. The general requirement is that the content of host cell proteins must be below the detection limit of the ELISA; usually this will be in the low parts per million range.

b) Related Impurities

Glucagon-related impurities are defined as impurities arising from covalent transformations of native glucagon. Although they are induced during all the different production stages, special concern is attached to the transformation products that arise in the glucagon-finished product during storage as these cannot be removed prior to administration to the patient.

To measure the overall purity of glucagon samples, RP-HPLC has become the method of choice. A chromatogram of an RP-HPLC analysis of a *degraded* sample of a finished product containing recombinant glucagon is shown in Fig. 4. It should be noticed that the impurities eluting just after the glucagon represent monodeamidated forms of glucagon. Apart from the methionine sulfoxide derivatives appearing rather early in the chromatogram, little is known about the other impurities.

α) *Deamidation.* Glucagon contains four potential deamidation sites (BROMER et al. 1972): glutamine in positions 3, 20 and 24 and asparagine in position 28. Stability studies have shown that deamidation is a major transformation pathway for glucagon in the finished product. A specific measurement of the deamidated glucagon content is therefore relevant.

Native polyacrylamide gel electrophoresis (PAGE) has been used extensively for the specific estimation of glucagon deamidation. However, an anion exchange HPLC method (IE-HPLC) has been developed, which is superior to gel electrophoresis in several aspects. It is faster, it is quantitative and it is able to separate the different monodeamidated forms of glucagon. Results obtained by the two methods are comparable. Figure 5 shows the chromatogram

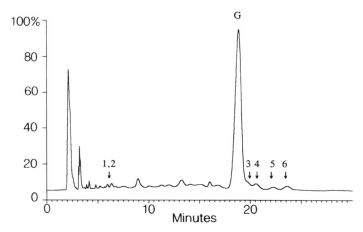

Fig. 4. Reversed-phase high-performance liquid chromatography (RP-HPLC) of a severely degraded sample of a freeze-dried preparation of recombinant glucagon (GlucaGen 1 mg). Column, Merck Lichrosorp C18, 250 × 4 mm; column temperature, 40°C; buffers, 200 mM sodium sulfate, 50 mM phosphoric acid, 10% v/v acetonitrile, pH adjusted to 2.5 with triethylamine (A); 50% v/v acetonitrile (B). Isocratic elution with A/B ~ 52/48. Flow, 1 ml/min; detection, 215 nm; injection load, 30 µg glucagon. *Peaks 1, 2* represent glucagon oxidized in Met-27. *Peaks 3, 4, 5 and 6* are monodeamidated glucagon derivatives. *Peak 3*, Asp-28; *peak 4*, Glu-3; *peak 5*, Glu-20; *peak 6*, Glu-24. *Peak G* is the pure glucagon peptide

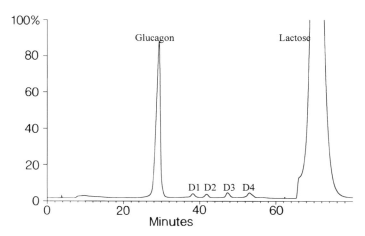

Fig. 5. Anion exchange chromatography of a severely degraded sample of a freeze-dried preparation of recombinant glucagon (GlucaGen 1 mg). Prior to the anion exchange step the sample is automatically desalted by gel filtration on a 10 × 1-cm column of Sephadex G25 superfine (Pharmacia Biosystems). Anion exchange column, Mono Q, HR 5/5 (Pharmacia Biosystems); buffers, 20 mM TRIS acetate, pH 8.3, 50% v/v ethanol (A); 20 mM TRIS, 200 mM sodium acetate, pH 8.3, 50% v/v ethanol (B). Elution, linear gradient from 0 to 54 mM acetate in 30 min, followed by a linear gradient to 160 mM acetate in 15 min. Flow, 1 ml/min; detection, 280 nm. Inj. load, 300 µg glucagon. *D1–D4*, monodeamidated glucagon derivatives; *D1*, Glu-3-glucagon; *D2*, Glu-20-glucagon; *D3*, Glu-24-glucagon; *D4*, Asp-28-glucagon

of a deamidated sample of glucagon-finished product. It is noteworthy that the four monodeamidated glucagon derivatives elute completely separated from each other and from glucagon itself. By PAGE all monodeamidated forms elute as one single band. Small but significant differences in the topological charge distribution are supposed to explain the better separation obtained by IE-HPLC. The monodeamidated glucagon forms are also separated by RP-HPLC (cf. above), but not as well as by IE-HPLC. Furthermore, they elute in a different sequence on RP-HPLC, indicating that hydrophobic interaction takes place. It should also be mentioned that the four monodeamidated glucagon forms are present in comparable amounts. This indicates that the high degree of flexibility of the glucagon peptide makes the four deamidation sites equally accessible to hydrolysis. Remarkably, glucagon derivatives being deamidated in more than one position have not yet been seen.

β) Aggregation. Glucagon aggregates covalently as well as noncovalently (BLUNDELL 1983). Noncovalent aggregation of glucagon, also known as fibrillation, is a phenomenon which only occurs in glucagon solutions, preferably at acidic pH. In solid glucagon including freeze-dried finished preparations small amounts of covalent aggregates of glucagon are formed. These impurities can be measured by RP-HPLC using gradient elution. High-performance size exclusion chromatography (HP-SEC) is, however, a more suitable method. Figure 6 shows an HP-SEC chromatogram of a glucagon sample containing covalent aggregates.

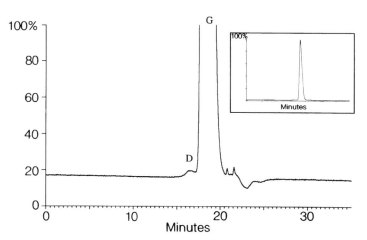

Fig. 6. High-performance size exclusion chromatography of glucagon. Column, Waters Protein Pak 125, 7.8 × 300 mm; mobile phase, 30% v/v acetic acid, 10% v/v 2-propanol; flow, 0.5 ml/min; detection, 280 nm; injection load, 100 μg; *peak G*, glucagon monomer; *peak D*, covalently aggregated glucagon. *Insert* is the same chromatogram scaled to show the entire glucagon monomer peak

IV. Assay

The potency of glucagon and glucagon drug products has so far been determined by biological assays using either rabbits or cats, where the glycogenolytic effect of glucagon on the liver is measured. However, these assays have several drawbacks.

Glucagon exerts several physiological and pharmacological actions. Among these are the glycogenolytic effect on the liver and the spasmolytic action on smooth musculature, which represent the two predominant clinical applications of glucagon.

With respect to the glycogenolytic effect, which is applied for the treatment of hypoglycemia, the above-mentioned official bioassays are most specific. For the spasmolytic effect of glucagon on the smooth musculature of the gastrointestinal tract, which is used extensively for several types of diagnostic examinations, no proper bioassay exists. Whereas the glycogenolytic effect of glucagon is exerted at physiological concentration levels, the spasmolytic effect requires pharmacological doses of glucagon. This indicates the presence of two different receptors. Therefore the glycogenolytic potency measured by the official bioassay does not necessarily reflect the potency with respect to the spasmolytic effect.

Although very specific for measuring the glycogenolytic effect of glucagon, the bioassay is very imprecise. Therefore glucagon may be transformed into less active forms without any significant reflection in the biological potency.

As an alternative to the bioassay, the RP-HPLC method used for purity testing has been introduced for the quantitation of recombinant glucagon. The high resolution obtained by RP-HPLC makes it possible to quantitate the content of pure glucagon in the presence of even substantial amounts of related substances. Furthermore the RP-HPLC assay exhibits a much better precision. The relative standard variation is typically around 1%. This is particularly valuable when performing stability studies on pharmaceutical preparations of glucagon. This is illustrated in Fig. 7, where the stability at 30°C of the glucagon content of a freeze-dried formulation of recombinant glucagon measured by both bioassay and RP-HPLC assay is shown. Although the bioassay data indicate a decrease in the glucagon content, the difference from the initial value is not statistically significant. The RP-HPLC values on the other hand show a constant and significant decrease with time. However, Fig. 7 also indicates that the two methods are in agreement with each other.

D. Stability

Stability of glucagon has been reviewed by PINGEL et al. 1983. As already mentioned in Sect. B, glucagon produced by recombinant DNA technology is identical to glucagon isolated from pancreas, whereby the basic stability issues for recombinant glucagon are covered in the review by PINGEL et al. (1983).

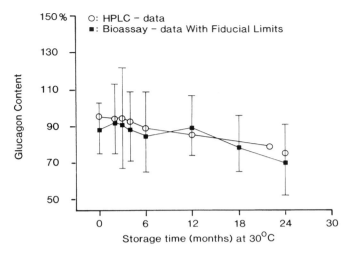

Fig. 7. Quantitation of stability samples of a freeze-dried preparation of recombinant glucagon (GlucaGen 1 mg Novo Nordisk) by bioassay (rabbits) and RP-HPLC. The bioassay is performed according to the Pharmacopeia Europea. RP-HPLC is performed as described in the legend to Fig. 2. The two curves describe the glucagon content relative to the declared content

The advances obtained since this review are mainly improvements of the analytical methods. Therefore, issues pertaining to stability are covered in the present chapter by Sect. C, e.g., C.III.3.b.α and C.IV.

References

Blundell TL (1983) The conformation of glucagon. In: Lefèbvre PJ (ed) Glucagon I. Springer, Berlin Heidelberg New York, pp 37–55
Bromer WW (1983) Chemical characteristics of glucagon. In: Lefèbvre PJ (ed) Glucagon I. Springer, Berlin Heidelberg New York, pp 1–19
Bromer WW, Boucher ME, Patterson JM, Pekar AH, Frank BH (1972) Glucagon structure and function I. Purification and properties of bovine glucagon and monodesamido-glucagon. J Biol Chem 247:2581–2585
Ishizaki J, Tamaki M, Shin M, Tsuzuki H, Yoshigawa K, Teraoka H, Yoshida N (1992) Production of recombinant human glucagon in the form of a fusion protein in Escherichia coli; recovery of glucagon by sequence-specific digestion. Appl Microbiol Biotechnol 36:483–486
Merrifield RB, Mosjov S (1983) The chemical synthesis of glucagon. In: Lefèbvre PJ (ed) Glucagon I. Springer, Berlin Heidelberg New York, pp 23–34
Moody AJ, Norris F, Norris K, Hansen MT, Thim L (1987) The secretion of glucagon by transformed yeast strains. FEBS Lett 212:302–306
Pingel M, Skelbaek-Pedersen B, Brange J (1983) Glucagon preparations. In: Lefèbvre PJ (ed) Glucagon I. Springer, Berlin Heidelberg New York, pp 175–186

CHAPTER 2
The Glucagon Gene and Its Expression

J. PHILIPPE

A. Introduction

A new era in the knowledge of the physiology of the proglucagon-derived peptides and its producing cell has been opened by the cloning of the glucagon gene. This major step has allowed a better understanding of the structure of the glucagon precursor, proglucagon, and the discovery of two unexpected peptides, glucagon-like peptides-1 and -2 (GLP-1 and GLP-2).

Since then, a wealth of information has accumulated on where and how proglucagon is produced and processed, the physiological roles of GLP-1 and the pancreatic A-cell-specific expression and regulation of the glucagon gene. This new information leaves us with many more questions than before but has given us a new perspective and attached new importance to glucagon and its related peptides.

The availability of powerful molecular biology techniques has allowed considerable advances in the knowledge of the structure, cell-specific expression and regulation of the glucagon gene. Data on the expression of the gene have come from experiments using fusion genes consisting of various parts of the rat gene 5′-flanking sequences linked to reporter genes introduced into transgenic mice or glucagon-producing cells. Since gene expression is mostly controlled at the level of transcription initiation, complementary studies have attempted to characterize the *cis*-acting DNA sequences localized within the glucagon gene and particularly its promoter and the transacting factors interacting with them. Characterization of the transacting factors involved in the cell-specific expression of genes has recently taught us that they have a much wider role than previously expected. Studies on growth hormone and insulin gene expression have revealed the importance of these proteins in processes such as cell proliferation, differentiation and maintenance of the differentiated state (Voss and ROSENFELD 1992; OHLSSON et al. 1993; PEERS et al. 1994 and JONSSON et al. 1994).

Identification of the factors responsible for the cell-specific expression of the glucagon gene will thus be a necessary step to understand its transcriptional regulation during development and in response to extracellular stimuli as well as to unravel the basis for islet cell differentiation. These studies should also shed some light on the dysregulation of glucagon biosynthesis in diabetes, as discussed in Chap. 7, this volume.

B. Structure

I. The Glucagon Gene

The human glucagon gene consist of six exons and five introns spanning 10 kb (WHITE and SAUNDERS 1986) and is located on chromosome 2 (TRICOLI et al. 1984). Nucleotide sequence analyses have indicated that glucagon is part of a large biosynthetic precursor, preproglucagon (see Chap. 3, this volume). The structural organization of the exon-intron junctions reveals that each functional domain of preproglucagon is encoded by a separate exon (WHITE and SAUNDERS 1986; BELL et al. 1983a; HEINRICH et al. 1984a). The 5'- and 3'-untranslated regions of glucagon mRNA are encoded by exons 1 and 6, respectively (Fig. 1); exon 2 encodes the signal peptide and part of the amino-terminal end of glucagon-related pancreatic peptide (GRPP), while exons 3, 4 and 5 contain the coding information for glucagon, GLP-1 and GLP-2, respectively. Introns B, C and D are located in connecting peptide sequences. The structural organization of the rat gene is similar to that of the human gene (HEINRICH et al. 1984a,b; BELL et al. 1983a).

The anglerfish, chicken, hamster, rat, guinea pig, degu, bovine and human glucagon cDNA sequences have been determined (WHITE and SAUNDERS 1986; BELL et al. 1983a,b; HEINRICH et al. 1984a; LUND et al. 1982, 1983; LOPEZ et al. 1983; SEINO et al. 1986; NISHI and STEINER 1990; HASEGAWA et al. 1990; BELL 1986). Interestingly, the anglerfish islets synthesize two glucagon mRNAs from two different genes (LUND et al. 1982, 1983); sequence analyses of the respective cloned cDNAs have revealed that glucagon is part of large biosynthetic precursors, preproglucagon I and II. Anglerfish preproglucagon I

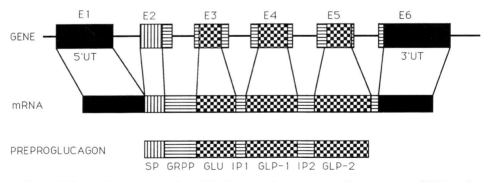

Fig. 1. Schematic representation of the human glucagon gene, the messenger RNA and the encoded precursor protein, preproglucagon. Correspondence between exons and functional domains of preproglucagon is indicated by *lines between gene and mRNA* and by *specific boxes*. Exons are represented by *boxes designated E1–E6* and introns by *solid lines*. *Black boxes* indicate 5' and 3'-untranslated sequences (*UT*); *SP*, signal peptide; *GRPP*, glucagon-related pancreatic peptide; *IP1, IP2*, intervening peptides 1 and 2; *GLP-1, GLP-2*, glucagon-like peptides 1 and 2; *GLU*, glucagon

and II contain 124 and 122 amino acids (aa), respectively, and have 62% identity. The glucagon sequence resides in the middle of each precursor with peptide extensions at both termini (Fig. 2). Glucagon I and II share 79% identity and possess 69% and 76% identity, respectively, with human glucagon. The amino-terminal end of each precursor contains a characteristic hydrophobic signal sequence, followed by a 30-aa peptide, the 29-aa glucagon and a glucagon-like peptide (GlP) at the carboxy terminus. GLP displays marked homology to glucagon and has been isolated and sequenced from the anglerfish and catfish pancreases (ANDREWS et al. 1986; ANDREWS and RONNER 1986). It represents the equivalent of the mammalian GLP-1. Chicken proglucagon, like anglerfish proglucagon, lacks a GLP-2 equivalent but is probably derived from a single gene (HASEGAWA et al. 1990). The cDNA encodes a 151-aa protein which contains a 22-aa signal peptide followed by a region of 30 aa, glucagon, a 24-aa linker region and a 37-aa carboxy-terminal peptide corresponding to GLP-1 (Fig. 2). Chicken glucagon and GLP-1 differ by only 1 and 5 aa, respectively, from their human equivalents. However, domains corresponding to intervening peptide 2 (IP2) and GLP-2 present in mammalian proglucagons are absent from chicken proglucagon.

The mammalian proglucagons are derived from a single gene but strong conservation is observed with anglerfish proglucagon. Rat and human preproglucagons are 180 aa long, have a molecular weight of 18 kDa and are longer than anglerfish preproglucagons I and II by 56 and 58 aa, respectively, and chicken preproglucagon by 29 aa. In contrast to anglerfish and chicken proglucagons, the mammalian proglucagons contain, in addition to glucagon, two glucagon-like peptides, GLP-1 and GLP-2. Human GLP-1 and GlP-2 share 35% identity and 68% homology and display 48% and 38% identity with glucagon, respectively. The close homologies between glucagon and GLPs

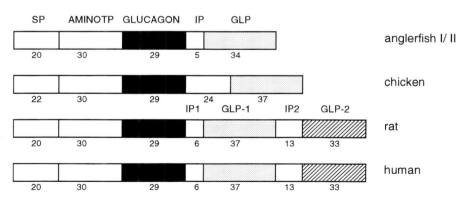

Fig. 2. Schematic representation of the glucagon precursor, preproglucagon, from anglerfish, chicken, rat and human. Anglerfish possess two nonallelic glucagon genes encoding two glucagon precursors. *Numbers* indicate the amino acid length of the respective peptides. *SP*, signal peptide; *AMINOTP* amino-terminal peptide; *IP*, intervening peptide; *GLP*, glucagon-like peptide

suggest that an ancestral glucagon-encoding exon duplicated early in evolution to give rise to GLP-1 in anglerfish and then triplicated (either from the glucagon- or the GLP-1-encoding exon) to result in two different GLPs. It is still unclear at what time during evolution the triplication occurred, since GLP-2 is not only found in mammalian pancreas but also in the amphibians (GAIL POLLOCK et al. 1988).

The aa sequences of human, bovine, hamster and rat preproglucagon are highly conserved. Human and rat glucagon and GLP-1 are 100% identical while GlP-2 displays 94% identity. The most divergent regions are GRPP and the signal peptide, which share only 69% and 85% identity, respectively. The marked degree of conservation of GLP-1 across evolution suggested at the time of its discovery an important biological function.

Despite the high degree of conservation of the glucagon aa sequence throughout evolution, it is interesting to note that within mammals there are exceptions (Fig. 3). Degu and guinea pig glucagons, which belong to the histricomorph rodents, differ from other mammalian glucagons mainly in their C-terminal regions by five amino acids (SEINO et al. 1986; NISHI and STEINER 1990). The C-terminal region of glucagon appears to be important for receptor-binding affinity and biological activity; consequently the changes found in guinea pig glucagon result in decreased biological potency, an observation likely to be valid for degu glucagon. In contrast to glucagon, degu and guinea pig GLP-1 and GLP-2 are highly homologous to other mammalian glucagon-like peptides.

The changes observed in the glucagon sequence of histricomorph rodents may have been to counterbalance the metabolically less active insulins found in these species. The insulin genes from these rodents have undergone multiple mutations, resulting in insulin molecules which have roughly 60% identity with rat and human insulins. They differ especially in residues that are involved in hexamer formation, suggesting that zinc-stabilized insulin hexamers may not form (NISHI and STEINER 1990).

II. Genes Encoding Peptide Hormones of the Glucagon Superfamily

Glucagon belongs to a family of peptides with similar primary structures that includes vasoactive intestinal peptide (VIP), gastric inhibitory polypeptide (GIP), growth hormone releasing factor (GHRF), secretin and pituitary adenylate cyclase activating protein (PACAP). These peptides are found in the gut and pancreas as well as in the central and peripheral nervous system; they may thus function as hormones and/or neurotransmitters.

VIP is a 28-aa peptide, synthesized in the intestine and the central and peripheral nervous system. PreproVIP is 170 aa long as deduced from the cloned rat and human cDNAs (NISHIZAWA et al. 1985; TSUKADA et al. 1985). It contains a signal peptide (21 aa), a 58-aa amino-terminal peptide, PHM-27 (a 27-aa peptide), a 12-residue spacer peptide, VIP itself and a 15-aa carboxy-terminal peptide. The VIP gene contains six exons and five introns spanning

The Glucagon Gene and Its Expression

HisSerGluGlyThrPheSerAsnAspTyrSerLysTyrLeuGluAspArgLysAlaGlnGluPheValArgTrpLeuMetAsnAsn	anglerfish I
HisSerGluGlyThrPheSerAsnAspTyrSerLysTyrLeuGluThrArgArgAlaGlnAspPheValGlnTrpLeuLysAsnSer	anglerfish II
HisSerGlnGlyThrPheThrSerAspTyrSerLysTyrLeuAspSerArgArgAlaGlnAspPheValGlnTrpLeuMetSerThr	chicken
HisSerGlnGlyThrPheThrSerAspTyrSerLysTyrLeuAspSerArgArgAlaGlnGlnPheLeuLysTrpLeuLeuAsnVal	guinea pig
HisSerGluGlyThrPheThrSerAspTyrSerLysTyrSerLysPheLeuAspThrArgArgAlaGlnAspPheLeuLysAsnThr	degu
HisSerGluGlyThrPheThrSerAspTyrSerLysTyrSerLysPheLeuAspThrArgArgAlaGlnAspPheValGlnTrpLeuMetAsnThr	hamster
HisSerGluGlyThrPheThrSerAspTyrSerLysTyrSerLysPheLeuAspThrArgArgAlaGlnAspPheValGlnTrpLeuMetAsnThr	rat
HisSerGluGlyThrPheThrSerAspTyrSerLysTyrSerLysPheLeuAspThrArgArgAlaGlnAspPheValGlnTrpLeuMetAsnThr	bovine
HisSerGluGlyThrPheThrSerAspTyrSerLysTyrSerLysPheLeuAspThrArgArgAlaGlnAspPheValGlnTrpLeuMetAsnThr	human

Fig. 3. Comparison of the amino acid sequence of anglerfish I, anglerfish II, chicken, guinea pig, degu, hamster, rat, bovine and human glucagons

about 9 kb and has a similar organization to the human glucagon gene (TSUKADA et al. 1985) (Fig. 4): exons 1 and 6 encode the 5'- and 3'-untranslated sequences, exon 2 the signal peptide and exons 3, 4 and 5 the spacer peptide, PHM-27 and VIP, respectively. VIP and PHM-27 share an N-terminal histidine, a phenylalanine-6 and a threonine-7 with glucagon.

Human GHRF is a 44-aa protein (43 aa in the rat) made in the hypothalamus. Human preproGHRF is 108 aa long (104 aa for the rat preprotein) and contains a signal peptide, an amino-terminal peptide, GHRF and a carboxy-terminal peptide. The human and rat genes are about 10 kb long and contain five exons which encode functionally distinct domains of preproGHRF (MAYO et al. 1983, 1985) (Fig. 4). In contrast to glucagon and VIP, GHRH is encoded by two exons.

GIP is a 42-aa hormone produced in the duodenum. PreproGIP is 153 aa long; its gene spans approximately 10 kb and contains six exons (INAGAKI et al. 1989). The structural organization of the human gene is similar to those of other members of the glucagon superfamily. Like GHRF, GIP is encoded by two exons.

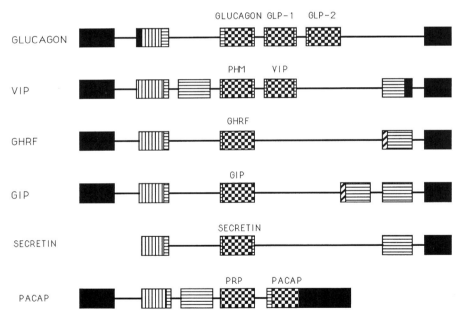

Fig. 4. Organization of the genes encoding the glucagon superfamily of peptides: glucagon, VIP, GHRF, GIP, secretin and PACAP. Exons are represented by *boxes* and introns by *solid lines*. Functional domains of the preprohormones are indicated by *specific box filling*. *Black boxes* represent untranslated sequences and *checked boxes* processed and bioactive peptides; *GLP*, glucagon-like peptides; *PHM*, peptide histidine-methionine; *VIP*, vasoactive intestinal peptide; *GHRF*, growth hormone releasing factor; *GIP*, gastric inhibitory polypeptide; *PRP*, PACAP-related peptide; *PACAP*, pituitary adenylate cyclase activating protein

Secretin is a 27-aa hormone produced by the S cells of the intestine and the brain. Preprosecretin consists of 134 aa which contain a signal peptide, a short N-terminal peptide, secretin and a long 72-aa carboxy-terminal extension (KOPIN et al. 1991). The gene includes four exons; the first exon encodes the 5'-untranslated region, the signal peptide and 8 aa of the N-terminal peptide; exon 2 codes for secretin and part of the amino-terminal peptide, and exon 3 and 4 code for the carboxy-terminal peptide. The secretin gene is the only member of the glucagon gene superfamily that does not have an intron separating the transcriptional and translational start site.

PACAPs which exist in two forms (PACAP38 and PACA27, the latter corresponding to the 27 amino-terminal residues of PACAP 38) were originally isolated from the hypothalamus and shown to stimulate adenylate cyclase in rat anterior pituitary cell cultures. Human preproPACAP is 172 aa long and contains a signal peptide, a long 55-aa amino-terminal peptide, a 29-aa PACAP-related peptide (PRP), a linker peptide and PACAP. The gene is composed of five exons and four introns (HOSOYA et al. 1992). PRP and PACAP are each encoded by a single exon arranged in tandem in the PACAP gene, suggesting exon duplication. The structural features of the PACAP gene are similar to those of the VIP gene, which is also characterized by the tandem arrangment of the exons encoding VIP and the VIP-related peptide, PMH-27.

Comparison of the genes encoding glucagon, VIP, GIP, GHRF, PACAP and secretin reveals organizational similarities but also significant differences. The exon boundaries relative to the biologically active peptides have not been highly conserved. In addition, comparison of the translated nucleotide seuqence of the different genes reveals similarities only at the N terminus of each hormone. If the genes encoding the glucagon superfamily of peptides are derived from a common ancestral gene, it is clear that they have diverged extensively to the point of displaying very few common features between, for instance, the secretin and the VIP or PACAP genes.

C. Expression

I. Tissue-Specific Expression

The glucagon gene is expressed in at least three different tissues, the endocrine pancreas, the intestine and the brain. In the brain, the glucagon gene is predominantly expressed in the brain stem and hypothalamus (DRUCKER and ASA 1988). However, important developmental changes occur within different regions of the mouse brain (LEE et al. 1993). During embryogenesis, at day 19 (E19), glucagon mRNA is found mostly in the cortex but also in the cerebellum and brain stem; glucagon mRNA is still detectable postnatally at 2 weeks of age in the cortex but is absent at 5 or 12 weeks. By contrast, mRNA levels increase in the cerebellum, hypothalamus and brain stem to become maximal at 12 weeks. The functional relevance of glucagon gene expression in the brain and of the region-specific developmental changes remains to be elucidated.

The glucagon gene is also expressed in the endocrine L cells of the stomach, and the small and large intestine (MOJSOV et al. 1986; NOVAK et al. 1987). Information related to the developmental and region-specific differences of glucagon gene expression in the intestine is lacking. Studies in transgenic mice expressing the simian virus (SV) 40 large T antigen under the control of 2000 bp of the glucagon gene 5'-flanking sequences suggest that glucagon gene expression may start first in the intestine and later, during the first 10 days post-natally, in the stomach (LEE et al. 1992).

Developmental expression of the glucagon gene in the pancreas has been studied in greater detail (ALPERT et al. 1988; GITTES and RUTTER 1992; TEITELMAN et al. 1993 and HERRERA et al. 1991, 1994). Glucagon mRNA is first detected in the mouse embryo at E8.5–9 at the same time as insulin mRNA and about 12 h before the condensation of the mesenchyme near the gastrointestinal junction, an event that announces the evagination of the pancreatic diverticulum. Glucagon mRNA is first localized to the dorsal wall of the duodenal anlage, in the precise area that gives rise to the pancreas. At that time, immediately adjacent tissues such as the dorsal and ventral duodenum, the stomach and the distal intestine do not contain glucagon mRNA transcripts. Glucagon immunoreactivity appears later at E10.5 in the dorsal buds; at that stage, glucagon-positive cells are five to ten times more abundant than insulin-containing cells. The relative rates of accumulation of glucagon- to PP- and insulin-containing cells indicate that the number of cells increases two to three times faster in dorsal buds and two to three times slower in ventral buds relative to PP cells and four and ten times slower in dorsal and ventral buds, respectively, relative to insulin cells.

Developmental studies indicate that the four islet hormone genes (encoding glucagon, insulin, somatostatin and pancreatic polypeptide) are expressed roughly at the same time rather than sequentially and importantly that the primitive islet cells appear in the epithelium of the foregut prior to morphogenesis of the pancreas. These observations suggest that the appearance of islet cells and thus of hormone gene expression initiates pancreatic development.

Age-related changes in pancreatic glucagon gene expression may also occur during adult life (GIDDINGS et al. 1995). Glucagon mRNA levels in the pancreas of Fisher rats between 2 and 12 months of age double while insulin mRNA remains stable. The relevance of this observation to the impaired glucose tolerance characteristic of ageing is unknown at present.

The proglucagon cDNAs isolated from pancreas, intestine and brain are identical in sequence, indicating that transcription of the glucagon gene in these three tissues generates a single and unique mRNA transcript (DRUCKER and ASA 1988; NOVAK et al. 1987). In addition, the transcriptional start site of the glucagon gene in the pancreas, intestine and brain is identical, implying that a common promoter region mediates transcription initiation. Major differences in the determinants that control glucagon gene expression in the pancreas and intestine have been revealed, however, by transgenic mouse studies (LEE et al. 1992; EFRAT et al. 1988). In vivo analyses of a glucagon-SV40 large T antigen transgene containing approximately 1300 bp of the rat glucagon 5′-flanking sequences demonstrated expression in the brain and pancreas but not in the gastrointestinal tract of transgenic mice. In contrast, transgenic mice containing additional glucagon gene 5′-flanking sequences (approx. 2.0 kb) express the transgene not only in the brain and pancreas but also in the stomach and the small and large intestines. These observations indicate that the DNA determinants for glucagon gene expression in the brain and pancreas lie within 1300 bp of the 5′-flanking sequence of the glucagon gene. These determinants are different or not sufficient for glucagon gene expression in the gastrointestinal tract which requires additional DNA regulatory sequences between 1300 and 2000 bp.

The precise determinants responsible for brain and gastro-intestinal glucagon gene expression are completely unknown. By contrast, gene transfer studies have identified several specific *cis*-acting enhancer and promoter sequences that control glucagon gene expression in pancreatic A cells.

II. A-Cell-Specific Expression

Information on the factors that promote cell-restricted expression of the glucagon gene has come from the technology of gene transfer into glucagon-producing cells in culture and mouse embryos to generate transgenic mice. To characterize the factors that mediate transcription of the glucagon gene, the 5′-flanking region of the rat gene has been analyzed by studying the expression of the rat gene has been analyzed by studying the expression of progressively shortened 5′ and 3′, and internally deleted sequences fused to the coding region of the bacterial enzyme, chloramphenocol acetyl transferase (CAT), transfected into glucagon- and non-glucagon-producing cells (DRUCKER et al. 1987a,b; PHILIPPE et al. 1988; KNEPEL et al. 1990a; PHILIPPE and ROCHAT 1991; MOREL et al. 1995).

Only limited information has been obtained on the vivo cell-specific expression of the glucagon gene. In transgenic mice carrying a fusion gene containing 1300 bp of of the rat glucagon gene promoter linked to the SV40 large T antigen-coding sequences, expression of the T antigen was limited to the A cells of the pancreas and the brain, indicating that 1300 bp is sufficient for pancreatic expression (EFRAT et al. 1988). More detailed analyses on the determinants that control islet and A-cell-specific expression have been performed in glucagon-producing cell lines. These determinants are contained within 300 bp upstream of the transcriptional start site; no difference in transcriptional activity is found between DNA constructs containing 1300 and 300 bp (PHILIPPE et al. 1988). The 300 bp of the glucagon promoter can be divided into two functionally different regions: a proximal promoter element from −1 to −150 bp containing the TATA box, G1 and G4 and a distal enhancer, localized between −150 and −300 bp, which contains two enhancer-like elements and a cAMP response element CRE (Fig. 5). These elements have been defined by deletional and mutational analyses as well as by binding assays.

Interestingly, 5-deletions of the two most distal elements, the CRE and G3, do not result in any significant loss of basal transcription, whereas deletion of G2 leads to an 80% decrease in activity (PHILIPPE et al. 1988). G2 and G3, however, are both capable of independently increasing transcription when directly upstream of the proximal glucagon promoter or a heterologous promoter. Both control elements are strongly distance dependent in their ability to activate transcription (PHILIPPE and ROCHAT 1991a). In the absence of G1 they need to be close to the TATA box to be active; when the distance that separates them from the TATA box exceeds 100 bp, they lose most of their activating properties.

G2, but not G3, functions in both orientations. This difference may be secondary to the architecture of their respective binding sites. G2 indeed interacts with two complexes, as analyzed by gel retardation assays, which bind to overlapping and probably mutually exclusive sites; one of these complexes (A2) is detected only in islet cells whereas the second complex represents the liver-enriched factor, hepatocyte nuclear factor $3\beta1$ (HNF-$3\beta1$) (PHILIPPE et al. 1994) (Fig. 6). Mutations that affect HNF-$3\beta1$ and not A2 binding result in increased transcriptional activity. In addition, overexpression of HNF-$3\beta1$ leads to a decrease in G2-mediated transcriptional activity. These results indicate that G2 enhancer activity is controlled both positively and negatively; HNF-$3\beta1$ acts as a repressor of glucagon gene expression whereas A2, an islet-specific complex, functions as an activator. The complexity of glucagon gene regulation at G2 is further enhanced by the presence in A cells of at least three different HNF-3β isoform proteins which result from alternative splicing events from the HNF-3β primary transcript. Although all three HNF-3β isoforms bind G2, two of them, HNF-$3\beta2$ and $\beta3$, do not affect glucagon gene transcription; rather they compete with the negatively acting HNF-$3\beta1$ isoform to decrease its repressing effects (PHILIPPE 1995).

The difference in the transcriptional properties between HNF-$3\beta1$ and the $\beta2$ and $\beta3$ isoforms may be explained by their respective amino-terminal ends. The amino-terminal end of HNF-$3\beta1$ indeed contains a transcriptional domain (PANI et al. 1992; QIAN and COSTA 1995). The properties of this domain may be modified by the presence of five additional aa at the amino-terminal end

Fig. 5. Schematic representation of the DNA control elements present in the first 300 bp of the 5′-flanking sequence of the glucagon gene. The 5′-flanking sequence can be divided into an A-cell-specific promoter and an islet-cell-specific enhancer. Arrow indicates the transcriptional start site. *CRE*, cyclic AMP-response element

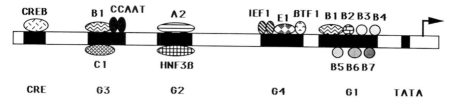

Fig. 6. Schematic representation of the *cis*-acting DNA elements and *trans*-acting factors believed to be involved in the expression and regulation of the glucagon gene. *Black boxes* indicate DNA control elements and *circles or ovals*, protein complexes binding to them. *Arrow* indicates the transcriptional start site CCAAT, CCAAT-binding protein; *HNF3β*, hepatocyte nuclear factor-3β; *IEF*, insulin enhancer factor; *βTF1*, B-cell transcription factor 1; *CRE, CREB*, cyclic AMP responsive element and CRE-binding protein, respectively. Information for this figure for the CRE is from KNEPEL et al. (1990a) and DRUCKER et al. (1991), for G3 from KNEPEL et al. (1990b), PHILIPPE (1991b) and PHILIPPE et al. (1995, for G2 from PHILIPPE et al. (1994) and PHILIPPE (1995), for G4 from CORDIER-BUSSAT et al. (1995) and for G1 from MOREL et al. (1995)

(HNE-3β2) or by the absence of the first 30 aa (HNF-3β3). Alternatively, the differences may be explained by the ability of the isoforms to generate protein-protein interactions that result in different activating potential. The three HNF-3β isoforms are present in phenotypically different islet cell lines and are also detected in normal islets. The functional significance of the presence of the isoforms in insulin- and somatostatin-producing cells as well as the dual regulation at the G2-binding site remains to be elucidated.

The structural organization of the G3 control element differs from that found for G2. G3 is divided into two separate domains, A and B (KNEPEL et al. 1990a; PHILIPPE 1991b; PHILIPPE et al. 1995). The B domain binds a ubiquitous DNA-binding protein that probably belongs to the family of CCAAT-box proteins (Fig. 7). Mutations of the B domain do not affect basal transcription significantly; the functional relevance of the B domain is thus unclear. The A domain is responsible for both enhancer activity and the mediation of the insulin effects on glucagon gene expression (PHILIPPE et al. 1995; PHILIPPE 1989, 1991a; CHEN et al. 1989a). The A domain binds two complexes that are present in all islet cell phenotypes but not in other cell types. One of the two complexes also interacts with the proximal upstream promoter element G1.

Both G2 and G3 function as islet cell-specific enhancers and, at least as assayed in transformed islet cell lines, are unable to restrict by themselves glucagon gene expression to the A cells relative to the other islet cell phenotypes (PHILIPPE et al. 1988; MOREL et al. 1995).

The proximal promoter of the rat glucagon gene contains in addition to the TATA box at least two control elements, the upstream promoter sequences (UPS) G1 and G4. Their functional relevance for glucagon gene expression is very different.

G4 is a composite DNA element localized upstream of G1 between nucleotides –100 and –140 (CORDIER-BUSSAT et al. 1995). It contains at least three

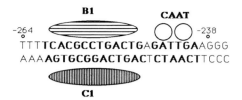

Fig. 7. Schematic representation of the G3 control element. *A, B* indicate the A and B domains, respectively; the DNA sequences of the respective binding sites are *in bold*. *IRE*, insulin-response element; *CCAAT*, CCAAT-binding protein; *HNF-3β1, HNF-3β2, HNF-3β3*, hepatocyte nuclear factor β1, β2 and β3, respectively

binding sites, two E box motifs and an intervening sequence (Fig. 6). E box motifs are characterized by the sequence CANNTG and interact with transcription factors of the helix-loop-helix protein family. Members of this family have been shown to be involved in cell-type specification through their activation of specific genes in muscle or neuronal cells (WRIGHT 1992; CAUDY et al. 1988). E boxes are present in the promoters of several genes transiently or permanently expressed in the B cells of the pancreatic islets and most notably the insulin gene, the expression of which appears critically dependent on the integrity of its E boxes (for review, PHILIPPE 1991b). The most distal binding site of G4 (E3) represents an E box motif and functions as such. E3 is identical to the FAR and NIR boxes of the rat insulin I gene and to the ICE/RIPE 3a sequence of the insulin II gene (for review, PHILIPPE 1991b); it also interacts with the same complex, IEF1, which is an islet-cell-specific activator involved in B-cell-specific expression of the rat insulin, gastrin and secretin genes (WANG and BRAND 1990; WHEELER et al. 1992). IEF1 is a heterodimeric complex composed of the ubiquitous HLH protein E12/E47 and an unidentified protein present in A and B cells. The functional relevance of IEF1 in glucagon gene expression is further enhanced by the observation that overexpression of the negative HLH regulator Id-1 inhibits transcriptional activity of DNA constructus containing G4 (CORDIER-BUSSAT et al. 1995).

The second E box motif, E2, interacts with two complexes, one being islet specific, which show no specificity for the E box motif itself. E2 is homologous to other DNA elements (called B elements) found in the rat insulin I, somatostatin and elastase I genes, respectively; complexes which bind E2 also bind these elements (KRUSE et al. 1993). The presence of potential B element equivalents in other pancreatic genes, both endocrine and exocrine specific, suggests that these elements and their transacting factors may play a role in pancreatic gene expression particularly in the early stages of development when the endocrine and exocrine compartments share a common precursor lineager.

E2 and E3 have no activating potential by themselves; they necessitate their intervening sequence to activate transcription. E2, E3 and the intervening sequence thus form a functional unit which can activate transcription in A, and also in B cells, thus conferring islet-specific activation to the glucagon gene.

G1, the most proximal element, is a large 48-bp protein-binding site localized between −52 and −100 bp (PHILIPPE et al. 1988; MOREL et al. 1995). It probably represents the major determinant that restricts glucagon gene expression to A cells as opposed to other islet cells. It is characterized by low intrinsic transcriptional activity.

At least seven protein complexes bind to G1. Four are detected by gel retardation assays using the whole G1 element (MOREL et al. 1995). Three of them appear to be related and may contain common proteins; their respective binding sites are overlapping and similar although not identical. Mutational analyses within G1 indicate that the complex which interacts with both G1 and G3 is critical for transcriptional activity. Three additional complexes are detected when using fractions of the G1 element (author's unpublished data). Most G1-binding complexes are islet specific, but some of them are ubiquitously distributed. None of them, however, has been found to be A cell specific. We are thus presently left with the paradoxical observation that G1 is a critical element which restricts glucagon gene expression to the A cells and that protein complexes which interact with G1 are found in all islet cells. It is still unclear, however, whether complexes detected in glucagon- and insulin-producing cells represent the same proteins. More detailed analyses of these complexes should help to resolve these discrepancies and unravel the molecular mechanisms of A-cell-specific expression.

III. Regulation of Glucagon Gene Expression

Many physiological stimuli, hormones and nutrients have been shown to regulate glucagon secretion (LEFÈBVRE 1995). Much less is known about the factors that control glucagon biosynthesis. Although the processes of hormone secretion and biosynthesis are often closely interrelated, specific studies on glucagon biosynthesis are needed to better understand the pathophysiology of diabetes.

Among the stimuli that physiologically regulate glucagon secretion, glucose, amino acids, insulin and catecholamines are the most relevant. There is no information on the effects of glucose and amino acids on glucagon biosynthesis or glucagon gene expression, whereas insulin and cAMP analogues (catecholamines act through the cAMP second messenger pathway) have been shown to modulate glucagon gene expression by changes in transcription rates. The molecular mechanisms of these changes have been the subject of recent studies.

1. Regulation by Insulin

Insulin inhibits glucagon gene expression both in vivo and in vitro (PHILIPPE 1989, 1991a; PHILIPPE et al. 1995; CHEN et al. 1989a,b). Chronic hyperglycemic clamping for 5 days in normal rats results in a 50% increase in insulin mRNA and an 81% decrease in glucagon mRNA levels. In insulin-dependent streptozotocin-treated diabetic rats, glucagon mRNA levels increase despite hyperglycemia and are rapidly reduced by insulin treatment. These results suggest that insulin is a major factor in the control of glucagon secretion and biosynthesis; the role of glucose in modulating insulin effects remains to be determined. Using a glucagon-producing cell line, we have shown that insulin regulates glucagon gene expression at the transcriptional level. The effects of insulin on transcription are rapid, dose and time dependent and do not require new protein synthesis (PHILIPPE 1989).

Regulation of glucagon gene expression by insulin is mediated by an insulin-response element (IRE) (PHILIPPE 1991a). The IRE can confer insulin responsiveness to a heterologous promoter and is orientation dependent; it corresponds to the previously identified enhancer-like element G3. G3 is a bipartite element which contains two domains, A and B. The A domain is transcriptionally active and mediates insulin effects while the B domain appears transcriptionally inactive (PHILIPPE 1991a; PHILIPPE et al. 1995) (Fig. 7).

The A domain interacts with two islet-specific complexes; these complexes bind to overlapping sequences within the A domain and display very similar binding specificities. Mutations that affect binding of both complexes result in decreased G3 enhancer activity and insulin-mediated inhibitory effects (PHILIPPE et al. 1995). The B domain interacts with a protein of the CCAAT box family of DNA-binding proteins. Experimental data suggest that protein binding to domains A and B may occur independently and may be mutually exclusive; competition between transcription factors for G3 binding may thus be a potential mechanism for the modulation of glucagon gene expression by insulin. The same proteins that bind the A domain of G3 have been proposed also to interact with the distal enhancer E1 of the rat insulin I gene and with the upstream element of the somatostatin gene, SMS-UE (KNEPEL et al. 1991). Interestingly, both genes have been shown to be regulated negatively by insulin (PAPACHRISTOU et al. 1989; KORANYI et al. 1992). Characterization of the transcription factors which bind G3 will allow a better understanding of the physiological regulation between insulin and glucagon.

2. Regulation by the Second Messenger cAMP

Analogues of the second messenger cAMP have been shown to regulate glucagon gene expression positively in pancreatic A cells, fetal intestinal L cells and immortalized cell lines from pancreatic and intestinal origins (KNEPEL et al. 1990b; DRUCKER and BRUBACKER 1989; DRUCKER et al. 1991, 1994; GAJIC and DRUCKER 1993). Cyclic AMP-dependent regulation is mediated at the transcriptional level through cyclic AMP-response element (CRE) located just upstream of G3. The rat glucagon CRE consists of a perfect palindromic consensus site TGACGTCA; it is not, however, as active, in response to CAMP analogues, as other similar sites, such as those found in the rat somatostatin and human gonadotropin α-subunit genes (MILLER et al. 1993). These observations may be explained by the nature of the nucleotides surrounding the CRE. Several uncharacterized proteins interact with the flanking motif (TCATT) of the glucagon CRE and mutations of this motif enhance the response of the cAMP analogs. These results suggest that the flanking motif-binding proteins negatively modulate the ability of the CRE-binding protein (CREB) to mediate protein kinase A-stimulated transcription.

The CRE may be the target for additional sgnals which regulate glucagon gene transcriptional (SCHWANINGER et al. 1993). The A cells exhibit spontaneous action potentials, and electrical activity is increased by glucagon secretagogues. Arginine increases spike frequency through the stimulation of calcium influx into the cells. Changes in cytosolic calcium concentrations are a major control point that regulates hormone secretion. Recent data suggest that increases in calcium levels induced by membrane depolarization can activate glucagon gene transcription through the CRE. Pharmacological experiments indicate that calcium influx and calcium/calmodulin-dependent protein kinase II (CaM kinase II) are critical for the membrane depolarization-induced effects on transcription. CREB may function as a depolarization factor after phosphorylation by CaM kinase II.

Increases in intracellular calcium levels and activation of the cAMP pathway result in synergistic effects on transcription. Both of these second messenger pathways appear to converge on CREB and the CRE; they may thus act by different mechanisms to activate CREB. Experimental data suggest that this is indeed the case. Serine-119 of CREB is required for intact depolarization-induced activity and is phosphorylated by CaM kinase II in vitro (SCHWANINGER et al. 1993).

To activate transcription, calcium may thus activate CaM kinase II, which in turn phosphorylates the CRE-bound CREB at serine-119. By contrast, protein kinase A will preferentially phosphorylate serine-133 of CREB. Phosphorylation of both serines could then activate transcription synergistically (Fig. 8).

Glucagon and various proglucagon-derived peptides have been shown to acutely suppress glucagon secretion from the pancreatic A-cell (KAWAI and UNGER 1982). An excess of these peptides may result in a marked decrease in their own biosynthesis and in glucagon mRNA levels in the pancreas and intestine (BRUBACKER et al. 1992; BANI et al. 1991; LOGOTHETOPOULOS et al. 1960). Transgenic mice containing a glucagon gene promoter linked to the SV40 large T antigen develop invasive neuroendocrine carcinoma of the large intestine; these tumors produce large amounts of glucagon and other peptides derived from proglucagon and are accompanied by decreases in glucagon mRNA and proglucagon-derived peptides in both pancreas and small intestine (BRUBACKER et al. 1992).

The precise mechanisms of these changes are not completely understood; one of the mechanisms involved may be a decrease in glucagon-producing cell

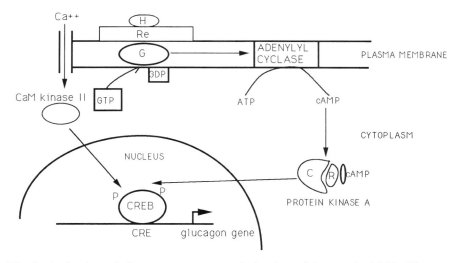

Fig. 8. Activation of glucagon gene transcription by calcium and cAMP. Glucagon secretagogues increase electrical activity and induce membrane depolarization. In this model, membrane depolarization leads to an increase in intracellular calcium concentration which activates calcium/calmodulin-dependent protein kinase II (*Ca M kinase II*). This enzyme will in turn phosphorylate the CRE-binding protein (*CREB*) at serine-119, which activates transcription catecholamines and other hormones (*H*), after binding to their receptors (*Re*) may activate the cAMP second messenger pathway through G proteins (*G*). Dissociation of the α- from the β- and γ-subunits results in the formation of cAMP from ATP after activation of the enzyme adenyl cyclase. cAMP in turn activates protein kinase A, leading to the dissociation of the catalytic from the regulatory subunit and the translocation of the catalytic subunit to the nucleus. The catalytic subunit phosphorylates the CRE-binding protein (*CREB*) at serine-133, an event activates transcription

number. In a patient with a glucagonoma, immunocytochemical and ultrastructural analyses have revealed virtual disappearance of islet A cells (Bani et al. 1991). Effects of exogenous glucagon on the rabbit endocrine pancreas in vivo were similar to the findings observed in this patient, suggesting that glucagon itself may be responsible, at least in part, for this effect (Logothetopoulos et al. 1960). Whether glucagon acts directly on the A and L cells is unknown. Its effects may be also mediated by an increase in intraislet insulin concentrations. More data are needed to elucidate the mechanisms by which glucagon inhibits its own biosynthesis.

D. Conclusions

The cloning of the glucagon gene has allowed a second explosion of knowledge, not particularly about glucagon itself but about the structure and physiology of the glucagon-related peptides and the expression and regulation of the glucagon gene. Much more needs to be learned, however. Studies on the regulation of glucagon gene expression by insulin and/or glucose should provide more information about the hormonal dysregulation characteristic of diabetes. Identification of the factors involved in the tissue-specific expression of the glucagon gene will permit a better understanding of the embryonic development of the endocrine pancreas and of islet cell differentiation. These studies along with those devoted to the regulation of the other pancreatic hormone genes promise to shed new light on the physiology and pathophysiology of the islets of Langerhans.

References

Alpert S, Hanahan D, Teitelman G (1988) Hybrid insulin genes reveal a developmental lineage for pancreatic endocrine cells and imply a relationship with neurons. Cell 53:295–308

Andrews PC, Ronner P (1986) Isolation and structures of glucagon and glucagon-like peptides from catfish pancreas. J Biol Chem 260:3910–3914

Andrews PC, Hawke DH, Lee TD, Legesse K, Noe BD, Schiverly JE (1986) Isolation and structure of the principal products of preproglucagon processing, including an amidated glucagon-like peptide. J Biol Chem 261:8128–8133

Bani D, Biliotti G, Sacchi TB (1991) Morphological changes in the human endocrine pancreas induced by chronic excess of endogenous glucagon. Virchows Arch [B] Cell Pathol 60:199–206

Bell GI (1986) The glucagon superfamily: precursor structure and gene organization. Peptides 7 [Suppl 1]:27–36

Bell GI, Sanchez-Pescador R, Laybourn PJ, Najarian RC (1983a) Exon duplication and divergence in the human glucagon gene. Nature 304:368–371

Bell GI, Santerre RF, Mullenbach GT (1983b) Hamster preproglucagon contains the sequence of glucagon and two related peptides. Nature 302:716–718

Brubaker PL, Lee YC, Drucker DJ (1992) Alterations in proglucagon processing and inhibition of proglucagon gene expression in transgenic mice which contain a chimeric proglucagon SV-40 T antigen gene. J Biol Chem 267:20728–20733

Caudy M, Vässin H, Brand M, Tuma R, Yan LY, Yan YN (1988) Daughterless, a drosophila gene essential for both neurogenesis and sex determination, has sequence similarities to myc and the achaete-scute complex. Cell 55:1061–1067

Chen L, Komiya I, Inman L, O'Neil J, Appel M, Alam T, Unger RH (1989a) Molecular and cellular responses of islets during perturbations of glucose homeostasis determined by in situ hybridization histochemistry. J Clin Invest 84:711–714

Chen L, Komiya I, Inman L, Mckorkle K, Alam T, Unger RH (1989b) Effects of hypoglycemia and prolonged fasting on insulin and glucagon gene expression. Proc Natl Acad Sci USA 86:1367–1371

Cordier-Bussat M, Morel C, Philippe J (1995) Homologous DNA sequences and cellular factors are implicated in the control of glucagon and insulin gene expression. Mol Cell Biol 15:3904–3916

Drucker DJ, Asa SL (1988) Glucagon gene expression in vertebrate brain. J Biol Chem 263:13475–13478

Drucker DJ, Brubacker PL (1989) Proglucagon gene expression is regulated by a cAMP-dependent pathway in rat intestine. Proc Natl Acad Sci USA 86:3953–3957

Drucker DJ, Philippe J, Jepeal L, Habener JF (1987a) Cis-acting DNA sequence controls glucagon gene expression in pancreatic islet cells. Trans Assoc Am Physicians 100:109–115

Drucker DJ, Philippe J, Jepeal L, Habener JF (1987b) Glucagon gene 5'-flanking sequences promote islet cell-specific glucagon gene transcription. J Biol Chem 262:15659–15665

Drucker DJ, Campos R, Reynolds R, Stobie K, Brubaker PL (1991) The rat glucagon gene is regulated by a protein kinase A-dependent pathway in pancreatic islet cells. Endocrinology 128:394–400

Drucker DJ, Jin T, Asa SL, Young TA, Brubaker PL (1994) Activation of proglucagon gene transcription by protein kinase-A in a novel mouse enteroendocrine cell line. Mol Endocrinol 8:1646–1655

Efrat S, Teitelman G, Anwar M, Ruggiero D, Hanahan D (1988) Glucagon gene regulatory region directs oncoprotein expression to neurons and pancreatic alpha cells. Neuron 1:605–613

Gail Pollock H, Hamilton JW, Rouse JB, Ebner KE, Rawitch KB (1988) Isolation of peptide hormones from the pancreas of the bullfrog. J Biol Chem 263:9746–9751

Gajic D, Drucker DJ (1993) Multiple cis-acting domains mediate basal and adenosine 3′, 5′-monophosphate-dependent glucagon gene transcription in a mouse neuroendocrine cell line. Endocrinology 132:1055–1323

Giddings SJ, Carnaghi LR, Mooradian AD (1995) Age-related changes in pancreatic islet cell gene expression. Metabolism 44:320–324

Gittes GK, Rutter WJ (1992) Onset of cell-specific gene expression in the developing mouse pancreas. Proc Natl Acad Sci 89:1128–1132

Hasegawa S, Terazono K, Nata K, Takada T, Yamamoto H, Okamoto H (1990) Nucleotide sequence determination of chicken glucagon precursor cDNA. FEBS Lett 264:117–120

Heinrich G, Gros P, Habener JF (1984a) Glucagon gene sequence: four or six exons separate functional domains of rat preproglucagon. J Biol Chem 259:14082–14084

Heinrich G, Gros P, Lund PK, Bentley RC, Habener JF (1984b) Preproglucagon mRNA: nucleotide and encoded amino acid sequences of the rat pancreatic cDNA. Endocrinology 115:2176–2181.

Herrera PL, Huarte J, Sanvito F, Meda P, Orci L, Vassalli JD (1991) Embryogenesis of the murine endocrine pancreas; early expression of pancreatic polypeptide gene. Development 113:1257–1265

Herrera PL, Huarte J, Zufferey R, Nichols A, Mermillod B, Philippe J, Muniesa P, Sanvito F, Orci L, Vassalli JD (1994) Ablation of islet endocrine cells by targeted expression of hormone-promoter-driven toxigenes. Proc Natl Acad Sci USA 91:12999–13003

Hosoya M, Kimura C, Ogi K, Ohkubo S, Miyamoto Y, Kugoh H, Shimizu M, Onda H, Oshimura M, Arimura A, Fujino M (1992) Structure of the human pituitary adenylate cyclase activating polypeptide (PACAP) gene. Biochim Biophys Acta 1129:199–206

Inagaki N, Seino Y, Takeda J, Yano H, Yamada Y, Bell GI, Eddy RL, Fukushima Y, Byers MG, Shows TB, Imura H (1989) Gastric inhibitory polypeptide: structure and chromosomal localization of the human gene. Mol Endocrinol 3:1014–1021

Jonsson J, Carlsson L, Edlung T, Edlund H (1994) Insulin-promoter-factor 1 is required for pancreas development in mice. Nature 371:606–609

Kawai K, Unger RH (1982) Inhibition of glucagon secretion by exogenous glucagon in the isolated, perfused dog pancreas. Diabetes 31:512–515

Knepel W, Jepeal L, Habener JF (1990a) A pancreatic islet cell-specific enhancer-like element in the glucagon gene contains two domains binding distinct cellular proteins. J Biol Chem 265:8725–8735

Knepel W, Chafitz J, Habener JF (1990b) Transcriptional activation of the rat glucagon gene by the cAMP-responsive element in pancreatic islet cells. Mol Cell Biol 10:6799–6804

Knepel W, Vallejo M, Chafitz JA, Habener JF (1991) The pancreatic islet-specific glucagon G3 transcription factors recognize control elements in the rat somatostatin and insulin I genes. Mol Endocrinol 5:1457–1466

Kopin A, Wheeler MB, Nishitani J, McBride EW, Chang T, Chey WY, Leiter AB (1991) The secretin gene: evolutionary history, alternative splicing, and developmental regulation. Proc Natl Acad Sci USA 88:5335–5339

Koranyi L, James DE, Kraegen EW, Permutt MA (1992) Feedback inhibition of insulin gene expression by insulin. J Clin Invest 89:432–436

Kruse F, Rose SD, Swif GH, Hammer RE, McDonal RJ (1993) An endocrine-specific element is an integral component of an exocrine-specific pancreatic enhancer. Genes Dev 7:774–786

Lee YC, Asa SL, Drucker DJ (1992) Glucagon gene 5′-flanking sequences direct expression of simian virus 40 large T antigen to the intestine, producing carcinoma of the large bowel in transgenic mice. J Biol Chem 267:10705–10708

Lee YC, Campos RV, Drucker DJ (1993) Region- and age-specific differences in proglucagon gene expression in the central nervous system of wild-type and glucagon-simian virus-40 T-antigen transgenic mice. Endocrinology 133:171–177

Lefèbvre PJ (1995) Glucagon and its family revisited. Diabetes Care 18:715–730

Logothetopoulos J, Sharma BB, Salter JM, Best CH (1960) Glucagon and metaglucagon diabetes in rabbits. Diabetes 9:278–282

Lopez LC, Frazier ML, Su C, Kumar A, Saunders GF (1983) Mammalian pancreatic preproglucagon contains three glucagon-related peptides. Proc Natl Acad Sci USA 80:5485–5489

Lund PK, Goodman RH, Dee PC, Habener JF (1982) Pancreatic preproglucagon cDNA contains two glucagon-related coding sequences arranged in tandem. Proc Natl Acad Sci USA 79:345–349

Lund PK, Goodman RH, Montminy MR, Dee PC, Habener JF (1983) Anglerfish islet preproglucagon II. Nucleotide and corresponding amino-acid sequence of the cDNA. J Biol Chem 258:3280–3284

Mayo KE, Vale W, Rivier J, Rosenfeld MG, Evans RM (1983) Expression-cloning and sequence of a cDNA encoding human growth hormone-releasing factor. Nature 306:86–88

Mayo KE, Cerelli GM, Rosenfeld MG, Evans RM (1985) Gene encoding human growth hormone releasing factor precursor: structure, sequence and chromosomal assignment. Proc Natl Acad Sci USA 82:63–67

Miller CP, Lin JC, Habener JF (1993) Transcription of the rat glucagon gene by the cyclic AMP response element-binding protein CREB is modulated by adjacent CREB-associated proteins. Mol Cell Biol 13:7080–7090

Mojsov S, Heinrich G, Wilson IB, Ravazzola M, Orci L, Habener JF (1986) Preproglucagon gene expression in pancreas and intestine diversifies at the level of post-translational processing. J Biol Chem 261:11880–11889

Morel C, Cordier-Bussat M, Philippe J (1995) The upstream promoter element of the glucagon gene, G1, confers pancreatic alpha cell-specific expression. J Biol Chem 270:3046–3055

Nishi M, Steiner DF (1990) Cloning of complementary DNAs encoding islet amyloid polypeptide, insulin, and glucagon precursors from a new world rodent, the degu, Octodon degus. Mol Endocrinol 4:1192–1198

Nishizawa M, Hayakawa Y, Yanaihara N, Okamoto H (1985) Nucleotide sequence divergence and functional constraint in vasoactive intestinal peptide precursor mRNA evolution between human and rat. FEBS Lett 183:55–59

Novak U, Wilks A, Buelln G, McEwen S (1987) Identical mRNA for preproglucagon in pancreas and gut. Eur J Biochem 164:553–558

Ohlsson H, Karlsson K, Edlund T (1993) IPF1, a homeodomain-containing transactivator of the insulin gene. EMBO J 12:4251–4259

Pani L, Overdier DG, Porcella A, Qian X, Lai E, Costa RH (1992) Hepatocyte nuclear factor 3 beta contains two transcriptional activation domains, one of which is novel and conserved with the Drosophila fork head protein. Mol Cell Biol 12:3723–3732

Papachristou DN, Pham K, Zingg HH, Patel YC (1989) Tissue-specific alterations in somatostatin mRNA accumulation in streptozotozotocin-induced diabetes. Diabetes 38:752–757

Peers B, Leonard J, Sharma S, Teitelman G, Montminy MR (1994) Insulin expression in pancreatic islet cells relies on cooperative interactions. Mol Endocrinol 8:1798–1806

Philippe J (1989) Glucagon gene transcription is negatively regulated by insulin in a hamster islet cell line. J Clin Invest 84:672–677

Philippe J (1991a) Insulin regulation of the glucagon gene is mediated by an insulin-responsive DNA element. Proc Natl Acad Sci USA 88:7224–7227

Philippe J (1991b) Structure and pancreatic expression of the insulin and glucagon genes. Endocrinol Rev 12:252–271

Philippe J (1995) Hepatocyte-nuclear factor 3 beta gene transcripts generate protein isoforms with different transactivation properties on the glucagon gene. Mol Endocrinol 9:368–374

Philippe J, Rochat S (1991) Strict distance requirements for transcriptional activation by two regulatory elements of the glucagon gene. DNA Cell Biol 10:119–124

Philippe J, Drucker DJ, Knepel W, Jepeal L, Misulovin Z, Habener JF (1988) Alpha cell-specific expression of the glucagon gene is conferred to the glucagon promoter element by the interactions of DNA-binding proteins. Mol Cell Biol 8:4877–4888

Philippe J, Morel C, Prezioso VR (1994) Glucagon gene expression is negatively regulated by hepatocyte nuclear factor 3 beta. Mol Cell Biol 14:3514–3523

Philippe J, Morel C, Cordier-Bussat M (1995) Islet-specific proteins interact with the insulin-response element of the glucagon gene. J Biol Chem 270:3039–3045

Qian X, Costa RH (1995) Analysis of hepatocyte nuclear factor 3 beta protein domains required for transcriptional activation and nuclear targeting. Nucleic Acids Res 23:1184–1191

Schwaninger M, Lux G, Blume R, Oetjen E, Hidaka H, Knepel W (1993) Membrane depolarizaton and calcium influx induce glucagon gene transcription in pancreatic islet cells through the cyclic AMP-responsive element. J Biol Chem 268:5168–5177

Seino S, Welsh M, Bell GI, Chan SJ, Steiner DF (1986) Mutations in the guinea pig preproglucagon gene are restricted to a specific portion of the prohormone sequence. FEBS Lett 203:25–30

Teitelman G, Alpert S, Polak JM, Martinez A, Hanahan D (1993) Precursor cells of mouse endocrine pancreas coexpress insulin, glucagon and the neuronal proteins tyrosine hydroxylase and neuropeptide Y, but not pancreatic polypeptide. Development 118:1031–1039

Tricoli JV, Bell GI, Shows TB (1984) The human glucagon gene is located on chromosome 2. Diabetes 33:200–202

Tsukada T, Horovitch SJ, Montminy MR, Mandel G, Goodman RH (1985) Structure of the human vasoactive intestinal polypeptide gene. DNA 4:293–300

Voss JW, Rosenfeld MG (1922) Anterior pituitary development: short tales from dwarf mice. Cell 70:527–530

Wang TC, Brand SJ (1990) Islet cell-specific regulatory domain in the gastrin promoter contains adjacent positive and negative DNA elements. J Biol Chem 265:8908–8914

Wheeler MB, Nishitani J, Buchan AM, Kopin AS, Chey WY, Chang TM, Leiter AB (1992) Identification of a transcriptional enhancer important for enteroendocrine and pancreatic islet cell-specific expression of the secretin gene. Mol Cell Biol 12:3531–3539

White JW, Saunders GF (1986) Structure of the human glucagon gene. Nucleic Acids Res 14:4719–4730

Wright WE (1992) Muscle basic helix-loop-helix proteins and the regulation of myogenesis. Curr Opin Genet Dev 2:243–248

CHAPTER 3
Preproglucagon and Its Processing

D. BATAILLE

A. Introduction

Many proteins become activated by post-translational events occurring inside a larger precursor (reviews in DOCHERTY and STEINER 1982; LAZURE et al. 1983; COHEN 1987). A major event in this process is endoproteolytic cleavage, which leads to one or several fragments that may undergo other post-translational modifications such as glycosylation, N-acylation, phosphorylation, sulfation or C-terminal amidation. It was recognized almost 30 years ago that peptide hormones are produced from precursors, such as proopiomelanocortin (CHRÉTIEN and LI 1967), leading to opiate peptides, melanocyte-stimulating hormone (MSH), γ-lipotropic hormone (LPH) and adrenocorticotropic hormone (ACTH), or proinsulin, from which insulin is produced (STEINER et al. 1967). Similarly, glucagon arises from post-translational processing of a large molecular weight precursor, proglucagon. The structure of this precursor has been deduced from the nucleotide sequence of the mRNA-derived cDNA from several species: ox (LOPEZ et al. 1983), hamster (BELL et al. 1983a), human (BELL et al. 1983b), anglerfish (LUND et al. 1983), rat (HEINRICH et al. 1984) and chicken (HASEGAWA et al. 1990). Further studies led to the knowledge of the structure of the genomic DNA coding for preproglucagon: The preproglucagon gene (BELL et al. 1983b; HEINRICH et al. 1984a,b; WHITE and SAUNDERS 1986), located on chromosome 2 in man (TRICOLI et al. 1984), contains four exons and five introns, three of which are inside the translated moiety (Fig. 1). The precursor, derived from this gene, is expressed in the pancreatic A cells, the intestinal L cells and some specialized neurons in the CNS. A single species of mRNA appears to exist, with no evidence for alternative splicing (NOVAK et al. 1987). From this mRNA, the 180-residue preproform which contains the 20-amino-acid signal peptide is translated. After elimination of the signal peptide, the 160-amino-acid proglucagon undergoes endoproteolytic processing, which differs according to the tissue considered, leading to the active fragments. Unexpectedly, two additional peptidic structures resembling glucagon clearly appeared in the mammalian proglucagon (LOPEZ et al. 1983; BELL et al. 1983a,b). This structure in tandem is considered to derive from gene duplication followed by progressive mutations (see PHILIPPE 1991). It is noteworthy that only one of these glucagon-like peptides is present in nonmammalian proglucagon (HASEGAWA

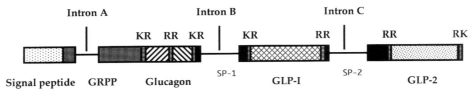

Fig. 1. General structure of the portion of the human preproglucagon gene which contains the translated sequences. Four exons are separated by three introns. Seven domains are separated by dibasic sites Lys-Arg (*KR*), Arg-Arg (*RR*) or Arg-Lys (*RK*). *GRPP*, glicentin-related pancreatic peptide; *SP-1*, *SP-2*, spacer peptides -1 and -2; *GLP-1*, *GLP-2*, glucagon-like peptides- 1 and -2. (From BELL et al. 1983b)

et al. 1990). The biological features of the peptides present in the C-terminal moiety of proglucagon, named "glucagon-like peptide I" and "glucagon-like peptide II" (GLP-1, GLP-2,) respectively, were subsequently studied. GLP-1, and more specifically a truncated form of this peptide (t-GLP-1), turned out to be of major importance in the regulation of insulin secretion (see below and Chap. 18 in this volume). In Fig. 2 is depicted the organizational structure of proglucagon as well as the amino acid sequences of rat and human proglucagons. Different domains, separated by dibasic sites, may be recognized within this structure. By proteolytic cleavage at these sites, many different fragments may be obtained. A limited number of possibilities have appeared, however, to be used in vivo. Figure 3 shows the proglucagon fragments, the existence of which was evidenced in various laboratories.

B. Tissue-Specific Post-translational Processing of Proglucagon

I. Glucagon-Containing Peptides

It was known for a long time that glucagon-like peptides exist in the intestinal mucosa (UNGER et al. 1966) and that these intestinal forms display immunological features which partially differ from that of their pancreatic counterparts (VALVERDE et al. 1970). Two peptides, containing the 29-amino-acid peptide glucagon, were isolated from porcine intestine, the 69-amino-acid peptide glicentin (THIM and MOODY 1981) and the 37-amino-acid peptide oxyntomodulin (BATAILLE et al. 1982a,b). Both contain a C-terminal octapeptide which is absent in glucagon. The presence of this octapeptide is the key to the original biological properties of these peptides (Bataille et al. 1988, 1989; BATAILLE 1989). It is also the main feature which differentiates the intestinal type from the pancreatic type of proglucagon processing when the observer's attention is focused on the N-terminal moiety of proglucagon. Indeed, all available data obtained using various approaches indicate that the final prod-

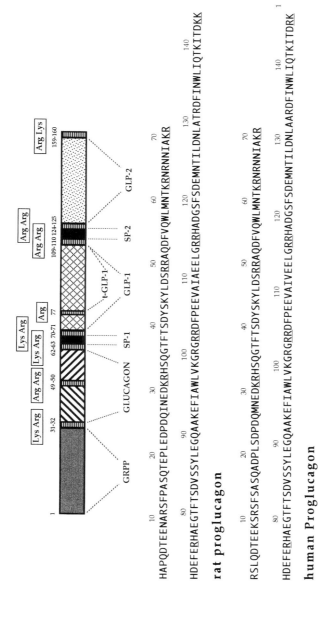

Fig. 2. General structure of proglucagon deduced from the mRNA-derived cDNA. Amino acid (one-letter code) sequences of rat and human proglucagon are given. *GRPP*, glicentin-related pancreatic peptide; *SP-1*, *SP-2*, spacer peptides -1 and -2; *GLP-1*, *GLP-2*, glucagon-like peptides -1 and -2; *t-GLP-1*, truncated glucagon-like peptide-1. (From BELL et al. 1983b; LOPEZ et al. 1983; HEINRICH et al. 1984a)

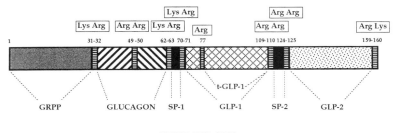

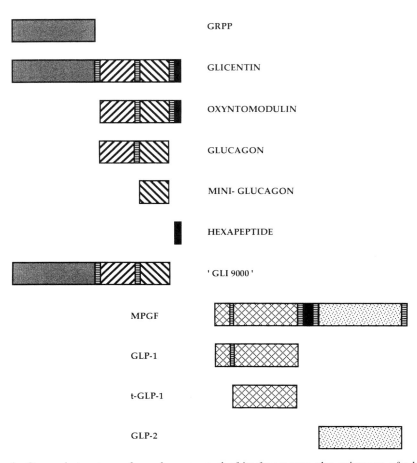

Fig. 3. General structure of proglucagon and of its fragments, the existence of which has been evidenced in proglucagon-containing tissues. *GRPP*, glicentin-related pancreatic peptide (Patzelt and Schiltz 1984); glicentin (Thim and Moody 1981; oxyntomodulin (Bataille et al. 1982a; 1982b); glucagon-(19-29) or miniglucagon (Blache et al. 1989a); hexapeptide (Yanaihara et al. 1985); *GLP-1*, *GLP-2*, glucagon-like peptides -1, -2; *t-GLP-1*, truncated glucagon-like peptide-1. (Mojsov et al. 1986)

ucts from the first 69 amino acids of proglucagon are essentially glucagon in the pancreatic A cells (see Fig. 4a,b) and a mixture of glicentin and oxyntomodulin in the intestinal L cells (see Fig. 5). Both glicentin and oxyntomodulin are released into the blood stream under appropriate physiological stimuli and, accordingly, are true hormones (KERVRAN et al. 1987). Only the proportion in blood between the longer and the shorter peptides differ slightly from one animal species to another; this proportion faithfully reflects what is observed in the intestinal tissue. As described in Chap. 19, this volume, which is specifically devoted to oxyntomodulin, this proportion is not of major importance since the final biologically competent peptide appears to be the C-terminal nonadecapeptide, which may be produced from either hormone. In the intestine, the scenario which leads to the production of glicentin and oxyntomodulin is clear: the first 71 amino acids of proglucagon (that is glicentin plus the basic doublet, see below) are detached from the rest of the molecule. It must be noted that, in contrast to what happens in pancreatic islets (see below), fully matured glicentin (proglucagon 1–69) is produced in intestine (see Fig. 5), indicating that the C-terminal basic amino acids are eliminated by carboxypeptidase before further processing. Glicentin is then partially cleaved into oxyntomodulin, leading to the storage of both peptides in the secretory granules (see Fig. 5). The proportion of cleavage of glicentin into oxyntomodulin drives the final proportion between the two peptides in the secretory granules and, eventually, in the circulation.

Some discrepancies appeared in the interpretation of the steps which, in the pancreatic A cells, lead to glucagon: Thanks to methods with both high sensitivity and high specificity, which use antibodies directed towards free epitopes that are unmasked following proteolytic cleavage in conjunction with high-performance liquid chromatography (HPLC), it was observed that glicentin is undetectable in either rat or human pancreas (BLACHE et al. 1988a,c), whereas oxyntomodulin is present at the level of 5%–10% of the glucagon content (BLACHE et al. 1988a,c) (see Fig. 4a,b). It must also be remembered that the "glicentin-like immunoreactivity" observed in pancreas by various groups (e.g., RAVAZZOLA et al. 1979; MOJSOV et al. 1986) may be any fragment containing glucagon-related pancreatic peptide (GRPP), including GRPP itself, which bears the "glicentin" epitope (see also MOODY and THIM 1983; ORCI et al. 1983). Using pulse-chase methods, however, a peptide having the appearance of glicentin was observed as an intermediate in a glucagon-producing cell line (ROUILLÉ et al. 1994). When we carry out careful investigations using the most precise immunological tools, a molecule recognized by classical antiglucagon antibodies is detected in HPLC which runs in front of oxyntomodulin at a retention time similar to that of standard glicentin (Fig. 4a,b). This molecule, however, is not recognized by our C-terminal antibody (Fig. 4a,b), which, as already pointed out, is directed towards a free epitope, which is masked in elongated molecules. It is very likely that this molecule, often interpreted as glicentin (e.g., ROUILLÉ et al. 1994; ROTHENBERG et al. 1995a,b) when a specific C-terminal assay is not available, is the 1–71 fragment

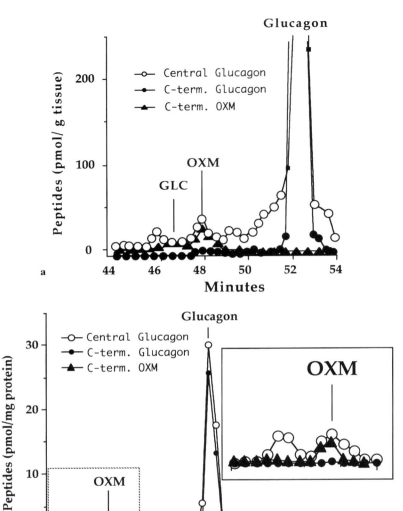

Fig. 4. a Analysis of the proglucagon-derived peptides in rat pancreas using high-performance liquid chromatography (*HPLC*) and radio-immunoassays recognizing specifically an "external" epitope common to all glucagon-containing peptides (*central glucagon*), a free C-terminal glucagon epitope (*C-term. glucagon*) and a free C-terminal epitope present in glicentin and oxyntomodulin which is masked in C-terminally elongated peptides (*C-term. OXM*). Retention times of standard glucagon, oxyntomodulin (*OXM*) and glicentin (*GLC*) are indicated (Data from BLACHE et al. 1989c). **b** Analysis, using the same tools as in Fig. 4, of the proglucagon-derived peptides in the α-TC cell line. (Data from BLACHE et al. 1994b)

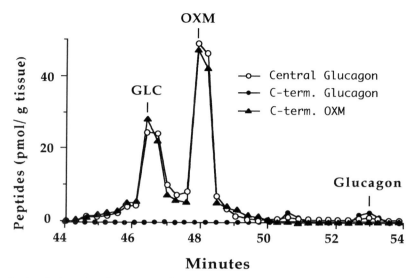

Fig. 5. Analysis, using the same tools as in Fig. 4, of the proglucagon-derived peptides in rat intestine. (Data from BLACHE et al. 1989c)

(which may be written glicentin-KR) resulting from the cleavage of proglucagon between Arg^{71} and His^{72} and leading to the release of major proglucagon fragment (MPGF) from the C-terminal moiety of proglucagon. It is logical that the 1–71 molecule appears during the course of the proglucagon processing, since it is now well established that the prohormone convertases (see below) cleave at the C-terminus of basic doublets, the amino acids remaining at the C-terminus being later eliminated by specific carboxypeptidases. Another possible explanation of the absence of glicentin in pancreas is that the cleavage occurring between residues 32 and 33 (between GRPP and glucagon) takes place earlier than or is contemporaneous with the one occurring at residues 62–63 (between glucagon and the octapeptide). This scenario, however, does not take the existence of the 1–71 peptide into account. Accordingly, the first hypothesis is the most likely. Since true oxyntomodulin (C-terminal free) is present in pancreas, the elimination by carboxypeptidase of the basic amino acids on its C-terminus should precede the maturation into glucagon. Oxyntomodulin thus appears as the last intermediate leading to glucagon, the passage from the former to the latter peptide giving rise to a hexapeptide which was isolated from pancreas and shown to be co-secreted with glucagon (YANAIHARA et al. 1985). Whether oxyntomodulin is released together with glucagon from pancreatic A cells and whether this possible release may be of physiological or pathological importance remains to be established.

II. Glucagon-Related Peptides

Within the C-terminal moiety of proglucagon, the existence of molecular structures corresponding to the glucagon-related peptides (GLP-1, GLP-2) stimulated the search for their biological significance. The biological role of GLP-2 is so far unknown. On the other hand, GLP-1 has acquired the status of an important hormone, the main biological role of which is the regulation of insulin release (MOJSOV et al. 1987). When looking carefully at the proglucagon structure corresponding to the GLP-1 moiety, it appears that the molecular structure which displays the best resemblance to the glucagon structure, in other words is the most "glucagon-like," is not the 72-108 sequence (GLP-1) but a truncated form (called truncated GLP-1 or t-GLP-1), corresponding to the sequence 78-108. Because of the presence in the C-terminal moiety of the GLP-1 sequence of a glycine residue followed by a basic doublet, a consensus sequence for the production of the amidated form of an active peptide (EIPPER and MAINS 1988), it was supposed that t-GLP-1 was C-terminally amidated. Amidated and nonamidated forms, truncated and not truncated, coexist in both pancreas and intestine (see HOLST et al. 1994 and Chap. 18, this volume). It is usually considered, however, that the most physiologically meaningful peptide is the truncated and amidated form, called GLP-1(7-36)amide. It was shown to be a potent stimulator of insulin release and most probably plays a large role in the "incretin" concept, a signal originating from intestine which potentiates the direct effects of glucose on the B cells in the islets of Langerhans (WEIR 1995).

Similarly to the differential processing between pancreas and intestine within the first 69 amino acids of proglucagon, the C-terminal moiety also undergoes specific processing which depends upon the tissue considered. Although small amounts of free GLP-1 (full length or truncated) and GLP-2 are detectable in porcine and human pancreas (HOLST et al. 1994), the C-terminal proglucagon moiety remains essentially untouched in pancreas, leading to the release of fragment 72-158 (HOLST et al. 1994), also called "major proglucagon fragment" (PATZELT and SCHILTZ 1984), the role of which, if any, is unknown. In contrast, intestinal L cells process the proglucagon C-terminus in such a way that biologically active fragments are produced and released. As already pointed out, the role of GLP-2 remains obscure. A reason for that is the difficulty of obtaining a high-quality radioactive tracer, impeding the development of reliable radioimmuno- and binding assays. The existence of this peptide only in mammals may suggest a specific role related to the development of the complex regulatory process appeared at a late stage of species evolution. On the other hand, GLP-1, and more specifically the 7-36-amide variant, is an important link between the nutritional signals present in intestine and the insulin response from the pancreatic islets (see Chap. 18, this volume). t-GLP-1 is stored in secretory granules within the intestinal L cells and released under appropriate stimuli. Whether this release is synchronized to that of the hormones deriving from the N-terminal of proglucagon (oxyntomodulin and glicentin) remains to be established (see Chap. 19, this volume).

C. Role of Prohormone Convertases in Proglucagon Processing

It is now well established that prohormone processing occurs mainly at pairs of basic residues, Lys-Arg, Arg-Lys or Arg-Arg, the Lys-Lys motif being rarely used (COHEN 1987). Some exceptions exist, such as cleavages at monobasic sites, one example being given by the transformation of GLP-1 into t-GLP-1 which occurs at a single Arg residue (see Fig. 2). Besides this exception, proglucagon processing has the full taste of classicism: The main proglucagon fragments described so far are the result of cleavages at dibasic residues (see Figs. 2, 3). From the search for endopeptidases involved in post-translational prohormone (and more generally proprotein) processing, a family of serine proteases implicated in these essential biological events evolved. These proteases belong to the group including the bacterial subtilisin. A protease, called Kex-2, responsible for the maturation of α-mating factor in yeast (JULIUS et al. 1984; MIZUNO et al. 1988; FULLER et al. 1989) turned out to be the head of a series of structurally related endoproteases in mammals, often referred to as 'kexin-like." Several of these proteases, displaying strong preferences for dibasic sites, named "prohormone-convertases" (PCs), were recognized by different laboratories as essential processing enzymes for prohormones (review in SEIDAH and CHRÉTIEN 1994). Up to now, the cDNA coding for PC1 (SEIDAH et al. 1991), also called PC3 (SMEEKENS et al. 1991), PC2 (SEIDAH et al. 1990; SMEEKENS and STEINER 1990; BENJANNET et al. 1991), PC4 (NAKAYAMA et al. 1992), PC5 (LUSSON et al. 1993) also called PC6 (NAKAGAWA et al. 1993) and, more recently PC7 (B. YOUNG, personal communication), were cloned. Other endoproteases of the same family, such as furin (ROEBROEK et al. 1986; VAN DEN OUWELAND et al. 1990), or proteases of the paired basic amino acid cleaving enzyme (PACE) family see (REHEMTULLA et al. 1993) were also structurally and functionally characterized. After cleavage of prohormones by prohormone convertases, the biologically active molecules are produced thanks to elimination of the C-terminal basic amino acids by specific carboxypeptidases such as carboxypeptidase E, also called carboxypeptidase H (see FRICKER et al. 1986) and, for the hormones the C-terminal of which is amidated, by the action of peptidyl glycine α-amidating monooxygenase or PAM (EIPPER and MAINS 1988). In this rapidly developing area, we will describe "state-of-the-art" knowledge on the relationship between the PCs and proglucagon processing.

The direct implication of PC1/3 and PC2 in the post-translational processing of proinsulin into insulin is now well established (see SMEEKENS et al. 1992). As far as proglucagon is concerned, the importance of the same endoproteases in its processing is still a matter of debate. The prohormone convertase PC2 is undoubtfully expressed at high levels in both authentic A cells (see NAGAMUNE et al. 1995) and in the α-TC cell line (BLACHE et al. 1994b; ROUILLÉ et al. 1994), which derives from microadenomas obtained in transgenic mice bearing a proglucagon promoter linked to SV-40 large T antigen (POWERS et al. 1990). This cell line displays characteristics which make it close to true A cells. In particular, a pancreatic-type proglucagon processing was observed in this cell line, an essential feature for the analysis of a possible involvement of prohormone convertases in proglucagon processing into glucagon. PC1/PC3 is barely detected in this cell line (BLACHE et al. 1994b; ROUILLÉ et al. 1994). Direct comparison of the PC1/3 and PC2 content of the α-TC cell line, which processes proglucagon into glucagon, with that of the S-TC cell line, which similarly processes proglucagon to intestinal L cells (BLACHE et al. 1994b), suggests that PC1 is implicated in the formation of oxyntomodulin and that PC2 may be responsible for the transformation of oxyntomodulin into glucagon. Furthermore, a partial suppression of PC2 expression in α-TC cells using antisense RNA resulted in a decrease in proglucagon processing into glucagon, suggesting a link between the two parameters (ROUILLÉ et al. 1994). Similarly, transfecting the proglucagon gene into a cell line (AtT-20) expressing large quantities of PC1 resulted in oxyntomodulin production with no appearance of glucagon, whereas transfecting proglucagon into a cell line (GH3) expressing PC2 resulted in the production of oxyntomodulin and glucagon (MINEO et al. 1995). At variance with these data, in vitro experiments measuring the effects of recombinant or purified PC1/3 and PC2 on HPLC-purified ^{3}H-tryptophan-labeled proglucagon (ROTHENBERG et al. 1995a,b) indicated that either

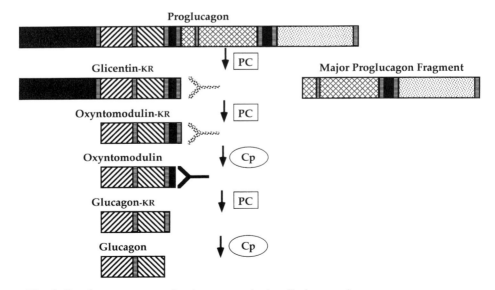

Fig. 6. Proglucagon processing in pancreatic A cells (see text)

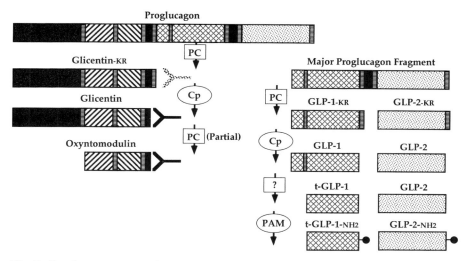

Fig. 7. Proglucagon processing in intestinal L cells (see text)

endoprotease was able to produce MPGF, glicentin (or more likely glicentin-KR) and oxyntomodulin but that glucagon production was not observed under these conditions, rendering doubtful that PC2 is the endoprotease which liberates glucagon and the hexapeptide from the precursor. No explanation is available at the present time to account for such discrepancies, but it is still possible that another PC (such as PC5/6 or other enzymes including furin and/or a PACE enzyme) should be taken into account to explain the differential proglucagon processing into the N-terminal-derived peptides. As far as the C-terminal peptides are concerned (GLP-1 and GLP-2), very little information is available, except that recombinant PC1 is able to release GLP-1 from MPGF (ROTHENBERG et al. 1995a) and that the AtT-20 cells, which, as already mentioned, express high levels of PC1/3, produce t-GLP-1 from proglucagon (ROUILLÉ and STEINER 1995). The release of t-GLP-1 from GLP-1, on the other hand, should implicate an as yet unidentified processing enzyme acting specifically at monobasic sites (Arg^{77} in proglucagon).

Figures 6 and 7 summarize the present knowledge on the proglucagon processing steps occurring in pancreas and in intestine, respectively.

D. Secondary, Postsecretory Processing of Proglucagon-Derived Peptides

All but one dibasic site are loci where proglucagon is cleaved by processing enzymes in proglucagon-containing endocrine cells, such as the pancreatic A cells or the intestinal L cells, leading to the production of biologically active peptides. The single exception is the Arg[49]-Arg[50] (amino acid numbers in proglucagon), referred to as Arg[17]-Arg[18] when the amino acid numbers in glucagon/oxyntomodulin are considered. This indicates that the prohormone convertases are unable to cleave proglucagon or processing intermediates at this site, for reasons that are, as yet, not understood. This may be related to some subtle conformation of the physiological substrates for endoproteases in the intracellular environment of the endocrine cells (trans-Golgi network, secretory granules, etc.), protecting them from being recognized and cleaved (see RHOLAM et al. 1986). Observations on the biological effects of C-terminal fragments from glucagon and oxyntomodulin made us suspect that this dibasic site is, in fact, a cleavage site for (an)other endoprotease(s) active in another environment.

I. Glucagon (19-29) or Miniglucagon

From the observation that the (19-29) fragment of glucagon was 3 orders of magnitude more potent (MALLAT et al. 1987) than glucagon (LOTERSZTAJN et al. 1984) in modulating the plasma membrane Ca^{2+} pump of the hepatocyte, it became apparent that this fragment, acting at nanomolar concentrations, was a biologically competent peptide. The probable implication of this peptide in biological regulation resulting from these observations was further substantiated by the data obtained on the modulation of the in vitro system by GTP or GTP analogues, leading to a 2 orders of magnitude increase in the potency of the fragment which acted under these conditions at picomolar concentrations (LOTERSZTAJN et al. 1990). This prompted us to direct our research in two directions:

1. To determine the consequences of this effect of the (19-29) fragment on intact cells that are known to be glucagon-sensitive cells: It was shown that at concentrations that are active on the Ca^{2+} pump (10^{-12} to $10^{-10} M$), the fragment, referred to as "miniglucagon," modulates negatively the biological processes regulated by glucagon in a positive way. This is true for cardiac myocytes in culture (PAVOINE et al. 1991) where miniglucagon displays a negative inotropic effect, opposite to the glucagon-positive inotropic effect, and in insulin-secreting pancreatic B cells in culture, where miniglucagon inhibits the glucose-induced and the glucagon-induced insulin release (BATAILLE et al. 1992). Details of this relationship between the effects of miniglucagon vs. that of glucagon and on the interplay between the intracellular cyclic AMP and the Ca^{2+} pathways is given in Chap. 7 of this volume.
2. To determine where and how the miniglucagon molecule is produced, an essential element of this novel regulatory process. Using the peptide chemical approaches which led us to obtain immunological tools allowing the analysis of peptide epitopes arising from endoproteolytic cleavage of hormone precursors (BLACHE et al. 1988a, see above), we developed a radioim-

munoassay recognizing specifically the N-terminal moiety of miniglucagon (BLACHE et al. 1989b, 1990). Using this tool, we observed that miniglucagon is undetectable in any tissue from the rat, except the pancreas, where it represents around 5% of the glucagon concentrations (BLACHE et al. 1989b). Furthermore, the half-life of exogenous miniglucagon both in vivo and in vitro (liver membranes) is extremely short, precluding an action using a hormonal pathway. The same site for miniglucagon production and action at the level of the glucagon target cells was thus very likely. We observed that miniglucagon is produced upon incubation of glucagon with purified rat liver plasma membranes, a strong indication for a local release of miniglucagon from its mother hormone, near the glucagon site of action (BLACHE et al. 1989a, 1990). We observed that miniglucagon production is very rapid and we calculated that, taken the dynamics of production and degradation of the fragment as well as the time allowed for its action, a few percent (less than 10%) of the glucagon concentration in blood is transformed into miniglucagon in one pass through the liver. The dynamics may be different in other organs, which do not seem to degrade miniglucagon as actively as the liver. We were also able to determine that miniglucagon is produced at a very significant rate when glucagon is incubated with intact cells in culture, such as hepatoma cells (BLACHE et al. 1994a), indicating that the release of the fragment occurs at the cell surface.

The next step was to determine which endopeptidase transforms glucagon into miniglucagon. It appeared (BLACHE et al. 1990, 1992, 1993) that the enzyme responsible for this cleavage displays both metalloprotease properties (e.g., the miniglucagon production was inhibited by metal chelating agents such as 1,10-phenantroline) and thiol-protease properties (inhibition by reagents of thiol groups such as *para*-chloro-mercuri-benzoate). In that sense, the enzyme, called "miniglucagon-generating endopeptidase" (MGE), appears to belong to a family of metalloproteases containing a free cysteine near the active site, a family which includes "insulin-degrading enzyme" or IDE (DUCKWORTH 1988) and "*N*-arginine dibasic convertase" or NRD convertase (PIEROTTI et al. 1994; CHESNEAU et al. 1994). MGE was extracted from purified liver plasma membranes using the detergent CHAPS (cholamidopropyl-dimethylammonio-propane sulfonate) (BLACHE et al. 1992, 1993) and purified to homogeneity by a series of chromatographic steps (BLACHE et al. 1993). It appeared that the MGE activity is borne by a single-chain 100-kDa protein, a partial sequence of which was obtained (BLACHE et al. 1993). This partial sequence may be aligned with the known sequences of proteases of the same family (BLACHE et al. 1994a). It is worth noting that one of the features of this family, referred to as the M16 family of proteases (see CHESNEAU et al. 1994), is the presence of an unusual, inverted consensus amino acid sequence (HXXEH) for zinc coordination when compared with that of "classical" metalloproteases (HEXXH). Cloning of the cDNA coding for MGE, underway in our laboratory, will allow the determination of whether this endoprotease indeed belongs to this new family of proteases.

From a physiological point of view, such secondary processing leading to the release from a circulating peptide of a fragment which modulates the action of a mother-hormone is still a matter of speculation. It may be seen as a refinement of a classical regulatory process leading to a fine-tuning of hormonal action necessary for highly sophisticated organisms to face various conditions. When more is known about MGE, it will be of major interest to determine whether the expression of miniglucagon-generating endopeptidase is regulated. On the other hand, it is highly likely that the mechanism described above is operative in vivo. Indeed, miniglucagon is produced at a rate which makes it present at the proper place at concentrations that are about 2 orders of magnitude lower than the mother hormone, whereas the fragment acts at concentrations (around $10^{-11} M$) that are also 2 orders of magnitude lower than the active doses of glucagon (around $10^{-9} M$). In the particular case of the islets of Langerhans, it must be remembered that miniglucagon is present there at several percent of the glucagon concentration, making it possible that miniglucagon acts physiologically as a modulator of the glucagon action on insulin release. It is now well established that pancreatic B cells contain authentic glucagon receptors coupled to insulin secretion (KAWAI et al. 1995). When glucagon is released from A cells every time an increase in blood glucose is necessary, an insulin release triggered by glucagon from the fringe of B cells that surround the A cells would be sufficient to abolish the hyperglycemic effect of glucagon. A release of miniglucagon together with glucagon might be a suppressive mechanism for this undesirable side effect of glucagon.

II. Oxyntomodulin (19-37)

MGE is able to release the (19-37) fragment from oxyntomodulin (BATAILLE et al. 1989; JARROUSSE et al. 1993), and, most probably, from glicentin too. This question and the biological consequences of this production is discussed in Chap. 19, this volume, which is entirely devoted to oxyntomodulin.

E. Conclusions

The tissue-specific proglucagon processing is, like that of proopiomelanocortin (POMC, see LAZURE et al. 1983), a good example of the production from a single precursor of a set of hormones with different biological roles. This differential processing is the key to the specificity of the system. For instance, it is clear that the release of glucagon into the blood is necessary under conditions that differ markedly from those which necessitate the release of oxyntomodulin and/or of t-GLP-1. The differential processing in pancreatic A cells and in intestinal L cells, making a completely different set of hormones possibly released independently from these two distant sites, allows this specificity. What happens in the CNS is still unclear, since no biological role has been assigned in the CNS to any of the proglucagon-derived peptides. The proglucagon processing is essentially of the intestinal type (glicentin and oxyntomodulin) but with a significant amount of glucagon, particularly in medulla oblongata (BLACHE et al. 1988b). It is not known whether this apparent intermediate form between intestinal and pancreatic types of processing is real or whether it is the result of the coexistence in a CNS structure of several sets of neurons expressing separately the two types of processing. The same question may be asked about human intestine, which expresses predominantly oxyntomodulin and glicentin but also contains a significant amount of glucagon (BLACHE et al. 1988a). For a review of extrapancreatic glucagon, see LEFEBVRE and LUYCKX (1983). On the other hand, the coexistence in the L cells of oxyntomodulin/glicentin and of t-GLP-1 makes it possible that these peptides are coreleased under common stimuli. This, however, has never been proved directly. If this was not the case, one should imagine that two sets of L cells exist, each with a differential sensitivity to stimuli; alternatively, oxyntomodulin/glicentin on one hand and t-GLP-1 on the other may be stored, within the same individual cell, in different secretory granules which may be secreted separately under selective stimuli. To our knowledge, such a mechanism has not been ascertained for any endocrine cell so far.

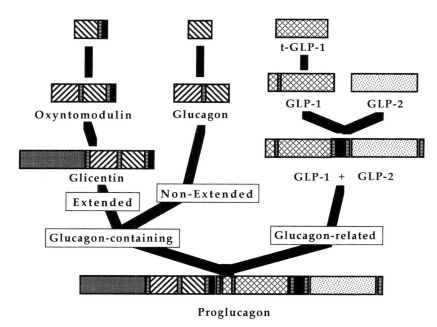

Fig. 8. The proglucagon tree (see text)

The proglucagon structure most probably derives from an ancestral gene which contained only one glucagon (or glucagon-like) sequence. By successive gene duplication followed by mutations, this structure has given three different sequences, and thus three different potential hormones. However, subtle cuts at certain places of this precursor by processing enzymes and using in a efficient way the "spacer peptides" led to a much greater variety of biologically competent peptides. If we disregard the fragments which have not been shown to possess an activity (such as the GRPP, the hexapeptide, the MPGF or the "GLI^{9000}") at least seven peptides are of physiological importance, either proved or very likely: glucagon, miniglucagon, glicentin, oxyntomodulin, the 19-37 fragment, t-GLP-1 and GLP-2. They may be placed on two branches of a tree which may be regarded as an evolutionary tree (Fig. 8). The right branch (with the glucagon-related peptides) bears the MPGF from which arise two further branches bearing the GLP-1 (eventually processed into t-GLP-1) and GLP-2. The left branch (which is likely to be the original one from an evolutionary point of view) is the branch bearing the glucagon-containing peptides. This branch is further separated into two branches: one bears the C-terminally extended peptides glicentin, oxyntomodulin and their C-terminal nonadecapeptide, which represents the ultimate "oxyntomodulin-like" peptide (see Chap. 19, this volume). The next branch, which may have appeared later in evolution; contains the nonelongated form glucagon and its C-terminal fragment, which may be seen as an additional sophistication to the regulatory processes induced by this hormone.

Such an ensemble looks more "coherent" than POMC, which contains peptides with extremely different biological specificities. Indeed, the proglucagon-derived peptides have their influence in the general context of nutrition (including digestion and metabolism). This highly structured and coordinated ensemble, which arose from an original single structure, is likely to represent a response to needs which appeared during evolution for more and more sophisticated regulation of this vital function.

Acknowledgements. I would like to thank Drs. Alain Kervran and Alain Le Quellec for helpful discussions during the preparation of the manuscript.

References

Bataille D (1989) Gut glucagon. In: Schulz SG, Makhlouf G, Rauner BB (eds) Handbook of physiology – section 6: the gastrointestinal system, vol II. Neural and endocrine biology neuroendocrinology of the gut. American Physiological Society, Bethesda, pp 455–474

Bataille D, Coudray AM, Carlqvist M, Rosselin G, Mutt V (1982a) Isolation of glucagon-37 (bioactive enteroglucagon/oxyntomodulin) from porcine jejunoileum. Isolation of the peptide. FEBS Lett 146:73–78

Bataille D, Tatemoto K, Gespach C, Jörnvall H, Rosselin G, Mutt V (1982b) Isolation of glucagon-37 (bioactive enteroglucagon/oxyntomodulin) from porcine jejunoileum. Characterisation of the peptide. FEBS Lett 146:79–86

Bataille D, Blache P, Mercier F, Jarrousse C, Kervran A, Dufour M, Mangeat P, Dubrasquet M, Mallat A, Lotersztajn S, Pavoine C, Pecker F (1988) Glucagon and

related peptides: molecular structure and biological specificity. Ann NY Acad Sci 527:168–185
Bataille D, Jarrousse C, Blache P, Kervran A, Dufour M, Mercier F, Le-Nguyen D, Martinez J, Bado A, Dubrasquet M, Mallat A, Pavoine C, Lotersztajn S, Pecker F (1989) Oxyntomodulin and glucagon: are the whole molecules and their C-terminal fragments different biological entities? Biomed Res 9 [Suppl 3]:169–179
Bataille D, Kuroki S, Blache P, Kervran A, Le-Nguyen D, Dufour M, Lotersztajn S, Mallat A, Brechler V, Pecker F (1992) Glucagon [19-29]: a daughter-molecule which modulates the action of the mother-molecule. Biomed Res 13 [Suppl 2]:137–142
Bell GI, Santerre RF, Mullenbach GT (1983a) Hamster preproglucagon contains the sequence of glucagon and two related peptides. Nature 302:716–718
Bell GI, Sanchez-Pescador R, Laybourn PJ, Najarian RC (1983b) Exon duplication and divergence in the human preproglucagon gene. Nature 304:368–371
Benjannet S, Rondeau N, Day R, Chrétien M, Seidah N (1991) PC1 and PC2 are proprotein convertases capable of cleaving proopiomelanocortin at distinct pairs of basic residues. Proc Natl Acad Sci USA 88:3564–3568
Blache P, Kervran A, Martinez J, Bataille D (1988a) Development of an oxyntomodulin/glicentin C-terminal radio-immunoassay using a "thiol-maleoyl" coupling method for preparing the immunogen. Anal Biochem 173:171–179
Blache P, Kervran A, Bataille D (1988b) Oxyntomodulin and glicentin: brain-gut peptides in the rat. Endocrinology 123:2782–2787
Blache P, Kervran A, Le-Nguyen D, Laur J, Cohen-Solal A, Devilliers G, Mangeat P, Martinez J, Bataille D (1988c) The glucagon-containing peptides and their fragments in the rat gastro-entero-pancreatic and central nervous systems. Biomed Res 9 [Suppl 3]:19–28
Blache P, Kervran A, Dufour M, Martinez J, Le-Nguyen D, Lotersztajn S, Pavoine C, Pecker F, Bataille D (1989a) Le glucagon est maturé en fragment (19-29) au niveau de la membrane hépatocytaire. CR Acad Sci [III] 17:467–472
Blache P, Kervran A, Laur J, Aumelas A, Le-Nguyen D, Martinez J, Bataille D (1989b) Antisera against preproglucagon fragments obtained by a "thiol-maleoyl" coupling method. Colloque INSERM/Libbey Eurotext 174:519–522
Blache P, Kervran A, Le-Nguyen D, Laur J, Cohen-Solal A, Devilliers J, Mangeat P, Martinez J, Bataille D (1989c) Glucagon-related peptides deriving from proglucagon in the gastroenteropancreatic and central nervous systems. Biomed Res 9 [Suppl 3]:19–28
Blache P, Kervran A, Dufour M, Martinez J, Le-Nguyen D, Lotersztajn S, Pavoine C, Pecker F, Bataille D (1990) Glucagon (19-29), a Ca^{2+} pump inhibitory peptide, is processed from glucagon in the rat liver plasma membrane by a thiol endopeptidase. J Biol Chem 265:21514–21519
Blache P, Kervran A, Duckworth W, Bataille D (1992) Purification and partial characterization of the liver membrane endopeptidase which transforms glucagon into glucagon (19-29). Biomed Res 13 [Suppl 2]:51–55
Blache P, Kervran A, Le-Nguyen D, Dufour M, Duckworth W, Bataille D (1993) Endopeptidase from rat liver membranes which generates miniglucagon from glucagon. J Biol Chem 268:21748–21753
Blache P, Kervran A, Le-Nguyen D, Bataille D (1994a) Miniglucagon production from glucagon: an extracellular processing of a hormone used as a prohormone. Biochimie 76:295–299
Blache P, Le-Nguyen D, Boegner-Lemoine C, Cohen-Solal A, Bataille D, Kervran A (1994b) Immunological detection of prohormone convertases in two different proglucagon-processing cell lines. FEBS Lett 344:65–68
Chesneau V, Pierotti AR, Prat A, Gaudoux F, Foulon T, Cohen P (1994) N-Arginine dibasic convertase (NRD convertase): a newcomer to the family of processing endopeptidases. An overview. Biochimie 76:234–240
Chrétien M, Li CH (1967) Isolation, purification and characterization of γ-lipotropic hormone from sheep pituitary glands. Can J Biochem 45:1163–1174

Cohen P (1987) Proteolytic events in the post-translational processing of polypeptide hormone precursors. Biochimie 69:87–89
Docherty K, Steiner DF (1982) Post-translational proteolysis in polypeptide hormone biosynthesis. Annu Rev Physiol 44:625–638
Duckworth WC (1988) Insulin degradation: mechanisms, products and significance. Endocr Rev 3:319–345
Eipper BA, Mains RE (1988) Peptide alpha-amidation. Annu Rev Physiol 50:333–344
Fricker L, Evans CJ, Esch FS, Herbert E (1986) Cloning and sequence analysis of cDNA for bovine carboxypeptidase E. Nature 323:461–464
Fuller RS, Brake AJ, Thorner J (1989) Intracellular targeting and structural conservation of a prohormone-processing endoprotease. Science 246:482–486
Hasegawa S, Terazono K, Nata K, Takuda T, Yamamoto H, Okamoto H (1990) Nucleotide sequence determination of chicken glucagon precursor cDNA. Chicken preproglucagon does not contain glucagon-like peptide II. FEBS Lett 264:117–120
Heinrich G, Gros P, Lund PK, Bentley RC, Habener JF (1984a) Preproglucagon messenger ribonucleic acid: nucleotide and encoded amino acid sequences of the rat pancreatic complementary deoxyribonucleic acid. Endocrinology 115:2176–2181
Heinrich G, Gros P, Habener JF (1984b) Glucagon gene sequence. Four of six exons encode separate functional domains of rat pre-proglucagon. J Biol Chem 259:14082–14087
Holst JJ, Bersani M, Johnsen AH, Kofod H, Hartmann B, Ørskov C (1994) Proglucagon processing in porcine and human pancreas. J Biol Chem 269:18827–18833
Jarrousse C, Carles-Bonnet C, Niel H, Sabatier R, Audousset-Puech M-P, Blache P, Kervran A, Martinez J, Bataille D (1993) Inhibition of gastric acid secretion by oxyntomodulin and its [19-37] fragment in the conscious rat. Am J Physiol 264 (Gastrointest Liver Physiol 27):G816–G823
Julius D, Brake AJ, Blair L, Kunisawa R, Thorner J (1984) Isolation of the putative structural gene for the Lys-Arg-cleaving endopeptidase required for processing of yeast prepro-α-factor. Cell 37:1075–1089
Kawai K, Yokota C, Ohashi S, Watanabe Y, Yamashita K (1995) Evidence that glucagon stimulates insulin secretion through its own receptor in rats. Diabetologia 38:274–276
Kervran A, Blache P, Bataille D (1987) Distribution of oxyntomodulin and glucagon in the gastro-intestinal tract and the plasma of the rat. Endocrinology 121:704–713
Lazure C, Seidah N, Pelaprat D, Chrétien M (1983) Proteases and post-translational processing of prohormones. Can J Biochem Cell Biol 61:501–515
Lefèbvre PJ, Luyckx AS (1983) Extrapancreatic glucagon and its regulation. In: Lefèbvre PJ (ed) Glucagon II. Springer, Berlin Heidelberg New York, pp 205–216 (Handbook of experimental pharmacology, vol 66/II)
Lopez LC, Frazier ML, Su CJ, Kumar A, Saunders GF (1983) Mammalian pancreatic preproglucagon contains three glucagon-related peptides. Proc Natl Acad Sci USA 80:5485–5489
Lotersztajn S, Epand RM, Mallat A, Pecker F (1984) Inhibition by glucagon of the calcium pump in liver plasma membranes. J Biol Chem 259:8195–8201
Lotersztajn S, Pavoine C, Brechler V, Roche B, Dufour M, Le-Nguyen D, Bataille D, Pecker F (1990) Glucagon (19-29) exerts a biphasic action on the liver plasma membrane Ca^{2+} pump which is mediated by G proteins. J Biol Chem 265:9876–9880
Lund PK, Goodman RH, Montminy MR, Dee PC, Habener JF (1983) Anglerfish islet preproglucagon II. Nucleotide and corresponding amino acid sequence of the cDNA. J Biol Chem 258:3280–3284

Lusson J, Vieau D, Hamelin J, Day R, Chrétien M, Seidah N (1993) Structure of the mouse and rat subtilisin/kexin-like PC5: a candidate proprotein convertase expressed in endocrine and nonendocrine cells. Proc Natl Acad Sci USA 90:6691–6695

Mallat A, Pavoine C, Dufour M, Lotersztajn S, Bataille D, Pecker F (1987) A glucagon fragment is responsible for the inhibition of the liver Ca^{2+} pump by glucagon. Nature 325:620–622

Mineo I, Matsumura T, Shingu R, Namba M, Kuwajima M, Matsuzawa Y (1995) The role of prohormone convertases PC1 (PC3) and PC2 in the cell-specific processing of proglucagon. Biochem Biophys Res Commun 207:646–651

Mizuno K, Nakamura T, Oshima T, Tanaka S, Matsuo H (1988) Yeast KEX2 gene encodes an endopeptidase homologous to subtilisin-like serine proteases. Biochem Biophys Res Commun 156:246–254

Mojsov S, Heinrich G, Wilson IB, Ravazzola M, Orci L, Habener JF (1986) Preproglucagon gene expression in pancreas and intestine diversifies at the level of post-translational processing. J Biol Chem 261:11880–11889

Mojsov S, Weir GC, Habener (1987) Insulinotropin: glucagon-like peptide-I (7-37), coencoded in the glucagon gene is a potent stimulator of insulin release in the perfused rat pancreas. J Clin Invest 79:616–619

Moody AJ, Thim L (1983) Glucagon, glicentin and related peptides. In: Lefebvre PJ (ed) Glucagon I. Springer, Berlin Heidelberg New York, pp 139–174 (Handbook of experimental pharmacology, vol 66/I)

Nagamune H, Muramatsu K, Akamatsu T, Tamai Y, Izumi K, Tsuji A, Matsuda Y (1995) Distribution of the kexin family proteases in pancreatic islets: PACE4 is specifically expressed in B cells of pancreatic islets. Endocrinology 136:357–360

Nakayama K, Kim W-S, Torii S, Hosaka M, Nakagawa T, Ikemizu J, Baba T, Murakami K (1992) Identification of the fourth member of the mammalian endoprotease family homologous to the yeast Kex2 protease. J Biol Chem 267:5897–5900

Nakagawa T, Hosaka M, Torii S, Watanabe T, Murakami K, Nakayama K (1993) Identification and functional expression of a new member of the mammalian kex2-like processing endopeptidase family: its striking similarity to PACE4. Biochem J 113:132–135

Novak U, Wilks A, Buell G, McEwen S (1987) Identical mRNA for preproglucagon in pancreas and gut. Eur J Biochem 164:553–558

Orci L, Bordi C, Unger RH, Perrelet A (1983) Glucagon- and glicentin-producing cells. In: Lefèbvre PJ (ed) Glucagon I. Springer, Berlin Heidelberg New York, pp 57–79 (Handbook of experimental pharmacology, vol 66/I)

Patzelt C, Schiltz E (1984) Conversion of proglucagon in pancreatic alpha cells: the major endproducts are glucagon and a single peptide, the major proglucagon fragment that contains two glucagon-like sequences. Proc Natl Acad Sci USA 81:5007–5011

Pavoine C, Brechler V, Kervran A, Blache P, Le-Nguyen D, Laurent S, Bataille D, Pecker F (1991) Miniglucagon "glucagon-(19-29)" is a component of the positive inotropic effect of glucagon. Am J Physiol 260:C993–C999

Philippe J (1991) Structure and pancreatic expression of the insulin and glucagon genes. Endocr Rev 12:252–271

Pierotti AR, Prat A, Chesneau V, Gaudoux F, Leseney AM, Foulon T, Cohen P (1994) N-Arginine dibasic convertase, a metalloendopeptidase as a prototype of a class of processing enzymes. Proc Natl Acad Sci USA 91:6078–6082

Powers AC, Efrat S, Mojsov S, Spector D, Habener JF, Hanahan D (1990) Proglucagon processing similar to normal islets in pancreatic α-like cell line from transgenic mouse tumor. Diabetes 39:406–414

Ravazzola M, Siperstein S, Moody AJ, Sundby F, Jacobsen H, Orci L (1979) Glicentin-immunoreactive cells and their relationship to glucagon-producing cells. Endocrinology 105:499–508

Rehemtulla A, Barr PJ, Rhodes CJ, Kaufman RJ (1993) PACE4 is a member of the mammalian propeptidase family that has overlapping but not identical substrate specificity to PACE. Biochemistry 32:11586–11590

Rholam M, Nicolas P, Cohen P (1986) Precursors for peptide hormones share common secondary structures forming features at the proteolytic processing sites. FEBS Lett 207:1–6

Roebroek AJM, Schalken JA, Leunissen JAM, Onnekink C, Bloemers HPJ, Van de Ven WJM (1986) Evolutionary conserved close linkage of the c-fes/fps proto-oncogene and genetic sequences encoding a receptor-like protein. EMBO J 5:2197–2202

Rothenberg ME, Eilertson CD, Zhou Y, Lindberg, I, McDonald JK, Noe BD (1995a) In vitro processing of mouse proglucagon by recombinant PC1 and PC2. J Cell Biochem 19B [Suppl]:248

Rothenberg ME, Eilertson CD, Klein K, Zhou Y, Lindberg I, McDonald JK, Mackin RB, Noe BD (1995b) Processing of mouse proglucagon by recombinant prohormone convertase I and immunopurified prohormone convertase 2 in vitro. J Biol Chem 270:10136–10146

Rouillé Y, Steiner DF (1995) Proglucagon is processed to glucagon-like peptide 1 in AtT-20 cells. J Cell Biochem 19B [Suppl]:248

Rouillé Y, Westermark G, Martin SK, Steiner DF (1994) Proglucagon is processed to glucagon by prohormone convertase PC2 in alpha TC1–6 cells. Proc Natl Acad Sci USA 91:3242–3246

Seidah N, Gaspar L, Mion P, Marcinkiewicz M, Mbikay M, Chrétien M (1990) cDNA sequence of two distinct pituitary proteins homologous to Kex2 and furin gene products: tissue-specific mRNAs encoding candidates for pro-hormone processing proteinases. DNA Cell Biol 9:415–424

Seidah N, Marcinkiewicz M, Benjannet S, Gaspar L, Beaubien G, Mattei MG, Lazure C, Mbikay M, Chrétien M (1991) Cloning and primary sequence of a mouse candidate prohormone convertase PC1 homologous to PC2, furin and Kex2: distinct chromosomal localization and messenger RNA distribution in brain and pituitary compared to PC2. Mol Endocrinol 5:111–122

Seidah NG, Chrétien M (1994) Pro-protein convertases of subtilisin/kexin family. Methods Enzymol 244:175–188

Smeekens SP, Steiner DF (1990) Identification of a human insulinoma cDNA encoding a novel mammalian protein structurally related to the yeast dibasic processing protease Kex2. J Biol Chem 265:2997–3000

Smeekens SP, Avruch AS, LaMendola J, Chan SJ, Steiner DF (1991) Identification of a cDNA encoding a second putative prohormone convertase related to PC2 in AtT20 cells and islets of Langerhans. Proc Natl Acad Sci USA 88:340–344

Smeekens SP, Montag AG, Thomas G, Albiges-Rizo G, Carroll R, Benig M, Phillips LA, Martin S, Ohagi S, Gardner P, Swift HH, Steiner DF (1992) Proinsulin processing by the subtilisin-related proprotein convertases PC2 and PC3. Proc Natl Acad Sci USA 89:8822–8826

Steiner DF, Cunningham D, Spiegelman L, Aten B (1967) Insulin biosynthesis: evidence for a precursor. Science 157:697–700

Thim L, Moody AJ (1981) The primary structure of porcine glicentin (proglucagon). Regul Pept 2:139–151

Tricoli JV, Bell GI, Shows TB (1984) The human glucagon gene is located on chromosome 2. Diabetes 33:200–202

Unger RH, Ketterer H, Dupré J, Eisentraut AM (1966) Distribution of immunoassayable glucagon in gastrointestinal tissues. Metabolism 15:865–867

Valverde I, Rigopoulo D, Marco J, Faloona GR, Unger RH (1970) Characterization of glucagon-like immunoreactivity. Diabetes 19:614–623

Van den Ouweland AMW, Van Duijinhoven HLP, Keizer GD, Dorssers LCJ, Van de Ven WJM (1990) Structural homology between the human fur gene and the subtilisin-like protease encoded by yeast KEX2. Nucleic Acids Res 18:664

Weir GC (1995) Glucagon-like peptide-1 (GLP-1): a piece of the incretin puzzle. J Clin Invest 95:1
White JW, Saunders GF (1986) Structure of the human glucagon gene. Nucleic Acid Res 14:4719–4730
Yanaihara C, Matsumoto T, Kadowaki M, Iguchi K, Yanaihara N (1985) Rat pancreas contains the proglucagon (64-69) fragment and arginine stimulates its release. FEBS Lett 187:307–310

CHAPTER 4
The Glucagon Receptor Gene: Organization and Tissue Distribution

V. IWANIJ

A. Introduction

The first step by which peptide hormones can initiate a cellular response is through the binding of hormone to specific receptor sites on the surface of target cells. The germane studies of RODBELL and coworkers (1971a,b) established that the glucagon receptor is involved in the activation of adenylyl cyclase, and this regulation is achieved through well-defined guanosine triphosphate (GTP)-binding heterotrimeric G-protein complex (see Chap. 5, this volume). RODBELL and coworkers have also established a reliable methodology to study glucagon-binding sites in target tissues. WAKELAM et al. (1986) have demonstrated that in addition to stimulation of adenylyl cyclase activity glucagon causes breakdown of phosphatidylinositol-4,5-bisphosphate. The report presented one of the first indications that a single peptide hormone may be capable of activation of two separate second messenger systems within the cell. Significant progress in the structural characterization of the glucagon receptor has been made due to the application of affinity-labeling approaches (JOHNSON et al. 1981; DEMOLIOU-MASON and EPAND 1982; IYENGAR and HERBERG 1984; IWANIJ and HUR 1985). This technique represents a useful tool in the gathering of information about minor components of plasma membrane, such as hormone receptors. Briefly the affinity-labeling approach allowed identification of the hepatic glucagon receptor as a 62-kDa polypeptide that contained four N-linked oligosaccharide chains (IYENGAR and HERBERG 1984; IWANIJ and HUR 1985) and intramolecular disulfide bonds (IYENGAR and HERBERG 1984; IWANIJ and HUR 1985). Detailed biochemical analysis of the glucagon receptor has been limited due to the scarcity of these receptors, the difficulty of solubilizing them from the membrane in their active form, and difficulties in purification. The success in cloning the glucagon receptor cDNA has provided new opportunities for the investigation of the structure and function of the glucagon receptor.

B. Cloning of the Glucagon Receptor

The development of expression cloning has been instrumental in the isolation of glucagon receptor cDNA. The cloning of glucagon receptor (GR) cDNA was simultaneously achieved by two groups, the Zymogenetics group of

Kindsvogel (JELINEK et al. 1993) and the Brussels group of SVOBODA and CHRISTOPHE (SVOBODA et al. 1993a,b). The Zymogenetics group has prepared an expression library from rat liver mRNA and has carried out transfections into the BHK cell line, which does not express endogenous glucagon receptor. Cells expressing glucagon receptor were identified subsequently by ^{125}I-glucagon binding to the tissue culture plates visualized by autoradiography. After several rounds of selection, a single cDNA was isolated. Transfection of this cDNA into BHK cells resulted in the ability of these cells to bind radiolabeled glucagon and increase intracellular cAMP levels or Ca^{2+} levels in response to glucagon. These functional assays supported the conclusion that the isolated cDNA indeed encoded the glucagon receptor. Furthermore, this experiment provided evidence that the glucagon receptor encoded by this cDNA was capable of elevating two different second messenger pathways within the same cell. The SVOBODA-CHRISTOPHE group used a separate approach in cloning the glucagon receptor. Reasoning that structural similarities exist among glucagon and the glucagon-like peptide receptors and secretin and calcitonin/PTH receptors, SVOBODA et al. (1993a) designed oligonucleotide primers derived from the reported sequence of the glucagon-like peptide receptor and secretin receptors. Using polymerase chain reaction (PCR) technology with synthetic oligoprimers, they have isolated a portion of the glucagon receptor gene. The sequence analysis indicated that the new gene encoded a receptor molecule that showed homology with the newly identified family of the secretin/calcitonin receptors. The expression studies that followed established that the new gene encoded the glucagon receptor (SVOBODA et al. 1993b). An analysis of the amino acid sequence indicated an open reading frame of 1455 nucleotides that encoded a protein of 485 amino acids (Fig. 1). A KYTE and DOOLITTLE (1982) analysis of this amino acid sequence indicated the presence of seven stretches of hydrophobic amino acids consistent with the seven transmembrane domains proposed for the structural model of a G-protein-linked receptor. The analysis also indicated the presence of a signal peptide of 19 amino acids that should be cleaved during maturation of the protein. The N-terminal domain is about 120 amino acids long, being of intermediate size relative to the short N-terminal of the β-adrenergic receptor group and the long N-terminal of the thyroid-stimulating hormone/luteinizing hormone (TSH/LH) receptor group. The N-terminal domain contains six cysteine residues and four N-linked glycosylation sites. This structural feature coincides with the biochemical characterization of affinity-labeled glucagon receptor, which indicated the presence of four N-linked oligosaccharides (IYENGAR and HERBERG 1984; IWANIJ and HUR 1985). The third intracellular loop, the proposed site of interaction with G proteins, is short and contains a C-terminal R-L-A-R sequence, a motif that was shown to be required for G-protein activation (OKAMOTO et al. 1991). The C-terminal is 80 amino acids long and contains Ser^{432}, which presents a potential site for kinase C activity. The C-terminal also contains three cysteine residues. These are of importance as they may be possible sites of modification by fatty acids as was observed in

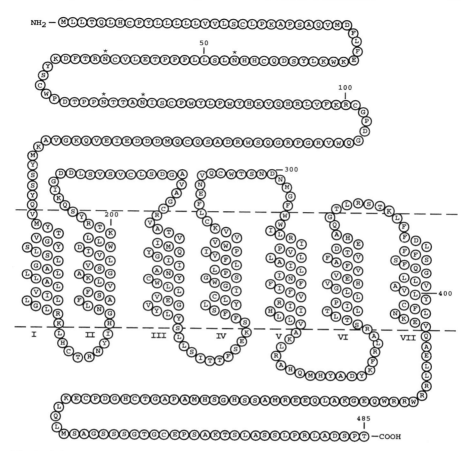

Fig. 1. Schematic representation of the rat glucagon receptor amino acid sequence and proposed topology of the seven transmembrane helices. The amino-terminal region and extracellular loops are oriented *toward the top*, while the carboxyl terminus and cytoplasmic loops are oriented *toward the bottom of the figure*. The four sites for N-linked glycosylation on the amino terminus are *labeled with asterisks*

the α- and β-adrenergic receptors (KENNEDY and LIMBIRD 1993; O'DOWD et al. 1989).

The human glucagon receptor cloned subsequently (LOK et al. 1994; MACNEIL et al. 1994) was shown to contain 477 amino acids with an 82% identity to the rat glucagon receptor (Fig. 2). The N-terminal domain and the transmembrane domains of the rat and human glucagon receptors showed the highest degree of similarity (97% and 84%, respectively), while the C-terminal regions showed the lowest degree of similarity. As glucagon belongs to the separate group of gut peptides that share structural similarities [glucagon/secretin vasoactive intestinal polypeptide (VIP)], the glucagon receptor also belongs to a separate subfamily of the G-protein-coupled hormone receptors

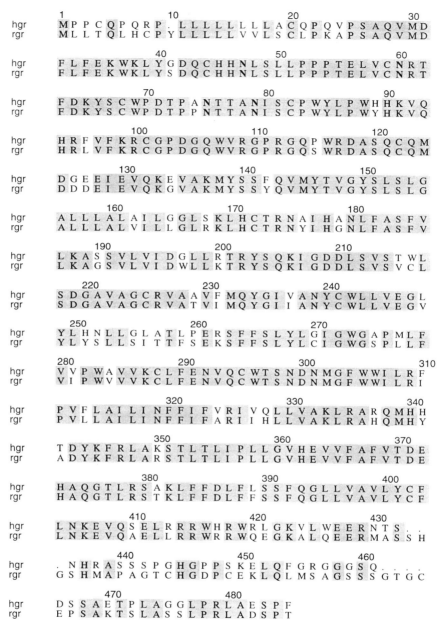

Fig. 2. Deduced amino acid sequence of the rat and human glucagon receptor cDNAs. Identical residues are *blocked in gray*. Sites for the N-linked glycosylation (asparagine residues) are *marked with bold letters*

(or serpentine receptors). To date this subfamily of receptors includes receptors for the ligands related to glucagon such as secretin (ISHIHARA et al. 1991), glucagon-like peptide-1 (GLP-1) (THORENS 1992), VIP-pituitary adenylate cyclase activating protein (PACAP) (ISHIHARA et al. 1992), and growth hormone releasing factor (GRF) (MAYO 1992). In addition, receptors for calcitonin (LIN et al. 1991), parathyroid hormone (JÜPPNER et al. 1991), corticotrophin-releasing factor (VITA et al. 1993), and invertebrate receptor for diuretic hormone (REAGAN 1994) all belong to the glucagon/secretin subfamily of the serpentine receptors. The sequences of these receptors shows a significant similarity between themselves but shows low (less than 20%) similarity with the β-adrenergic subgroup of receptors or the TSH/LH group of receptors. A comparison of multiple alignments (Fig. 3) indicates a great deal of structural similarity among these receptors. The order of sequence alignment shown in Fig. 3 indicates that GLP-1 receptor is most similar in sequence to the glucagon receptor while the calcitonin receptor is the least similar. Overall, the greatest degree of similarity could be observed within the transmembrane domains (TM), where the sequence is 40%–60% identical, with TM domains 1, 3, 4, and 7 being the most similar. A high degree of homology among the transmembrane domains has been observed in the case of other subgroups of serpentine receptors, thus indicating the importance of the TM domains in the maintenance of the structure and the bilayer arrangement of these receptors, and is required for each subfamily of receptors. It has been observed that the primary sequence in the TM domains of serpentine receptors ranges from 85%–95% identity for species homologues of a given receptor to 60%–80% identity for related subtypes of that same receptor, to 35%–45% identity for other members of the same family. Unrelated G-protein-coupled receptors show 20% or less sequence similarity.

The N-terminal domains of the glucagon receptor subfamily of receptors, although of similar size with respect to the number of amino acids (120–140), show 25% or less identity in amino acid sequence. Nevertheless, several amino acid residues are located in the same conserved position among these receptors. For example, the cysteine residues are located in conserved positions in all of the receptors of this subgroup (see Fig. 3), and may indicate that three-dimensional folding of the N-terminal is maintained by the intermolecular disulfide bridges. Furthermore, the aspartic acid residue (Asp^{64} in rat GR and Asp^{63} in human GR, highlighted in Fig. 3) found in the same position in all the receptors of this subgroup was shown to be necessary for the hormone binding of the growth hormone releasing factor (GHRF) receptor (LIN et al. 1993; GODFREY et al. 1993) and of glucagon receptor (CARRUTHERS et al. 1994), respectively. These data again indicate that the basic features of the N-terminal domain that may be needed to accommodate interaction with a ligand are present in all members of this subfamily of receptors despite the low overall similarity at an amino acid level.

The third intracellular loop, which has been shown to link serpentine receptors to G proteins, is relatively short, consisting of 28–30 amino acids.

```
         1         10        20        30        40        50
rgr      .MLLTQLHCPYLLLLVVLSCLPKAPSAQV.MDFLFEKWKLYSDQC..HH
rglp-1r  MAVTPSLLRLALLLLGAVGRAGPRPQGATVSLSETVQKWREYRHQC..QR
rsr      .MLSTMRPRLSLLLLRLLL...........LTKAAHTVGVPPRLCDVRR
rvipr    ....MRASVVLTCYCWLL......V......RVSSIHPECRFHL
rphr     MGAARIAPSLALLLCCPVLSSAYALVDADDVFTKEEQIFLLHRAQAQCDK
rghrfr   ...................MDSLLWATWVLCLLNLWGVALGHLHLECDFIT
rcr      ...MRFLLLNRFTLLLLLLVSPTPVLQAPTNLTDSGLDQEPFLYLVG.RK

                   60        70        80        90        100
rgr      NLSLLP.PPTELV.....................................
rglp-1r  FLTEAPLLATGLF.....................................
rsr      VLLEERAHCLQQLSKEK.................KGALGPET........
rvipr    EIQEEETKCAELLSSQM.......................EN........
rphr     LLKEVLHTAANIMESDKGWTPASTSGKPRKEKASGKFYPESKENKDVPTG
rghrfr   QLRDDELACLQAAEGTNNS...............................
rcr      KLLDAQYKCYDRIQQLPPYEGE............................

                   110       120       130       140       150
rgr      .......CNRTFDKYSCWPDTPPNTTANISCPWYLPWYHKVQHRLVFKRC
rglp-1r  .......CNRTFDDYACWPDGPPGSFVNVSCPWYLPWASSVLQGHVYRFC
rsr      ....ASGCEGLWDNMSCWPSSAPARTVEVQCPKFL.LMLSNKNGSLFRNC
rvipr    ....HRACSGVWDNITCWRPADIGETVTVPCPKVF.SNFYSRPGNISKNC
rphr     SRRRGRPCLPEWDNIVCWPLGAPGEVVAVPCPDYI.YDF.NHKGHAYRRC
rghrfr   ....SMGCPGTWDGLLCWPPTGSGQWVSLPCPEFF.SHFGSDPGAVKRDC
rcr      ....GPYCNRTWDGWMCWDDTPAGVMSYQHCPDYFPDF..DPTEKVSKYC

                   160       170       180       190       200
rgr      GPDGQWV.RGPRGQSWRDASQCQMDDDEIEVQKGVAKMYSSYQVMYTVGY
rglp-1r  TAEGIWLHKDNSSLPWRDLSECE.ESKQGERNSPEEQLLSLY.IIYTVGY
rsr      TQDGWSETFP.....RPDLACGVNINNSFNERRHA..YLLKLKVMYTVGY
rvipr    TSDGWSETF.......PDFIDACGYNDPEDESKIT..FYILVKAIYTLGY
rphr     DRNGSWEVVPGHNRTWANYSECLKFMTN.ETRERE..VFDRLGMIYTVGY
rghrfr   TITGWSDPFPPY.........PVACPVPLELLTEEKSYFSTVKIIYTTGH
rcr      DENGEWFRHPDSNRTWSNYTLCNAFTPDKLHNAYVSYYLAL......VGH

                   210       220       230       240       250
rgr      SLSLGALLLALVILLGLRKLHCTRNYIHGNLFASFVLKAGSVLVIDWLLK
rglp-1r  ALSFSALVIASAILVSFRHLHCTRNYIHLNLFASFILRALSVFIKDAALK
rsr      SSSLAMLLVALSILCSFRRLHCTRNYIHMHLFVSFILRALSNFIKDAVLF
rvipr    SVSLMSLTTGSIIICLFRKLHCTRNYIHLNLFLSFMLRAISVLVKDSVLY
rphr     SMSLASLTVAVLILAYFRRLHCTRNYIHMHMFLSFMLRAASIFVKDAVLY
rghrfr   SISIVALCVAIAILVALRRLHCPRNYIHTQLFATFILKASAVFLKDAAVF
rcr      SMSIAALIASMGIFLFFKNLSCQRVTLHKNMFLTYILNSIIIIIHLVEVV

                   260       270       280       290       300
rgr      TRYSQKIGDDLSVSVCLSDGAV...........AGCRVATVIMQYGIIAN
rglp-1r  WMYSTAAQQHQWDGL.LSYQDS...........LGCRLVFLLMQYCVAAN
rsr      SSDDVTYCD..............AH.....KVGCKLVMIFFQYCIMAN
rvipr    SSSGTLRCH...............DQPGSWVGCKLSLVFFQYCIMAN
rphr     SGFTLDEAERLTEEELHIIAQVPPPPAAAAVGYAGCRVAVTFFLYFLATN
rghrfr   QGDSTDHC.................SMSTILCKVSVAVSHFATMTN
rcr      PNGDLVRRDP................ISCKILHFFHQYMMACN
```

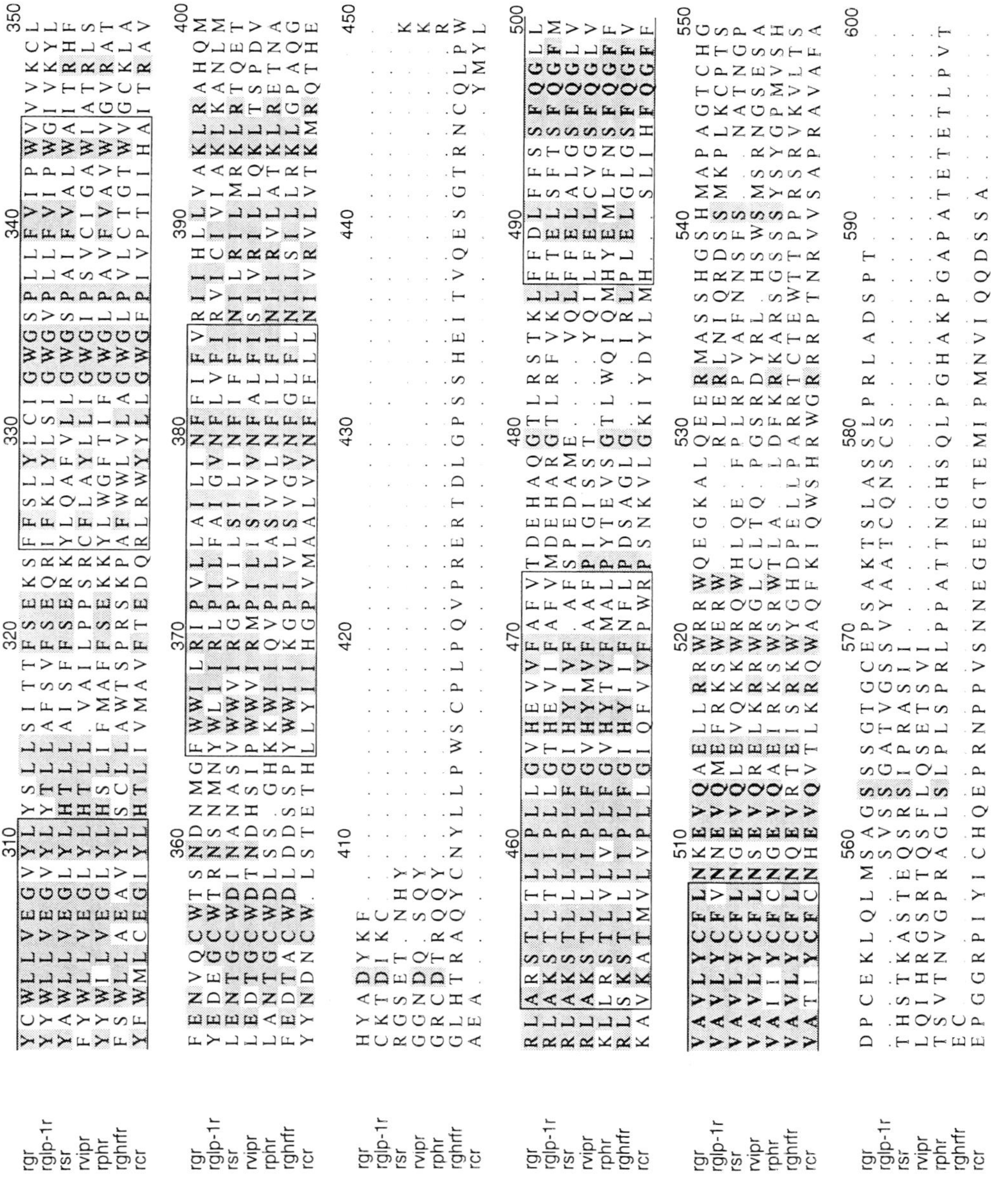

Fig. 3. Alignment of the amino acid sequence of the rat glucagon receptor (*rgr*), with rat receptors for GLP-1 (*rglp-1r*), secretin (*rsr*), VIP (*rvipr*), PTH (*rphr*), GHRF (*rghrfr*), and calcitonin (*rcr*). Identical residues are *blocked in gray*. Sequences for the putative transmembrane domains are *blocked*, conserved cysteine residues and conserved aspartic acid within N-terminal domain are *marked with bold letters*; the C-terminal consensus sequence is *also marked with bold letters*

The loop contains two sequence motifs: the K-L-R/K sequence close to the N-terminal portion of the loop, and the R-L-A-R/K-S-T-L sequence on the C-terminal portion of the loop present in all receptors of this subfamily. This R-L-A-R/K-S-T-L consensus sequence matches the K-A-L-K motif present at the same location in the β_2-adrenergic receptor that was shown to be important for G-protein coupling (OKAMOTO et al. 1991).

All of the receptors within the glucagon/secretin subfamily of receptors contain a highly conserved region spanning the C-terminal portion of the seventh putative transmembrane domain and the beginning of the C-terminal domain. This region was proposed to form a specific sequence motif for this receptor subfamily (LOK et al. 1994). This pattern, F-Q-G-Hydr-Hydr-V-A-X-Hydr-Y-C-F-X-E-V-Q, selectively detects all sequences in the subfamily (see Fig. 3). The C-terminal domains of the GR subfamily are relatively long (about 75–80 amino acids) and, except for the region adjacent to the seventh transmembrane domain, show a low level of amino acid sequence similarity.

C. Organization of the Glucagon Receptor Gene

A Southern analysis of human and rat genomic DNA indicates that the GR gene is a single copy gene (JELINEK et al. 1993; LOK et al. 1994). The genomic clones, isolated and sequenced from human and rat genomic libraries, respectively, and the complete gene sequence encoding the glucagon receptor from these two species is now available (LOK et al. 1994; MAGET et al. 1994). Sequence analysis of the rat genomic DNA indicates that the rat GR gene spans about 4 kb and contains 12 exons and 11 introns interrupting the coding sequence (MAGET et al. 1994), while the human gene spans approximately 5.5 kb and is interrupted by 12 introns (LOK et al. 1994). All splice junctions contain the consensus acceptor/donor sequences as described by MOUNT (1982), and the overall location of the introns within the human and the rat genes is very similar; four introns are present within the coding area of N-terminal domain including the signal peptide (see Fig. 4). In the rat gene, separate exons encode the first, second, third, and parts of the fourth, sixth, and seventh transmembrane domains. The human gene was reported to contain an additional intron that interrupts the seventh transmembrane domain. In both of these genes, regions encoding the first and second extracellular loops are interrupted by introns. Also, the first and third intracellular loops, and the beginning of the C-terminal domain, are interrupted by introns.

Mapping of the untranslated 5'-end of the human gene revealed the presence of an additional intron within this region (BUGGY et al. 1995a). It is possible that further analysis of the rat genomic sequence may reveal the presence of an intron of similar location in the 5'-end of the gene.

The organization of the gene with respect to intron/exon boundaries is of particular interest as it allows the evolutionary relatedness of the genes belonging to a particular family to be inferred. Of the receptors in the glucagon/secretin receptor subfamily, the genomic structure of the parathyroid hor-

The Glucagon Receptor Gene: Organization and Tissue Distribution

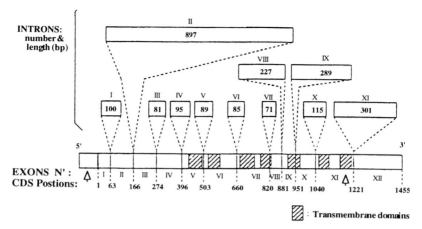

Fig. 4. Organization of the rat glucagon receptor gene (redrawn after MAGET et al. 1994). Introns are represented by *open boxes and numbered with roman numerals*. Two *arrowheads* represent approximate positions of two additional introns identified within the human glucagon receptor gene. Exon numbers are listed *on the bottom of the coding region model*. Coding sequence starts (*CDS*) with ATG where A is #1 nucleotide. Putative transmembrane domains are *marked with hatched boxes*

mone receptor (McCUAIG et al. 1994) and the growth hormone-releasing factor receptor (LIN et al. 1993) have been reported. In both cases the coding sequences of these receptors are interrupted by 12 introns at a position similar to the human glucagon receptor, suggesting that these genes diverged from a common ancestor. In contrast, genes encoding the serpentine receptors for other types of ligands vary significantly in their organization. For example, genes encoding the LH/FSH subfamily of receptors were found to have 9–10 introns interrupting the coding region for the large N-terminal domain, while the region encoding the seven transmembrane domains of the receptor is located within a single exon (SEGALOFF and ASCOLI 1993). At the other end of the range, the β-adrenergic receptor subfamily of genes is usually devoid of any introns (DIXON et al. 1986).

D. Tissue Distribution of the Glucagon Receptors

I. Adipose Tissue Glucagon Receptors

Although the liver is recognized as the major target site for the glucagon action, several other tissues have been shown to express the glucagon receptors. Since the major known physiological role of glucagon is to mobilize stored nutrients during fasting, adipose tissue responds to glucagon by an increase of lipolysis with a concomitant release of glycerol and non-estrified fatty acids (see Chap. 7, this volume). BIRNBAUMER and RODBELL (1969) have shown that the biological effects of glucagon in adipose tissue result from an increase in adenylyl cyclase activity. Furthermore, in the rat, glucagon-binding sites have been characterized in dissociated adipocytes and isolated adipocyte plasma membranes (DESBUQUOIS and LAUDAT 1974; LIVINGSTON et al. 1974). The affinity cross-linking approach resulted in the characterization of the glucagon receptor complex as about 62 kDa, identical in size to the rat liver glucagon receptor (IWANIJ et al. 1994). This study also established that in rat, in addition to epididymal fat, brown fat expresses the glucagon receptor in agreement with the physiological results of KUROSHIMA and YAHATA (1979). Therefore glucagon may also contribute to the regulation of thermogenesis.

The presence of the glucagon receptors has been documented in human adipose tissue (MERIDA et al. 1993) and malignant liposarcoma (CORRANZA et al. 1993), supporting the evidence for the glucagon-dependent stimulation of adenylyl cyclase in normal human adipose tissue (see Chap. 7, this volume).

II. Kidney Glucagon Receptors

In the kidney, glucagon is believed to regulate fluid and electrolyte secretion (KATZ and LINDHEIMER 1977; AHLOULAY et al. 1992). A glucagon-dependent increase in the adenylyl cyclase activity in kidney membrane preparations was observed to be less pronounced, especially at lower hormone concentrations, than in liver cells (MELSON et al. 1970). These results were explained by BAILLY et al. (1980), who showed that, in the rat, nephron glucagon-responsive adenylyl cyclase is restricted to the distal and collecting tubules. Autoradiographic studies on the nephron using ^{125}I-glucagon have provided independent evidence for the presence of the glucagon receptors in these well-defined regions of the kidney tubules (BUTLEN and MOREL 1985; MOREL and BUTLEN 1990). The biological effects of glucagon in kidney result from an increase in the cAMP levels, although Ca^{2+} may act as an additional messenger since it was detected in the glucagon-treated MDCK cells (KURSTJENS et al. 1990). Biochemical characterization using the affinity cross-linking approach indicates that in rat kidney membranes the glucagon receptor has an identical size to the receptor found in liver. However, in the canine kidney both the glucagon receptor and the MDCK receptor are larger in size, as judged from their electrophoretic mobility in sodium dodecyl sulfate (SDS) gel, than the canine liver receptor, and may represent a unique form of the glucagon receptor (IWANIJ 1995).

III. Pancreatic Islets Glucagon Receptors

It is well established that glucagon potentiates glucose-induced insulin secretion in pancreatic B cells during fasting (SAMOLS et al. 1965; KAWAI et al. 1995). A biochemical study using insulinoma (GOLDFINE et al. 1972) as a model system for the islets provided evidence for glucagon-binding sites and the glucagon-dependent elevation of the cAMP levels in this system. In addition, binding of ^{125}I-glucagon and autoradiography of intact and cultured islets provided further evidence for the glucagon receptors present on the B cells and other islet cells (PATEL et al. 1982; AMHERDT et al. 1989). The sequence of the human glucagon receptor expressed in pancreatic islets was shown to be identical to the human liver glucagon receptor (LOK et al. 1994).

IV. Brain Glucagon Receptors

Based on their physiological experiments with glucagon, GEARY et al. (1993) proposed that glucagon may be involved in the regulation of satiety (see Chap. 14, this volume). They reported that, upon infusion of glucagon into the portal vein of rat, a neural sensory signal is transmitted by the hepatic branch of the abdominal vagus to the nucleus tractus solitarii in the brain, causing reduction of meal size. Whether glucagon affects the brain in a direct or indirect manner and the exact mechanism of evoking satiety remain to be determined. Nevertheless, hormone-binding analysis of membrane preparations from different regions of the rat brain does indicate the presence of the glucagon receptors therein (HOOSEIN and GURD 1984). Interestingly, hypothalamus was shown to express glucagon (DRUCKER and ASA 1988). In addition to brain glucagon, receptors were detected in retina (FERNANDEZ-DURANGO et al. 1990).

V. Heart Glucagon Receptors

It has been well established that glucagon exerts several actions on heart (LUCCHESI 1968). Glucagon induces positive ionotropic and chronotropic effects on ventricular muscle. Phosphorylase activity and glycogenolysis are stimulated by glucagon in perfused heart (LAUGHLIN et al. 1992). Also glucagon was shown to activate the L-type Ca^{2+} channel (MÉRY et al. 1990). Overall, as in the case of the glucagon receptors in other target tissues, receptors found in the heart are present at a significantly lower density than are the liver receptors. Despite the lower density of the receptors, glucagon is capable of generating a physiological response such as an increase in the strength and rate of contraction in vivo and in vitro (see Chap. 11, this volume).

VI. Intestinal Tract Glucagon Receptors

Glucagon exerts several actions on the gastrointestinal tract (see Chap. 12, this volume). Glucagon receptors on the intestinal smooth muscle cell were reported to regulate contraction/relaxation of the intestinal wall (SANTAMARIA and DE MIGUEL 1985). In addition, glucagon at physiological levels inhibits gastric acid secretion presumably indirectly via somatostatin-secreting cells (LOUD et al. 1988).

E. Tissue Distribution of Glucagon Receptor Transcripts

The cloning of the cDNA encoding the glucagon receptor has allowed the study of glucagon receptor transcripts in different tissues. Analysis of glucagon receptor transcripts in the human, rat, and mouse liver by RNA blot indicates the presence of a single band of about 2.3 kb (YOO-WARREN et al. 1994; CAMPOS et al. 1994). When an RNA blot was applied to visualize transcripts in tissues other than liver, only the kidney transcript of a similar size to liver was observed, and at a significantly lower intensity in agreement with the lower density of the glucagon receptors in this tissue. Other tissues, such as brain or heart, did not produce a positive band on an RNA blot, indicating that the amounts of transcript were too low for the sensitivity of this technique. In order to study the distribution of the glucagon receptor transcripts, other more sensitive approaches have had to be applied. SVOBODA et al. (1994) utilized the RNAse protection procedure that provides a quantitative approach and the reverse transcriptase (RT)-PCR procedure that is semiquantitative, while CAMPOS et al. (1994) predominantly used RT-PCR. SVOBODA et al. (1994) reported the presence of the glucagon receptor transcripts in liver, kidney, heart, adipose tissue, duodenum, stomach, and brain in ever-decreasing concentrations. CAMPOS et al. (1994), using a similar approach, screened several mouse tissues and reported the presence of glucagon receptor transcripts in liver, kidney, pancreas, jejunum, and ileum. Also, they studied several regions of the mouse brain and found expression of glucagon receptor in cortex, cerebellum, hypothalamus, and brain stem. Overall, tissue distribution of glucagon receptor transcripts obtained by these two laboratories is in agreement with previously described target tissues for glucagon action. In addition, SVOBODA reported the presence of glucagon receptors in rat adrenal gland. Lung was found to express low levels of the glucagon receptor in mouse; however, no transcripts were detected in the rat lung. This difference may reflect species differences or technique differences since PCR may detect levels of transcripts at the lower end of resolution.

The presence of several intervening sequences as revealed by the detailed structure of the glucagon receptor gene raises the possibility of alternative splicing that could generate various receptor molecules from the single copy gene. It is believed that alternative splicing may be a general physiological mechanism for modulation of the function of the receptor molecules. In the case of the somatostatin receptor, the longer splice form was found to be less efficient in coupling to the G_i and more susceptible to agonist-induced desensitization (VANETTI et al. 1993), while prostaglandin E receptors that differ in their C-terminal tails couple to different G proteins (NAMBA et al. 1993). Other examples of alternatively spliced G-protein-linked receptors include the PACAP receptors (SPENGLER et al. 1993), which are structurally related to the glucagon receptor subfamily, in which a variation in the length of the third intracellular loop results in an altered pattern of adenylyl cyclase and phospholipase C regulation.

The RT-PCR analysis by SVOBODA et al. (1994) indicated a polymorphism of the glucagon receptor mRNA since several size bands were generated by the same set of primers. Interestingly, the pattern of bands obtained as a result of RT-PCR was dependent on the origin of tissue. This observation may indicate that the glucagon receptor gene may undergo splicing in a tissue-specific manner. While studying glucagon receptor transcripts in liver, MAGET et al. (1994) detected a longer transcript that contained an extension within the N-terminal domain due to unspliced intron 3. The sequence of the gene also indicated that there could be an alternatively spliced intron present within the C-terminal portion of the receptor molecule. IWANIJ (1995) has described the presence of a larger than liver-sized glucagon receptor in canine kidney. This difference in size could not be explained by glycosylation; thus alternative splicing remains a possibility. Whether indeed alternatively spliced receptor molecules are being expressed in the tissue remains to be established.

F. Regulation of Glucagon Receptor Gene Expression

During embryonic development, the A cells of the islets of Langerhans differentiate first followed by the insulin-secreting, pancreatic polypeptide, and somatostatin-containing cells. GITTES and RUTTER (1992) showed that in embryonic mouse glucagon is expressed early in development, while the density of the glucagon receptors in embryonic liver is low (BLAZQUES et al. 1987). Prior to birth (about day 20 of gestation), cells show large deposits of stored glycogen, the level of glucagon receptors is low, and the glucagon-dependent increase in cAMP levels is low. Postnatally the receptor density increases, reaching adult levels within 2–3 weeks (BLAZQUES et al. 1987). The experiments carried out by CAMPOS et al. (1994) using RT-PCR analysis in fetal and postnatal mouse support the biochemical data acquired from the binding of radiolabeled hormone to liver membranes. These results indicated a six fold increase in the GR mRNA from fetal day 19 to postpartum day 14 in the mouse. Thus the GR gene belongs to the group of hepatic genes that are being expressed late in liver development.

Ontogenesis of the glucagon receptors has been studied in chick heart (IWANIJ and HUR 1987). This study noted that early in the development of the heart cardiac cells do not respond to hormones such as glucagon, epinephrine, and acetylcholine. The onset of responsiveness to the hormone correlates with the increase in glucagon receptor density and the concomitant appearance of G proteins that link the receptors to adenylyl cyclase (IWANIJ and HUR 1987).

CAMPOS et al. (1994) have followed the levels of glucagon receptor transcripts in fetal and adult pancreas and lung mouse tissue, detecting a small increase in the levels of transcripts in the 2-week postpartum period over that seen in the day 19 embryo. The availability of the GR gene will enable the developmentally regulated onset of GR gene transcription to be further studied.

The regulation of glucagon receptor mRNA was studied in cultured primary hepatocytes as a model system for regulation in normal hepatic tissue. ABRAHAMSEN et al. (1995) reported that elevated glucagon levels in tissue culture media decreased the glucagon receptor's mRNA in a dose-dependent manner. Also, inclusion of reagents that elevate intracellular cAMP levels, such as IBMX and forskolin, resulted in a decrease in the glucagon receptor's mRNA levels in cultured hepatocytes. The down-regulation of glucagon receptor levels upon stimulation of the cell with glucagon has been reported in liver (SANTOS and BLAZQUEZ 1982), primary hepatocytes (PREMONT and IYENGAR 1988), and the MDCK cell line (RICH et al. 1984). Clearly in parallel with the loss of the glucagon receptors from the cell surface, the levels of the GR transcript also decrease. This study indicates that cAMP serves as a negative regulator of the glucagon receptor mRNA levels and as a signal for downregulation of glucagon receptor number. Since primary function of glucagon is to regulate circulating glucose levels, ABRAHAMSEN et al. (1995) investigated the influence of glucose concentration in media on the levels of glucagon receptor's mRNA expressed by cultured hepatocytes. Unexpectedly, ABRAHAMSEN et al. (1995) observed an increase in the glucagon receptor mRNA levels in response to the high glucose concentration present in tissue culture. The physiological significance of this observation is not clear, but this glucose-dependent upregulation of the glucagon receptor mRNA levels merits further investigation.

In addition to the study of receptor regulation under normal physiological conditions, it is of importance to investigate the levels of glucagon receptor and its mRNA under pathological conditions. YOO-WARREN et al. (1994) studied glucagon receptor mRNA in genetically diabetic (*db/db*) mice. These animals show high circulating levels of insulin, hyperglycemia (COLEMAN and HUMMEL 1967; TUMAN and DOISY 1977), and a high level of gluconeogenesis (COLEMAN and HUMMEL 1967). YOO-WARREN et al. (1994) observed that there were no detectable changes in the levels of the glucagon receptor mRNA when diabetic mice were compared with their normal littermates. However, the number of glucagon receptors was elevated in the diabetic animals. Further studies in this important area will be required to assess the correlation between the pathological state and the regulation of the glucagon receptor gene expression.

Because glucagon and thus its receptor affects insulin secretion from the pancreatic islets, the regulation of GR gene expression in cultured islet cells was studied by ABRAHAMSEN and NISHIMURA (1995), who reported that treatment of the cells with IBMX and forskolin, which effectively increase cAMP levels in the cells, resulted in a three fold decrease in the level of GR mRNA, while there was no significant change in the levels of GLP-1 receptor mRNA. Treatment of the cultured cells with somatostatin, a hormone known to inhibit adenylyl cyclase and consequently to reduce intracellular cAMP levels, resulted in a detectable increase in GR mRNA. A decrease in the glucagon receptor mRNA was also induced by treatment of the cell cultures with dexamethasone. The mRNA for the GLP-1 receptor was similarly affected by inclusion of dexamethasone. Of interest is that the levels of GR mRNA were upregulated in the presence of elevated concentrations of glucose in the tissue culture media, while mRNA levels for GLP-1 were not so affected. Furthermore, the extent of this increase in GR mRNA was dose dependent, increasing with the glucose concentration. This study indicated that, at least with respect to cultured islets, cAMP and glucocorticoids such as dexamethasone are negative modulators of GR mRNA levels, while elevated glucose concentrations are positive modulators. Certainly, the expression of the GR gene is highly regulated, being controlled in a tissue and a developmentally specific manner. In addition, the GR gene is subject to regulation in response to hormone and metabolite level fluctuations. Overall, the regulatory mechanism that governs the expression of serpentine receptors is complex and not fully understood. Further work on the detailed analysis of the promoter region of the GR gene in conjunction with the reporter gene would be necessary to elucidate the mechanism of this gene expression. The sequencing of approximately 0.5 kb of the 5'-end of the human glucagon receptor gene has been reported by BUGGY et al. (1995a). Unlike many typical eukaryotic gene promoters, human GR gene lacks the conserved TATA and CAAT boxes usually positioned at 25–30 base pairs and 100 base pairs upstream, respectively, from the transcription start point. The 5'-end region, however, contains several GC-rich consensus sequences for the transcription factor Sp-1, which is often present in housekeeping genes. It is interesting to note that the 5'-end promoter of the parathyroid receptor was reported to lack TATA and CAAT boxes and include several GC-rich sequences (MCCUAIG et al. 1994), resembling the features of the glucagon receptor gene's promoter. One putative AP-2-binding site was found approximately 85 base pairs from the transcription initiation site. As AP-2 responds to both kinase A and kinase C, and is abundant in liver (a tissue where GR is primarily expressed), AP-2 may be involved in the regulation of GR gene expression. A putative liver-specific transcription factor binding site, LF-A1, was identified upstream from the AP-2-binding motif. The LF-A1 may be involved in the tissue-specific and time-dependent regulation of the GR gene on a transcriptional level.

G. Structure/Function Analysis of the Glucagon Receptor

The agonists that bind and activate G-protein-coupled receptors vary widely in size, from glycoprotein hormones to a single photon. Large glycoproteins such as LH or TSH bind to the externally located amino-terminal region of the receptor, while small agonists such as epinephrine or serotonin interact with the transmembrane domains of the receptor (SEGALOFF and ASCOLI 1993; STRADER et al. 1994). The precise regions of the receptor that interact with the intermediate-size bioactive peptides such as glucagon have not yet been defined. The size of the peptide (3.5 kDa) suggests multiple sites of receptor ligand interactions with the possible involvement of the N-terminal domain and extracellular loops.

Construction of chimeric and point-mutated receptor proteins has provided insight into the structure-function relationship within the other types of serpentine receptors (STRADER et al. 1994). As a first step toward elucidating the role of the N-terminal domain in the binding of the glucagon receptor site, directed mutagenesis has been carried out by CARRUTHERS et al. (1994). Conserved Asp^{64} residue within the N-terminal domain was changed to Glu, Asn, Lys, or Gly. In each case the replacement of Asp^{64} residue resulted in a failure to bind glucagon. Since a similar loss of hormone binding was observed for the GHRF receptor with a mutation of Asp^{60} to Gly (LIN et al. 1993; GODFREY et al. 1993), the data are consistent with the interpretation that Asp substitution results in a change of protein domain conformation rather than the identification of an amino acid residue that is responsible for the specific peptide binding. A recently described missense mutation in the N-terminal region of human glucagon receptor, with Gly^{40} to Ser substitution, resulted in a decrease in binding affinity and a concomitant decrease in target tissue sensitivity to glucagon (HAGER et al. 1995).

BUGGY et al. (1995) carried out a detailed study of glucagon binding by several chimeras constructed between the glucagon receptor and the GLP-1 receptor. These two receptors are most closely related (66% similarity and 47% amino acid identity) in the members of the glucagon/secretin subfamily of receptors; nevertheless, they show a high level of selective peptide recognition. This study demonstrated that the region of the N-terminal between Ser^{80} and Gln^{142} is crucial for glucagon-binding specificity. Among the extracellular loops, the first and third were found to be required for glucagon binding. In addition, the third, fourth, and sixth transmembrane domains also contributed to the specificity of glucagon binding. A biochemical approach that utilized affinity cross-linking of radiolabeled glucagon to its receptor followed by a digest with proteolytic enzymes (IYENGAR and HERBERG 1984; IWANIJ and VINCENT 1990) has identified a 20 to 25-kDa fragment containing the covalently attached glucagon, but devoid of the carbohydrate moieties indicative of the N-terminal domain. The presence of the cross-linked glucagon to the carboxyl-terminal two-thirds of the receptor reveals the direct contact region of the receptor and its ligand therein. Overall, the biochemical data coincide well with the results of the receptor chimera approach outlined by BUGGY et al. (1995).

The chimera study of BUGGY et al. (1995) focused on an analysis of the ability of various chimeric glucagon/GLP-1 receptors to bind glucagon. Clearly, a similar set of chimeric receptors would be useful in the study of regions of molecule required for the signaling in response to glucagon binding.

STROOP et al. (1995) have investigated ligand-binding properties of chimeric receptors constructed between the glucagon and calcitonin receptors. Since these two receptors show a significantly lower level of similarity, most constructed chimeras showed little capacity to bind glucagon. Nevertheless, the results of this study support the conclusion that in addition to the N-terminal region of the glucagon receptor other regions of the receptor are required for hormone binding. Clearly, the investigation of the structure-function relationship of the glucagon receptor is in its initial stages. Nevertheless, the outline of a general model for the glucagon-binding domains can be generated. The N-terminal domain most probably generates a "fold" or a "pocket" for the ligand, promoting and stabilizing the interaction of the peptide with other regions of the receptor such as first and third extracellular loops. The Asp^{64} residue is implicated in the maintenance of this region's conformation. The receptor regions located within the proximal portion of the N-terminal and within the extracellular loops may constitute sites of specific ligand receptor interaction. Of special interest would be the identification of the contact point between the N-terminal His of glucagon and the receptor. Investigations by BREGMAN et al. (1980) and UNSON et al. (1991, 1993) indicate that glucagon analogs without the N-terminal His (des-His^1-glucagon) or with modified His^1 are capable of binding to the receptor but do not activate adenylyl cyclase. Therefore, glucagon interaction with receptor may occur in two distinct steps, fitting the ligand into binding space, followed by the subsequent activation upon His^1 interaction with specific residue(s) of the receptor (see also Chap. 8, this volume).

H. Human Glucagon Receptor

The implication that elevated glucagon levels (UNGER and ORCI 1981; see review in Chap. 7, this volume) in diabetic patients contribute to elevated glucose levels sparked an interest in the study of the glucagon receptor as a potential focus for the development of new pharmacological agents that could serve as a glucagon antagonist (see Chap. 8, this volume). Therefore, shortly after the isolation of the rat glucagon receptor cDNA, the human glucagon receptor sequence was reported by two laboratories (LOK et al. 1994; MACNEIL et al. 1994). The sequence of the human glucagon receptor is 80% identical to the rat glucagon receptor (see Fig. 2). The human receptor contains 13 exons spanning over 5.5 kilobases (LOK et al. 1994) and is located on chromosome 17q25 (LOK et al. 1994; MENZEL et al. 1994).

HAGER et al. (1995) have found an association between a missense mutation in the glucagon gene and late-onset non-insulin-dependent diabetes mellitus (NIDDM) in French and Sardinian patients. Sequence analysis established that the mutant form of the glucagon receptor has a Gly^{40} to Ser substitution, and this receptor binds glucagon with a threefold lower affinity than the wild-type human receptor. The authors propose that the decrease in the glucagon receptor ability to signal may affect glucagon-dependent insulin secretion, and consequently lead to an elevation of blood glucose levels. Alternatively, the glucagon receptor mutation might be in linkage disequilibrium with another mutational event within or near the glucagon receptor gene, as another polymorphism found in intron 8 of the receptor cosegregates with the Gly^{40}Ser mutation in all individuals tested. Overall these data describe a possible new and useful marker for NIDDM, and provide the first correlation between a mutated glucagon receptor and late-onset diabetes mellitus.

I. Conclusions

Cloning the glucagon receptor gene has provided an important means for studies of the receptor's structural features and the physiological role of its hormone. Future research will focus on several major issues: structure and function of the receptor will be analyzed further using site-directed mutagenesis and the chimera construction approach. This would be necessary for the precise mapping of the ligand-binding site and the elucidation of ligand-induced conformational changes of the receptor. The establishment of requirements for glucagon binding to the receptor will open possibilities for designing effective analogs that would be of therapeutic value (see Chap. 8, this volume). Although intensively investigated, the question of how these receptors activate G proteins remains unsolved. Uncovering the mechanism of the receptor-dependent G protein activation remains a major goal in this field.

Furthermore, the question of regulation of glucagon receptor gene expression in a tissue-specific and a developmental stage-dependent manner needs to be addressed, such as by using promoter/reporter gene analysis in cell culture and in transgenic animals. The application of homologous recombination for generating glucagon receptor "knockout" mouse may provide additional ways to assess this gene's function.

Acknowledgments. Our research has been supported by funds from NIH, American Diabetes Association, American Heart Association, and University of Minnesota Graduate School. I thank Kurt LaBresh for help in preparation of figures.

References

Abrahamsen N, Nishimura E (1995) Regulation of glucagon and glucagon-like peptide-1 receptor messenger ribonucleic acid expression in cultured rat pancreatic islets by glucose, cyclic adenosine 3′,5′-monophosphate, and glucocorticoids. Endocrinology 136:1572–1578

Abrahamsen N, Lungreen K, Nishimura E (1995) Regulation of glucagon receptor mRNA in cultured primary rat hepatocytes by glucose and cAMP. J Biol Chem 270:15853–15857

Ahloulay M, Bouby N, Machet F, Kubrusly M, Coutaud C, Bankir L (1992) Effects of glucagon on glomerular filtration rate and urea and water excretion. Am J Physiol 263:F24–F36

Amherdt M, Patel Y, Orci L (1989) Binding and internalization of somatostatin, insulin, and glucagon by cultured rat islet cells. J Clin Invest 84:412–417

Bailly C, Imbert-Teboul M, Chabardes D, Hus-Citharel A, Montegut M, Clique A, Morel F (1980) The distal nephron of rat kidney: a target site for glucagon. Proc Natl Acad Sci USA 77:3422–3424

Birnbaumer L, Rodbell M (1969) Adenyl cyclase in fat cells. II. Hormone receptors. J Biol Chem 244:3477–3482

Blazques E, Perez Castillo A, de Diego JG (1987) Characterization of glucagon receptors in liver membranes and isolated hepatocytes during rat ontogenic development. Mol Cell Endocrinol 49:149–157

Bregman MD, Trivedi D, Hruby VJ (1980) Glucagon amino group, evaluation of modifications leading to antagonism and agonism. J Biol Chem 255:11725–11733

Buggy J, Hull J, Yoo-Warren H (1995a) Isolation and structural analysis of the 5′ flanking region of the gene encoding the human glucagon receptor. Biochem Biophys Res Commun 208:339–344

Buggy JJ, Livingston JN, Rabin DU, Yoo-Warren H (1995b) Glucagon/glucagon-like peptide I receptor chimeras reveal domains that determine specificity of glucagon binding. J Biol Chem 270:7474–7478

Butlen D, Morel F (1985) Glucagon receptors along the nephron: [^{125}I]glucagon binding in rat tubules. Pflugers Arch 404:348–353

Campos RV, Lee YC, Drucker DJ (1994) Divergent tissue-specific and developmental expression of receptors for glucagon and glucagon-like peptide-1 in the mouse. Endocrinology 134:2156–2164

Carruthers CJL, Unson CG, Kim HN, Sakmar TP (1994) Synthesis and expression of a gene for the rat glucagon receptor. J Biol Chem 269:29321–29328

Coleman DL, Hummel KP (1967) Studies with the mutation, diabetes, in the mouse. Diabetologia 3:238–248

Corranza MC, Simon MA, Torres A, Romero B, Calle C (1993) Identification of glucagon receptors in human adipocytes from a liposarcoma. J Endocrinol Invest 16:439–442

Demoliou-Mason C, Epand RM (1982) Identification of the glucagon receptor by covalent labeling with a radiolabeled photoreactive glucagon analogue. Biochemistry 21:1996–2004

Desbuquois B, Laudat MH (1974) Glucagon-receptor interactions in fat cell membranes. Mol Cell Endocrinol 1:355–370

Dixon RA, Kobilka BK, Strader DJ, Benovic JL, Dohlman HG, Frielle T, Bolanowski MA, Bennet CD, Rands E, Diehl RE et al (1986) Cloning of the gene and cDNA for mammalian beta-adrenergic receptor and homology with rhodopsin. Nature 321:75–79

Drucker DJ, Asa S (1988) Glucagon gene expression in vertebrate brain. J Biol Chem 263:13475–13478

Fernandez-Durango R, Sanchez D, Fernandez-Cruz A (1990) Idntification of glucagon receptors in rat retina. J Neurochem 54:1233–1237

Geary N, Sauter J, Noh U (1993) Glucagon acts in the liver to control spontaneous meal size in rats. Am J Physiol 264:R116–R122

Gittes GK, Rutter WJ (1992) Onset of cell-specific gene expression in the developing mouse pancreas. Proc Natl Acad Sci USA 89:1128–1132

Godfrey P, Rahal JO, Beamer WG, Copeland NG, Jenkins NA, Mayo KE (1993) GHRH receptor of little mice contains a missense mutation in the extracellular domain that disrupts receptor function. Nat Genet 4:227–232

Goldfine ID, Roth J, Birnbaumer L (1972) Glucagon receptors in β-cells. Binding of ^{125}I-glucagon and activation of adenylate cyclase. J Biol Chem 247:1211–1218

Hager J, Hansen L, Vaisse C, Vionnet N, Philippi A, Poller W, Velho G, Carcassi C, Contu L, Julier C, Cambien F, Passa P, Lathrop M, Kindsvogel W, Demenais F, Nishimura E, Froguel P (1995) A missense mutation in the glucagon receptor gene is associated with non-insulin-dependent diabetes mellitus. Nat Genet 9:299–304

Hoosein NM, Gurd RS (1984) Identification of glucagon receptors in rat brain. Proc Natl Acad Sci USA 81:4368–4372

Ishihara T, Nakamura S, Kaziro Y, Takahashi T, Takahashi K, Nagata S (1991) Molecular cloning and expression of a cDNA encoding the secretin receptor. EMBO J 10:1635–1641

Ishihara T, Shigemoto R, Mori K, Takahashi K, Nagata S (1992) Functional expression and tissue distribution of a novel receptor for vasoactive intestinal polypeptide. Neuron 8:811–819

Iwanij V (1995) Canine kidney glucagon receptor: evidence for the tissue-specific variant of the glucagon receptor. Mol Cell Endocrinol 115:21–28

Iwanij V, Hur KC (1985) Direct cross-linking of ^{125}I-labeled glucagon to its membrane receptor by UV irradiation. Proc Natl Acad Sci USA 82:325–329

Iwanij V, Hur KC (1987) Development of physiological responsiveness to glucagon during embryogenesis of avian heart. Dev Biol 122:146–152

Iwanij V, Vincent AC (1990) Characterization of the glucagon receptor and its functional domains using monoclonal antibody. J Biol Chem 265:21302–21308

Iwanij V, Amos TM, Billington CJ (1994) Identification and characterization of the glucagon receptor from adipose tissue. Mol Cell Endocrinol 101:257–261

Iyengar R, Herberg JT (1984) Structural analysis of the hepatic glucagon receptor. J Biol Chem 259:5222–5229

Jelinek LJ, Lok S, Rosenberg GB, Smith RA, Grant FJ, Biggs S, Bensch PA, Kuijper JL, Sheppard PO, Sprecher CA, O'Hara PJ, Foster D, Walker KM, Chen LHJ, Mckernan PA, Kindsvogel W (1993) Expression cloning and signaling properties of the rat glucagon receptor. Science 259:1614–1616

Johnson GL, MacAndrew VI, Pilch PF (1981) Identification of the glucagon receptor in rat liver membranes by photoaffinity cross-linking. Proc Natl Acad Sci USA 78:875–878

Jüppner H, Abou-Samra AB, Freeman M, Kong XF, Schipani E, Richards J, Kolakowski LFJr, Hock J, Potts JT, Kronenberg HM, Segre GV (1991) A G protein-linked receptor for parathyroid hormone and parathyroid hormone-related peptide. Science 254:1024–1026

Katz AI, Lindheimer MD (1977) Action of hormones on the kidney. Annu Rev Physiol 39:97–134

Kawai K, Yakota C, Ohashi S, Watanabe Y, Yamashita K (1995) Evidence that glucagon stimulates insulin secretion through its own receptors in rats. Diabetologia 38:274–276

Kennedy ME, Limbird LE (1993) Mutation of the α2A-adrenergic receptor that eliminates detectable palmitoilation do not perturb receptor-G-protein coupling. J Biol Chem 268:8003–8011

Kuroshima A, Yahata T (1979) Thermogenic responses of brown adipocytes to noradrenaline and glucagon in heat-acclimated and cold-acclimated rats. Jpn J Physiol 29:683–690

Kurstjens NP, Heithier H, Cantrill RC, Hahn M, Boege F (1990) Multiple hormone actions: the rises in cAMP and Ca^{2+} in MDCK-cells treated with glucagon and prostaglandin E_1 are independent processes. Biochem Biophys Res Commun 167:1162–1169

Kyte J, Doolittle RF (1982) A simple method for displaying the hydropathic character of protein. J Mol Biol 157:105–132

Laughlin MR, Taylor JF, Chesnick AS, Balaban RS (1992) Regulation of glycogen metabolism in canine myocardium; effects of insulin and epinephrine in vivo. Am J Physiol 262:E875–E883

Lin HY, Harris TL, Flannery MS, Aruffo A, Kaji EH, Gorn A, Kolakowski LF Jr, Lodish HF, Goldring SR (1991) Expression cloning of an adenylate cyclase-coupled calcitonin receptor. Science 254:1022–1024

Lin S-C, Lin CR, Gukovski I, Luis AJ, Sawchenko PE, Rosenfeld MG (1993) Molecular basis of the little mouse phenotype and implications for cell type-specific growth. Nature 364:208–213

Livingston JN, Cuatrecasas P, Lockwood DH (1974) Studies of glucagon resistance in large rat adipocytes: ^{125}I-labeled glucagon binding and lipolytic capacity. J Lipid Res 15:26–32

Lok S, Kuijper JL, Jelinek LJ, Kramer JM, Whitmore TE, Sprecher CA, Mathewes S, Grant FJ, Biggs SH, Rosenberg GB, Sheppard PO, O'Harra PJ, Foster DC, Kindsvogel W (1994) The human glucagon receptor encoding gene: structure, cDNA sequence and chromosomal localization. Gene 140:203–209

Loud FB, Holst JJ, Christiansen J, Rehfeld JF (1988) Effect of glucagon on vagally induced gastric acid secretion in humans. Dig Dis Sci 33:405–408

Lucchesi BR (1968) Cardiac actions of glucagon receptors. Circ Res 22:777–787

MacNeil DJ, Occi JL, Hey PJ, Strader CD, Graziano MP (1994) Cloning and expression of a human glucagon receptor. Biochem Biophys Res Commun 198:328–334

Maget B, Tastenoy M, Svoboda M (1994) Sequencing of eleven introns in genomic DNA encoding rat glucagon receptor and multiple alternative splicing of its mRNA. FEBS Lett 351:271–275

Mayo KE (1992) Molecular cloning and expression of a pituitary-specific receptor for growth hormone-releasing hormone. Mol Endocrinol 6:1734–1744

McCuaig KA, Clarke JC, White JH (1994) Molecular cloning of the gene encoding the mouse parathyroid hormone/parathyroid-related peptide receptor. Proc Natl Acad Sci USA 91:5051–5055

Melson GL, Chase LR, Aurbach GD (1970) Parathyroid hormone-sensitive adenyl cyclase in isolated renal tubules. Endocrinology 86:511–518

Menzel S, Stoffel M, Espinosa R III, Fernald AA, Le Beau MM, Bell GI (1994) Localization of the glucagon receptor gene to human chromosome band 17q25. Genomics 20:327–328

Merida E, Delgado E, Molina LM, Villanueva-Penacarrillo ML, Valverde I (1993) Presence of glucagon and glucagon-like peptide-1-(7-36)amide receptors in solubilized membranes of human adipose tissue. J Clin Endocrinol Metab 77:1654–1657

Méry PF, Brechler V, Pavoine C, Pecker F, Fischmeister R (1990) Glucagon stimulates the cardiac Ca^{2+} currents by activation of adenylyl cyclase and inhibition of phosphodiesterase. Nature 345:158–161

Morel F, Butlen D (1990) Hormonal receptors in the isolated tubule. Methods Enzymol 191:303–325

Mount SM (1982) A catalogue of splice junction sequences. Nucleic Acids Res 10:459–472

Namba T, Sugimoto Y, Negishi M, Irie A, Ushikubi F, Kakizuka A, Ito S, Ichikawa A, Narumiya S (1993) Alternative splicing of C-terminal tail of prostaglandin E receptor subtype EP3 determines G-protein specificity. Nature 365:166–170

O'Dowd BF, Hnatowich M, Caron MG, Lefkowitz RJ, Bouvier M (1989) Palmitoilation of the human β_2-adrenergic receptor. J Biol Chem 264:7564–7469

Okamoto T, Murayama Y, Hayashi Y, Inagaki M, Ogata E, Nishimoto I (1991) Identification of a Gs activator region of the β_2-adrenergicreceptor that is autoregulated via protein kinase A-dependent phosphorylation. Cell 67:723–770

Patel Y, Amherdt M, Orci L (1982) Quantitative electron microscopic autoradiography of insulin, glucagon and somatostatin binding sites on islets. Science 217:1155–1156

Premont RT, Iyengar R (1988) Glucagon-induced desensitization of adenylyl cyclase in primary cultures of chick hepatocytes. J Biol Chem 263:16087–16095

Reagan J (1994) Expression cloning of an insect diuretic hormone receptor. J Biol Chem 269:9–12

Rich KA, Codina J, Floyd G, Secura R, Hildebrandt JD, Iyengar R (1984) Glucagon-induced heterologous desensitization of the MDCK adenylyl cyclase. J Biol Chem 259:7893–7901

Rodbell M, Birbbaumer L, Pohl S, Krans HMJ (1971a) The glucagon-sensitive adenyl cyclase system in plasma membranes of rat liver, V. An obligatory role of gluanyl nucleotides in glucagon activation. J Biol Chem 246:1877–1882

Rodbell M, Krans HM, Pohl S, Birnbaumer L (1971b) The glucagon-sensitive adenyl cyclase system in plasma membranes of rat liver. III. Binding of glucagon: method of assay. J Biol Chem 246:1861–1871

Samols E, Marri G, Marks V (1965) Promotion of insulin secretion by glucagon. Lancet 2:415–416

Santamaria L, de Miguel E (1985) Localization of glucagon receptors in intestinal and vascular smooth muscle fibers in dog. Use of autoradiographic technics. Rev Esp Enferm Apar Dig 67:301–308

Santos A, Blazquez E (1982) Regulatory effect of glucagon on its own receptor concentrations and target-cell sensitivity in the rat. Diabetologia 22:362–371

Segaloff DL, Ascoli M (1993) The lutropin/choriogonadotropin receptor... 4 years later. Endocr Rev 14:324–346

Spengler D, Waeber C, Pantaloni C, Holsboer F, Bockaert J, Seeburg PH, Journot L (1993) Differential signal transduction by five splice variants of the PACAP receptor. Nature 365:170–174

Strader CD, Fong TM, Tota MR, Underwood D, Dixon RAF (1994) Structure and function of G protein-coupled receptors. Annu Rev Biochem 63:101–132

Stroop SD, Kuestner RE, Serwold TF, Chen L, Moore EE (1995) Chimeric human calcitonin and glucagon receptors reveal two dissociable calcitonin interaction sites. Biochemistry 43:1050–1057

Svoboda M, Ciccarelli E, Tastenoy M, Cauvin A, Stièvenart M, Christophe J (1993a) Small introns in a hepatic cDNA encoding a new glucagon-like peptide 1-type receptor. Biochem Biophys Res Commun 191:479–486

Svoboda M, Ciccarelli E, Tastenoy M, Robberecht P, Christophe J (1993b) A cDNA construct allowing the expression of rat hepatic glucagon receptor. Biochem Biophys Res Commun 192:135–142

Svoboda M, Tastenoy M, Vertongen P, Robberecht P (1994) Relative quantitative analysis of glucagon receptor mRNA in rat tissues. Mol Cell Endocrin 105:131–137

Thorens B (1992) Expression cloning of the pancreatic beta cell receptor for the glucoincretin hormone glucagon-like peptide 1. Proc Natl Acad Sci USA 89:8641–8645

Tuman RW, Doisy RJ (1977) The influence of age on the development of hypertriglycerimia and hypercholesterolemia in genetically diabetic mice. Diabetologia 13:7–11

Unger RH, Orci L (1981) Glucagon and the A cell. N Engl J Med 304:1518–1524

Unson CG, Macdonald D, Ray K, Durrah TL, Merrifield RB (1991) Position 9 replacement analogs of glucagon uncouple biological activity and receptor binding. J Biol Chem 266:2763–2766

Unson CG, Macdonald D, Merrifield RB (1993) The role of histidine-1 in glucagon action. Arch Biochem Biophys 300:747–750

Vanetti M, Vogt G, Hollt V (1993) The two isoforms of the mouse somatostatin receptor (mSSTR2A and mSSTR2B) differ in coupling efficiency to adenylate cyclase and in agonist induced receptor desesitization. FEBS Lett 331:260–266

Vita N, Laurent P, Lefort S, Chalon P, Lelias JM, Kaghad M, Le Fur G, Caput D, Ferrara P (1993) Primary structure and functional expression of mouse pituitary and human brain corticotrophin releasing actor receptor. FEBS Lett 355:1–5

Wakelam MJO, Murphy GJ, Hruby VJ, Houslay MD (1986) Activation of two signaltransduction system in hepatocytes by glucagon. Nature 323:68–71

Yoo-Warren H, Willse AG, Hancock N, Hull J, McCaleb M, Livingston JN (1994) Regulation of rat glucagon receptor expression. Biochem Biophys Res Commun 205:347–353

CHAPTER 5
Mode of Action of Glucagon Revisited

F. PECKER and C. PAVOINE

A. Introduction

It has long been accepted that glucagon action is linked to a unique pathway, the stimulation of adenylyl cyclase activity and cyclic AMP synthesis, initiated by the binding of glucagon to its receptor (Fig. 1). However, this view has been disputed and there is now evidence that the physiological effects of glucagon may occur through other mechanisms namely: (a) actions mediated through the glucagon receptor, but relying on transduction signals other than adenylyl cyclase activation and (b) effects of glucagon, secondary to the processing of the hormone into its carboxy-terminal fragment (19-29).

These parallel or additional pathways of glucagon action have been described both in the liver and in the heart. In the liver, this is relevant when one considers that the major metabolic effect of glucagon is the activation of the hepatic glycogenolysis and gluconeogenesis. In the heart, in addition to its metabolic effects, glucagon is an important therapeutic agent in critical care medicine due to its positive inotropic properties (FARAH 1983). Thus, intravenous administration of high doses of glucagon is proven effective in the treatment of cardiogenic shock and heart failure (see Chap. 11, this volume, and HALL-BOYER et al. 1984), in the face of β-adrenergic blockade (ROBSON 1980; ZALOGA et al. 1986), Ca^{2+} channel blocker-induced myocardial dysfunction (ZARITSKY et al. 1988) or poisoning by tricyclic antidepressants (SENER et al. 1995).

This chapter will not cover the classical, and probably still essential action of glucagon via the stimulation of adenylyl cyclase, but will deal with the newly identified mechanisms of glucagon action, such as the possible involvement of the phosphoinositide metabolism in the hepatic action of glucagon, the cyclic AMP phosphodiesterase pathway in cardiac cells, the role of glucagon as a prohormone and the effects of its metabolite glucagon (19-29), referred to as mini-glucagon (UNGER and ORCI 1990).

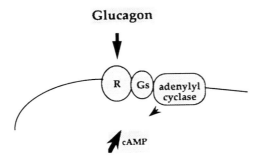

Fig. 1. Early 1980s concept of glucagon action. Adenylyl cyclase is the unique effector of glucagon action

B. Glucagon Actions Mediated Through Glucagon Receptors

I. Glucagon Receptor

The recent cloning and expression of the rat (JELINEK et al. 1993) and the human (MACNEIL et al. 1994) glucagon receptor demonstrated that this receptor, because of structural homologies, belonged to a distinct family of G-protein-coupled seven transmembrane domain receptors. This includes the receptors for calcitonin, parathyroid hormone and parathyroid hormone-related peptide, vasoactive intestinal peptide, pituitary adenylyl cyclase-activating polypeptide (PACAP), glucagon-like peptide 1, growth hormone-releasing hormone and secretin (see Chap. 4, this volume, and SEGRE and GOLDRING 1993).

Expression of several of these receptors in transfected cells showed that a single receptor stimulated intracellular accumulation of cAMP and increased inositol 1,4,5-triphosphate (InsP$_3$) and/or [Ca^{2+}]$_i$ (cytosolic free Ca^{2+} concentration) (JELINEK et al. 1993; SEGRE and GOLDRING 1993). So does the rat glucagon receptor expressed in either BHK cells or COS-7 cells (JELINEK et al. 1993); the PTH receptor expressed in COS-7 cells (ABOU-SAMRA et al. 1992); the PACAP receptor expressed in LLCPK1 (SPENGLER et al. 1993); and the calcitonin receptor expressed in MC 3T3 cells (CHABRE et al. 1992) and in M18 cells (FORCE et al. 1992).

These observations led to the question as to the cause and effect relationship between cAMP increase and activation of the InsP$_3$/Ca pathway. That activation of the InsP$_3$/Ca pathway was not simply a consequence of adenylyl cyclase stimulation was supported by the observation that, in COS-7 cells expressing the β$_2$-adrenergic receptor, isoproterenol increased cAMP accumulation but did not raise either [Ca^{2+}]$_i$ or InsP$_3$ (ABOU-SAMRA et al. 1992). Also, in the kidney, calcitonin activated both the adenylyl cyclase and the phospholipase C pathways, but in distinctly different regions of the nephron and in a cell cycle-dependent manner in LLC-PK1 cells, which demonstrated that both events were completely independent (HORNE et al. 1994). Finally, in pancreatic acinar tissue a structure function-relationship study with secretin analogs indicated that the mechanisms by which secretin stimulated adenylyl cyclase and activated phospholipase C were independent (TRIMBLE et al. 1987).

With regard to glucagon, the link between the Ca^{2+} and the cyclic AMP pathways, and the triggering signal for glucagon-induced $[Ca^{2+}]_i$ mobilization, remain to be clarified, and are documented in the following sections.

II. Glucagon Action in Liver

1. Glucagon Mobilizes Ca^{2+} in Hepatocytes

Hepatic glycogenolysis may be stimulated either by cAMP or by $[Ca^{2+}]_i$ elevations (STALMANS 1983). Glucagon has long been cited as a classical example of glycogenolytic hormone acting via a unique second messenger, cyclic AMP (ROBISON et al. 1967; EXTON et al. 1971; POHL et al. 1971).

However, there have been persistent claims that part of the glycogenolytic effect of glucagon in liver occurred via a cAMP-independent mechanism. In particular, subnanomolar concentrations of glucagon were reported to activate glycogen phosphorylase without a measurable rise in cyclic AMP or in protein kinase A activity (STALMANS 1983). Different authors have investigated the regulation of $[Ca^{2+}]_i$ by glucagon and the possible role of Ca^{2+} in the hepatic action of the hormone (for reviews see BYGRAVE and BENEDETTI 1993; KRAUS-FRIEDMANN 1995). It was found that glucagon evoked $[Ca^{2+}]_i$ increase in the hepatocyte (CHAREST et al. 1983; SISTARE et al. 1985; MAUGER et al. 1985; BLACKMORE and EXTON 1986; STADDON and HANSFORD 1986, 1989; COMBETTES et al. 1986; MAUGER and CLARET 1986; POGGIOLI et al. 1986; ALTIN and BYGRAVE 1986; CONNELLY et al. 1987; MINE et al. 1988; BENEDETTI et al. 1989; GARCIA-SAINZ et al. 1990), due to a Ca^{2+} release from intracellular stores into the cytosol (BLACKMORE and EXTON 1986; STADDON and HANSFORD 1986; MAUGER and CLARET 1986; BENEDETTI et al. 1989) and to an increase in net Ca^{2+} inflow across the plasma membrane (MAUGER et al. 1985; STADDON and HANSFORD 1986, 1989). MAUGER and CLARET (1986) showed that glucagon released Ca^{2+} from the same intracellular stores as did Ca^{2+}-mobilizing agonists. However, Ca^{2+} movements elicited by glucagon were less pronounced, slower in onset and their physiological significance was less certain than those induced by Ca^{2+}-mobilizing agonists, namely α-adrenergic agonists, vasopressin and angiotensin. Stimulation of plasma-membrane Ca^{2+} inflow was an early action of glucagon or dibutyryl cAMP and was likely to occur before Ca^{2+} mobilization from internal stores (BYGRAVE et al. 1993).

A major unresolved question about the mechanism by which glucagon affected Ca^{2+} metabolism was whether its effects were primary or secondary to the increase in cAMP which followed its administration. Many reports pointed out that the effects of glucagon on Ca^{2+} mobilization were mimicked by forskolin or cyclic AMP analogues (MAUGER et al. 1985; SISTARE et al. 1985; MAUGER and CLARET 1986; BLACKMORE and EXTON 1986; CONNELLY et al. 1987;

STADDON and HANSFORD 1986, 1989; BENEDETTI et al. 1989; GARCIA-SAINZ et al. 1990; CAPIOD et al. 1991) or β-adrenergic agonists (CAPIOD et al. 1991) and potentiated by the cAMP phosphodiesterase inhibitor, isobutylmethylxanthine (STADDON and HANSFORD 1989; BLACKMORE and EXTON 1986). These results suggested that the effects of glucagon on Ca^{2+} mobilization were derived from the activation of the cyclic AMP pathway.

However, it has been reported that low concentrations of glucagon ($\leq 10\,nM$), but not higher, produced a small increase in $InsP_3$ release in hepatocytes (WAKELAM et al. 1986; BLACKMORE and EXTON 1986; WHIPPS et al. 1987). Additionally, these authors demonstrated that the glycogenolytic analogue of glucagon, [1-N^α-trinitrophenylhistidine, 12-homoarginine] glucagon (TH-glucagon), caused a transient increase in $InsP_3$, while it did not increase cellular cAMP, suggesting that glucagon could act independently of cyclic AMP. These results are still a matter of debate, since several other authors have been unable to observe any activation of the phospholipase C by glucagon (POGGIOLI et al. 1986; PITTNER and FAIN 1989, 1991). Furthermore, CORVERA et al. (1984) and GARCIA-SAINZ et al. (1990) found that TH-glucagon, above $1\,\mu M$, did cause an increase in the cAMP content of hepatocytes. Such a partial agonist activity of TH-glucagon was also found in isolated rat fat cells (GARCIA-SAINZ et al. 1986).

It may be concluded that there is strong evidence suggesting that glucagon mobilizes Ca^{2+} in the hepatocyte. This seems to rely on the activation of the cyclic AMP pathway, but whether it is also directly or indirectly linked to phospholipase C activation has not been convincingly proved.

2. Glucagon Potentiates the Effect of Ca^{2+}-Mobilizing Agonists

Glucagon has been reported to potentiate the effects of Ca^{2+}-mobilizing hormones. Coadministration of glucagon with vasopressin, phenylephrine or angiotensin in hepatocytes led to rates of Ca^{2+} inflow much greater than those observed with either glucagon or Ca^{2+}-mobilizing agonist alone (MAUGER et al. 1985; MAUGER and CLARET 1986; POGGIOLI et al. 1986; BLACKMORE and EXTON 1986; KASS et al. 1990; SOMOGYI et al. 1992; and see the review by BYGRAVE and BENEDETTI 1993). Similar synergism was also observed with agents other than glucagon which elevated cellular cyclic AMP, such as isoproterenol (POGGIOLI et al. 1986; BURGESS et al. 1986) or exogenously applied dibutyryl cyclic AMP (MAUGER et al. 1985; MAUGER and CLARET 1986; KASS et al. 1990; BURGESS et al. 1986, 1991; SOMOGYI et al. 1992), suggesting that the interaction occurred downstream of cyclic AMP formation. One possibility was that potentiation occurred at/or before hormone-induced phosphoinositide metabolism, and data have been provided (BLACKMORE and EXTON 1986; BOCCKINO et al. 1985; PITTNER and FAIN 1989) showing two fold enhanced levels of inositol phosphates when glucagon was applied along with Ca^{2+}-mobilizing hormones to hepatocytes. However, other investigators (BURGESS et al. 1986; POGGIOLI et al. 1986) have been unable to detect any effect of cAMP-linked hormones (glucagon and isoproterenol) on Ca^{2+}-mobilizing hormone-mediated breakdown of phosphoinositides or formation of inositol phosphates at times when marked potentiation of Ca^{2+}-dependent responses could be observed. These authors suggested that interaction was likely to occur after $InsP_3$ formation and that the sensitivity of either the $InsP_3$ receptor or the hormone-sensitive Ca^{2+} store might be modulated by cyclic AMP. In an elegant report, using permeabilized hepatocytes and by microinjecting inositol phosphates into single cells, BURGESS et al. (1991) provided evidence that cyclic AMP was able to increase the sensitivity of the Ca^{2+}-releasing organelles to $InsP_3$, and that potentiation was not due to an increase in the cellular concentration of inositol phosphates.

In conclusion, potentiation of the effects of Ca^{2+}-mobilizing hormones by glucagon in liver is likely to be initiated by cAMP.

3. Regulation of the Cyclic-GMP-Inhibited Phosphodiesterase (CGI-PDE) by cAMP-Dependent Phosphorylation in Liver

In rat liver, glucagon has been shown to activate a particulate CGI-PDE (type III PDE), which has a low K_m for cAMP and is inhibited by milrinone and micromolar concentrations of cGMP (LOTEN et al. 1978; HEYWORTH et al. 1983; and for ref. see CONTI et al. 1991). PDE activity was also increased upon cholera-toxin treatment or by dibutyryl-cAMP (HEYWORTH et al. 1983). The indication that this was related to cAMP-dependent activation of the enzyme came from studies using the cAMP analogue 8-p-chlorophenyl-thio-cAMP (GETTYS et al. 1988). Degradation of this compound was enhanced when intact hepatocytes were challenged with glucagon (GETTYS et al. 1988). Thus, there is now general agreement that a phosphorylation catalyzed by a cAMP-dependent protein kinase is responsible for glucagon activation of the hepatic CGI-PDE.

Intracellular cyclic nucleotide level is the result of a steady state of synthesis and degradation, since extrusion of the cyclic nucleotides accounts only for a minor component in the control of the intracellular concentration of cyclic nucleotides. Thus, simultaneous increase in synthesis and degradation of cAMP elicited by glucagon in liver would result in an accelerated turnover which may increase the dynamics of the cyclic AMP-dependent systems by producing more rapid changes in PKA activation/deactivation.

In summary, to date, most of the data in the literature argue for cAMP as the mediator of glucagon action in liver.

III. Glucagon Action in Heart

1. Adenylyl Cyclase Activation Versus cAMP-Phosphodiesterase Inhibition in Heart Cells

Like its glycogenolytic effect in liver, the cardiac effects of glucagon are often correlated with an increase in intracellular cyclic AMP (FARAH 1983). How-

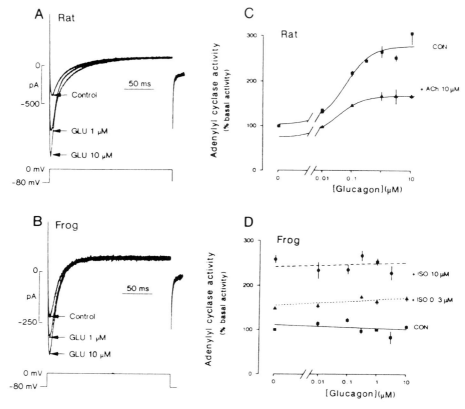

Fig. 2A–D. Cyclic AMP mediates the activation of I_{Ca} by glucagon in frog and rat heart ventricles. Glucagon activates the L-type Ca^{2+} current (I_{Ca}) in rat heart cells (**A**) and in frog heart cells (**B**). In rat heart cells, I_{Ca} activation is correlated with an increase in cAMP synthesis via adenylyl cyclase (**C**). In frog heart cells, I_{Ca} activation is not correlated with adenylyl cyclase activation (**D**). [Reprinted with permission from *Nature* (MERY et al. 1990), Macmillan Magazines Ltd.]

ever, the inability of the peptide to stimulate adenylyl cyclase was reported in various animals, such as guinea pig (CHATELAIN et al. 1983), mouse (CLARK et al. 1976), monkey (WILDENTHAL et al. 1976) and frog (MÉRY et al. 1990), where the peptide nevertheless elicits positive inotropic effects.

Since the activity of cardiac Ca^{2+} channels is strongly regulated by cAMP-dependent phosphorylation, Ca^{2+} channel current (I_{Ca}) measured in isolated cardiac myocytes may be used as a probe of cAMP metabolism. In rat and in frog ventricular heart cells, glucagon stimulated I_{Ca} (Fig. 2A,B). In rat heart, acetylcholine inhibited adenylyl cyclase activation by glucagon (Fig. 2C), and hence reduced the activation of I_{Ca} by glucagon (MÉRY et al. 1990). In contrast, the stimulatory effect of glucagon on I_{Ca} in frog heart was not reduced by acetylcholine, suggesting that this effect of glucagon was not due to adenylyl cyclase activation (MÉRY et al. 1990). Indeed, glucagon did not activate

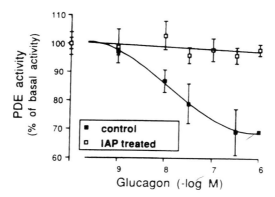

Fig. 3. Glucagon inhibits the low-K_m-cyclic AMP phosphodiesterase activity (low-K_m-cAMP PDE) in frog heart ventricles. In frog heart cells, I_{Ca} activation is correlated with a decrease in cAMP degradation via phosphodiesterase inhibition. Pertussis toxin (*IAP*) treatment abolishes the inhibition of the low-K_m-cAMP PDE by glucagon, suggesting the involvement of a Gi-like protein. (From BRECHLER et al. 1992)

adenylyl cyclase activity in frog heart (Fig. 2D), in contrast to isoproterenol (MÉRY et al. 1990).

Since both adenylyl cyclase and cyclic nucleotide phosphodiesterases (PDEs) control the cellular level of cAMP, the effects of glucagon on the cAMP PDE activity in frog heart cell have been examined. It has been demonstrated that glucagon elicited a dose-dependent inhibition of a membrane-bound low-K_m cAMP PDE, which correlated with an activation of I_{Ca} (Figs. 3, 2B). It should be noted that glucagon had no effect on the cytosolic PDE activity. Also, inhibition of PDE activity by glucagon did not appear to be restricted to frog heart: it also occurred in mouse and guinea pig heart (Fig. 4A,B; BRECHLER et al. 1992). In those tissues glucagon did not activate adenylyl cyclase (Fig. 4C,D).

It thus appears that glucagon, at concentrations which correlate with the concentrations at which it exerts its cardiac inotropic effects in heart, activates I_{Ca} in heart cells by increasing intracellular cAMP. This effect of glucagon may be mediated through either adenylyl cyclase stimulation or a low-K_m cAMP PDE inhibition, depending on the species (MÉRY et al. 1990; BRECHLER et al. 1992).

These results further emphasized the crucial role of PDEs in the control of cardiac contraction (HARTZELL and FISCHMEISTER 1986; FISCHMEISTER and HARTZELL 1987, 1990; SIMMONS and HARTZELL 1988; MÉRY et al. 1993), and raised the questions: (a) the isozyme family of the glucagon-sensitive PDE and (b) the mechanism involved in the inhibition process. In the heart four families of cAMP-PDEs have been defined: (a) the Ca^{2+}-calmodulin-regulated PDEs that hydrolyze both cAMP and cGMP (PDE I); (b) a cGMP-stimulated PDE that hydrolyzes cAMP in a relatively high range of concentrations ($1-100\,\mu M$) and is inhibited by EHNA (erythro-9-[2-hydroxy-3-nonyl]adenine) (PDE II); (c) the cGMP-inhibited PDE which has a high affinity for cAMP ($K_m < 1\,\mu M$)

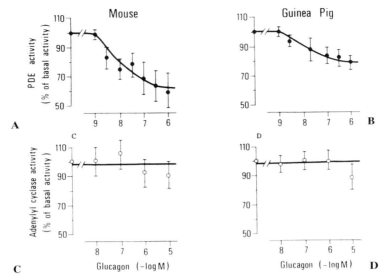

Fig. 4A–D. Inhibition by glucagon of the low-K_m cAMP phosphodiesterase in membrane fractions obtained from mouse and guinea pig heart ventricles. In mouse (**A**) and guinea pig (**B**) hearts, glucagon inhibits the membrane, low-K_m cAMP PDE, whereas it has no effect on the adenylyl cyclase activity (**C,D**). (From BRECHLER et al. 1992)

and is inhibited by low concentrations of cGMP and bipyridine PDE inhibitors such as milrinone (CGI-PDE or PDE III); and (d) the cAMP-specific PDE which has a high selectivity for cAMP as substrate ($K_m < 1\,\mu M$) and is inhibited by rolipram and Ro 20-1724 (PDE IV) (for reviews see: BEAVO and REIFSNYDER 1990; MULLER et al. 1993; MÉRY et al. 1995). Characterization of the PDE isozyme sensitive to glucagon has been done by examining the additivity of the effect of glucagon with that of the inhibitors of the two low-K_m PDEs: milrinone and Ro-20-1724. It appeared that glucagon still elicited a dose-dependent inhibition of the PDE activity, in the presence of Ro-20-1724. In contrast, its ability to inhibit PDE activity was abolished in the presence of milrinone (BRECHLER et al. 1992). Taken together, these results suggested that the milrinone-sensitive CGI-PDE was the target of glucagon action.

It is noteworthy that specific pharmacological inhibitors of the CGI-PDE have been used in cardiac medicine, due to their potent cardiotonic action (MULLER et al. 1990). In this regard, a singular role has been attributed to the CGI-PDE, as compared to the other PDE subtypes, in the regulation of cardiac contractility. To date, glucagon is the unique substance of physiological origin identified as a cardiac CGI-PDE inhibitor.

Inhibition of the CGI-PDE by glucagon was potentiated by guanine nucleotides, and abolished after pertussis toxin treatment of the membrane fraction. These results suggested that a Gi protein controlled the inhibition of the CGi

PDE by glucagon. The coupling of glucagon receptors to Gs, which mediates adenylyl cyclase activation, is well known. The possibility that the glucagon receptor could couple to Gi was first supported by the early observation that glucagon inhibited adenylyl cyclase in rabbit heart (KISS and TKATCHUK 1984). However, the latter was seen only in extreme conditions (high guanine nucleotide concentration) (KISS and TKATCHUK 1984). It may be added that the dual coupling to Gs and another G protein seems to be a common feature to the receptors in the glucagon-related receptor family (SEGRE and GOLDRING 1993).

In conclusion, data concerning glucagon action in heart demonstrate that cAMP is its second messenger, constitute the first evidence for hormonal inhibition of the cardiac CGi PDE and support the dual coupling of the glucagon receptor to both Gs and Gi.

C. Glucagon is Processed by its Target Cells

Pairs of basic amino acids are recognized as classical cleavage sites in post-translational processing of peptide hormones (GLUSCHANKOF and COHEN 1987; SEIDAH et al. 1993; MARTINEZ and POTIER 1986). The glucagon molecule includes a dibasic doublet (Arg^{17}-Arg^{18}) which has long been considered as unprocessed (Fig. 5A).

The use of a specific radioimmunoassay (BLACHE et al. 1988a) in association with the separation and identification of peptides by high-performance liquid chromatography has been used to detect the presence of the C-terminal fragment of glucagon, glucagon (19-29) (BLACHE et al. 1988b,c). Upon incubation at 37°C with liver plasma membranes, glucagon was processed into its (19-29) C-terminal fragment, as confirmed by amino acid sequencing (BLACHE et al. 1990). The accumulation of glucagon (19-29) reached a maximum at 2 min (1% of initial glucagon), followed by a slow decline. Incubation of $1\mu M$ glucagon with embryonic chick heart cells also resulted in a rapid release of glucagon (19-29) in the cell medium (PAVOINE et al. 1991). Accumulation of glucagon (19-29) was maximal after 8 min and reached 60 nM, corresponding to the conversion of 6% of initial glucagon (Fig. 5B). The level of glucagon (19-29) was stable for 15 min, suggesting that the degradation of the peptide by heart cell was slow as compared with that in liver membranes (BLACHE et al. 1990). In the presence of bacitracin, glucagon processing into glucagon (19-29) was inhibited by 60% (PAVOINE et al. 1991) (Fig. 5B).

From a physiological point of view, it must be pointed out that, whereas the glucagon concentration in the portal blood is roughly 1 nM, a 1%–6% processing of the hormone into its (19-29) fragment provides a 10–60 pM concentration of this fragment. As it will be discussed below, glucagon (19-29) is quite active, in vitro, at these concentrations (PAVOINE et al. 1991). This is a good indication that the processing of glucagon into glucagon (19-29) may have physiological meaning.

Glucagon (19-29) has also been detected in rat pancreas and stomach, its tissue concentration corresponding to about 3% of that of glucagon (BLACHE

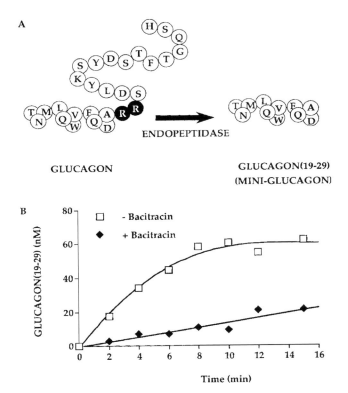

Fig. 5A,B. Processing of glucagon into glucagon (19-29), referred to as mini-glucagon. **A** Mini-glucagon is generated by cleavage of glucagon at the dibasic doublet R_{17}-R_{18}. **B** Time course of production of mini-glucagon upon incubation of glucagon with embryonic chick heart cells. Processing of glucagon to mini-glucagon is inhibited by bacitracine. (From PAVOINE et al. 1991)

et al. 1988c). In contrast, glucagon (19-29) has not been detected in rat plasma.

Glucagon processing to glucagon (19-29) did not appear to be associated with the binding of glucagon to its receptors since GTPγS, which is known to displace glucagon from its receptors (IYENGAR et al. 1988), did not reduce glucagon processing. This was further substantiated by the observations that [D-Gln³]glucagon, an analogue with a decreased affinity for the receptor (ROBBERECHT et al. 1988), was processed similarly to glucagon. Also, the hepatoma cell line (HepG2 cells), which did not contain high-affinity glucagon receptors, did process glucagon into glucagon (19-29).

An endopeptidase that cleaves glucagon-producing glucagon (19-29)-has been isolated from rat liver membranes (BLACHE et al. 1993) (Fig. 5A). The purified enzyme had a molecular weight of approximately 100kDa. Immunoreactivity assayed using an antiserum raised against a synthetic octapeptide corresponding to the N-terminal sequence was found in liver, pancreas, kidney and heart, all glucagon target tissues, and in gastric mucosa. Low levels were detected in spleen, and no immunoreactivity was detected in skeletal muscle and intestinal mucosa (BLACHE et al. 1993). The endopeptidase was inhibited by both sulfhydryl-blocking reagents and metal chelating reagents and activated by thiol compounds.

Enzymes belonging to the subtilisin family of serine proteinases have been shown to selectively cleave precursors at pairs of Lys-Arg and Arg-Arg residues to generate biologically active proteins and peptides (FULLER et al. 1989; SMEEKENS and STEINER 1990). In contrast to prohormone convertases, the endopeptidase that cleaved glucagon to glucagon (19-29) did not display strict selectivity toward basic doublets. Its activity was inhibited by insulin, glucagon-like peptide-1 or its truncated form glucagon-like peptide-1 (7-36) amide, none of which did contain any dibasic doublets, whereas it was not affected by other peptides such as vasoactive intestinal peptide (VIP), gastric inhibitory polypeptide (GIP) and peptide histidine isoleucine (PHI), which contained dibasic sites (BLACHE et al. 1990).

Glucagon metabolism in liver has been the subject of many other studies. Tager's group showed that glucagon was degraded by an enzyme from canine hepatic plasma membrane into several fragments including fragments (1-13), (14-29), (14-22), (23-29), (14-25) and (26-29) (SHEETZ and TAGER 1988). In contrast to the endopeptidase-producing glucagon (19-29) described above, this enzyme appeared to be linked to the glucagon receptors, as demonstrated by the inhibitory effect of GTP, or GTPγS (SHEETZ and TAGER 1988; IYENGAR et al. 1988).

Another study by ROSE et al. (1988) showed that an insulin-degrading enzyme cleaved glucagon between residues Arg^{17} and Arg^{18}, leading to fragments (1-17) and (18-29). In their set of experiments, glucagon (19-29) was not detected. They suggested that the formation of glucagon (19-29) from glucagon (18-29) might involve the removal of Arg^{18} by another protease such as aminopeptidase B. However, BLACHE et al. (1990) indicated that incubation of glucagon (18-29) with liver membranes did not lead to the formation of glucagon (19-29) immunoreactivity.

In conclusion, glucagon is a 29-amino-acid circulating hormone which can be processed into glucagon (19-29) upon interaction with its target tissues, in particular liver and heart. This maturation process seems independent of the binding of glucagon to its receptor and is ensured by an ectomembranous endopeptidase which has been purified.

D. Action of Mini-glucagon [Glucagon (19-29)] in Liver

I. Pharmacological Concentrations of Glucagon Inhibit the Liver Plasma Membrane Ca^{2+} Pump

A high-affinity (Ca^{2+}-Mg^{2+})ATPase activity, which ensures Ca^{2+} extrusion out of the cell (PAVOINE et al. 1987), has been identified in rat liver plasma membranes (LOTERSZTAJN et al. 1981; LOTERSZTAJN and PECKER 1982). It appeared that glucagon inhibited the Ca^{2+} pump in liver plasma membranes (Fig. 6) (LOTERSZTAJN et al. 1984). This effect was mimicked neither by cyclic AMP, nor by dibutyryl cyclic AMP, and a study of the structure-activity relationships of six glucagon derivatives demonstrated the total absence of correlation between adenylyl cyclase activation and (Ca^{2+}-Mg^{2+})-ATPase inhibition (LOTERSZTAJN et al. 1984).

However, there was a difference of 2 orders of magnitude between the glucagon concentrations effective for adenylyl cyclase activation and those required for calcium pump inhibition, the latter being observed only at high concentrations of glucagon ($IC_{50} = 0.7\,\mu M$). Moreover, in the presence of bacitracin, an inhibitor of glucagon degradation, the Ca^{2+} pump was no longer sensitive to glucagon (MALLAT et al. 1985), while adenylyl cyclase activation was potentiated (DESBUQUOIS et al. 1974). These findings suggested that glucagon processing might be required to observe inhibition of the liver Ca^{2+} pump.

II. Mini-glucagon is the True Effector of the Liver Plasma Membrane Ca^{2+} Pump

Glucagon (19-29), referred to as mini-glucagon, proved to be 1000-fold more potent than the whole 29-amino-acid hormone, glucagon, in inhibiting the two components that defined the Ca^{2+} pump: (a) the Ca^{2+}-activated and Mg^{2+}-dependent ATPase activity [(Ca^{2+}-Mg^{2+})ATPase] and (b) the ATP-dependent Ca^{2+} transport (Fig. 6; MALLAT et al. 1987). Nanomolar concentrations of mini-glucagon elicited a 40% maximal inhibition (MALLAT et al. 1987; LOTERSZTAJN et al. 1990; JOUNEAUX et al. 1993). The inhibitory action of mini-glucagon was

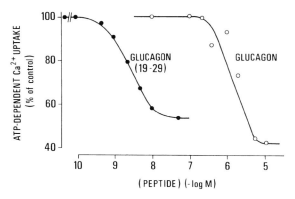

Fig. 6. Inhibition by glucagon and mini-glucagon of the ATP-dependent Ca^{2+} uptake by liver plasma membrane vesicles. Mini-glucagon is 2 orders of magnitude more potent than glucagon in inhibiting the ATP-dependent Ca^{2+} uptake in liver plasma membrane vesicles. [Reprinted with permission from *Nature* (MALLAT et al. 1987), Macmillan Magazines Ltd.]

specific on the basis of the order of potency of a series of glucagon fragments (LOTERSZTAJN et al. 1984). Glucagon (18-29) and glucagon (22-29) acted only as partial agonists. The amino-terminal (1-21) fragment of glucagon was completely inactive (MALLAT et al. 1987). It is noteworthy that mini-glucagon was totally ineffective in activating adenylyl cyclase in liver plasma membranes. However, inhibition of the plasma membrane Ca^{2+} pump paralleled an increase in phosphorylase a activity (LOTERSZTAJN et al., unpublished results).

Taken together, these results demonstrated that mini-glucagon is the true effector of the plasma membrane Ca^{2+} pump and that this peptide possesses its own biological activity, different from that of glucagon.

III. αs- and $\beta\gamma$-Subunits of G Protein Mediate Inhibition of the Liver Ca^{2+} Pump by Mini-glucagon

The regulation of the Ca^{2+} pump by mini-glucagon was dependent on guanine nucleotides. In the absence of guanine nucleotides, mini-glucagon caused a monophasic, dose-dependent inhibition of ATP-dependent Ca^{2+} uptake in inside-out oriented vesicles. The addition of GTP or GTPγS increased the affinity of the Ca^{2+} pump for mini-glucagon and revealed a biphasic regulation by the peptide (LOTERSZTAJN et al. 1990). In the presence of micromolar concentrations of either one nucleotide, low concentrations of mini-glucagon (10 pM to 1 nM) caused inhibition of the Ca^{2+} pump, while higher concentrations of the peptide (10–100 nM) reversed the inhibition caused by lower ones.

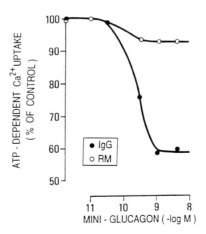

Fig. 7. A Gs protein mediates inhibition of the liver plasma membrane Ca^{2+} pump by mini-glucagon. The antibody raised against the carboxyl-terminal decapeptide, RMHLRQYELL, representing the conserved COOH-terminal sequence of the two forms of Gsα proteins (RM antibody), impairs inhibition of the Ca^{2+} pump in the liver plasma membranes by mini-glucagon. *IgG*, control immunoglobin G. (From JOUNEAUX et al. 1993)

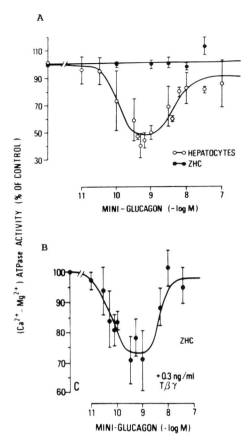

Fig. 8A,B. Effect of mini-glucagon on (Ca^{2+}-Mg^{2+})-ATPase activity in plasma membranes obtained from hepatocytes or Zajdela hepatoma cells. **A** Mini-glucagon exerts a biphasic effect on the plasma membrane Ca^{2+} pump in hepatocytes. In contrast, it has no effect on the system in Zajdela hepatoma cell membranes (*ZHC*). **B** Supplementation with T$\beta\gamma$-subunits reveals inhibition of the plasma membrane Ca^{2+} pump in ZHC. This result gives evidence for the control of the plasma membrane Ca^{2+} pump by $\beta\gamma$-subunits. (From LOTERSZTAJN et al. 1992)

Cholera toxin treatment resulted in a loss of sensitivity of the Ca^{2+} pump to mini-glucagon (LOTERSZTAJN et al. 1990). The G protein involved was further characterized using the antibody raised against the carboxyl-terminal decapeptide, RMHLRQYELL, representing the conserved COOH-terminal sequence of the two forms of Gsα proteins (RM antibody; JOUNEAUX et al. 1993). As shown in Fig. 7, the RM antibody totally blocked Ca^{2+} pump inhibition by mini-glucagon.

These observations led us to conclude that a Gs protein mediated mini-glucagon action. Interestingly, mini-glucagon had no effect on adenylyl cyclase, while isoproterenol which activated adenylyl cyclase had no effect on the Ca^{2+} pump (JOUNEAUX et al. 1993). These results further emphasized the pleiotypic action of Gs and also showed that Gs may be functionally specialized, raising the possibility of its compartmentalization.

The inhibition of the Ca^{2+} pump by mini-glucagon was also controlled by G$\beta\gamma$-subunits. This has been demonstrated in a cloned line of transformed rat hepatocytes, Zajdela hepatoma cells (ZH cells), with purified $\beta\gamma$-subunits from transducin (T$\beta\gamma$). As compared to hepatocytes, ZH cells were unresponsive to mini-glucagon (Fig. 8A). However, complementation experiments with T$\beta\gamma$-subunits revealed sensitivity of the system to mini-glucagon (Fig. 8B) (LOTERSZTAJN et al. 1992).

After cholera toxin treatment of Zajdela hepatoma cells, T$\beta\gamma$ no longer reconstituted the response of the Ca^{2+} pump to mini-glucagon, suggesting that the mechanism of $\beta\gamma$ action was dependent on an association with the Gsα-subunit.

In conclusion, the finding that Gsα- and $\beta\gamma$-subunits mediate Ca^{2+} pump inhibition by mini-glucagon would argue for the existence of mini-glucagon receptors. One would expect the existence of receptors specific to mini-glucagon, distinct from the glucagon receptors, since mini-glucagon is ineffective in antagonizing adenylyl cyclase activation by glucagon (PAVOINE et al. 1991). However, these receptors have as yet neither been isolated nor physically characterized.

E. Mini-glucagon Action in Heart

I. Mini-glucagon is a Component of the Positive Inotropic Effect of Glucagon

Since the report by Farah and Tuttle which described the positive inotropic effect of glucagon in a dog heart-lung preparation (FARAH and TUTTLE 1960), there has been a general agreement that glucagon increased cardiac contractile force and heart rate. However, discrepancies existed between the contractile responses to glucagon and the increase in the cyclic AMP levels (CHATELAIN et al. 1983; WILDENTHAL et al. 1976). The conversion of glucagon to mini-glucagon could open a new pathway of glucagon action in the heart.

The inotropic response of electrically stimulated, primary monolayer culture heart cells has been examined in response to glucagon and mini-glucagon. In this study, the cells were perfused with serum-free medium continuously renewed to minimize the degradation of the peptides and the accumulation of metabolites. Under these conditions, $30 nM$ glucagon, added alone, had no immediate effect on the amplitude of cell contraction (Fig. 9A), but evoked a 15%–20% increase over the control amplitude after a 15-min delay (PAVOINE et al. 1991). This could be related to a slow accumulation of mini-glucagon in the cell environment, despite the continuous renewal of the cell medium. At

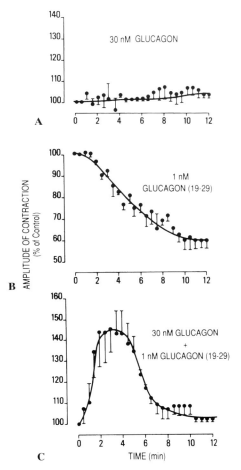

Fig. 9A–C. Time course of amplitude of contraction of cultured embryonic chick heart cells exposed to glucagon, mini-glucagon or both peptides. **A** Glucagon added alone, in conditions in which its degradation is minimized, has no effect on the amplitude of contraction of the myocytes. **B** At 1 nM, mini-glucagon, added alone, exerts a negative effect. **C** Both peptides added in combination produce a marked increase in the amplitude of contraction of the myocytes. This action reproduces the positive inotropic effect observed in vivo after i.v. administration of glucagon. (From Pavoine et al. 1991)

1 nM, mini-glucagon evoked a pronounced decrease in cell contractility (40% maximal decrease under the control level), which persisted beyond 10 min (Fig. 9B)(Pavoine et al. 1991). Higher concentrations of mini-glucagon severely impaired the cell integrity and led to cell death (Pavoine et al. 1991). Interestingly, the combined addition of 30 nM glucagon plus 1 nM mini-glucagon resulted in a positive inotropic effect (Fig. 9C). A 45% increase in the amplitude of contraction over the control level was reached, correspond-

ing to the maximal positive inotropic response of these cells determined using 3.6mM $CaCl_2$. The increase elicited was rapid in onset, reached a maximum at 2-4min, and returned to control level after 8-10min (PAVOINE et al. 1991). This mimicked the cardiac-positive inotropic response observed after in vivo intravenous injection of glucagon, which is characterized by a rapid increase in the amplitude of contraction and a maximal effect in the first few minutes after injection (FARAH 1983; KONES and PHILLIPS 1971).

Importantly, 8-bromo-cyclic AMP could substitute for glucagon in the positive inotropic effect elicited by the combination of glucagon and mini-glucagon (PAVOINE et al. 1991), which argued for cyclic AMP-mediating glucagon action.

The specificity of mini-glucagon action on cardiac cell contractility has been assessed by the study of two other glucagon peptides: a COOH-terminal peptide, glucagon (22-29), which lacks the three Ala^{19}-Gln^{20}-Asp^{21} residues of mini-glucagon, and the NH_2-terminal peptide, glucagon (1-21). Glucagon (22-29) appeared to act as a partial agonist of mini-glucagon. In contrast, glucagon (1-21) had no effect when added alone or in combination with glucagon.

Taken together, these results gave evidence that the effect of glucagon on cardiac contraction depends on both (a) an increase in intracellular cyclic AMP elicited by glucagon and (b) the conversion of glucagon to mini-glucagon, which possesses its own biological activity.

The mechanisms underlying the effects of mini-glucagon on heart contraction do not rely on classical pathways. Mini-glucagon did not evoke any detectable change in either cyclic AMP or cyclic GMP levels (PAVOINE et al. 1991), or IP_3 production, in embryonic chick heart cells (SAUVADET et al. 1996).

II. The Sarcolemmal Ca^{2+} Pump, in Heart, Is a Target for Mini-glucagon Action

In the search for the biochemical mechanism supporting mini-glucagon action in heart, the participation of the Ca^{2+} pump in the sarcolemma was considered, since this system had been previously identified as a target for this peptide in hepatocytes.

The biochemical and molecular characterization of the cardiac SL Ca^{2+} pump is well documented (for reviews see CARAFOLI 1994; GROVER and KHAN 1992). However, both its possible participation in physiological responses and its regulation by hormones have been overlooked. The reasons may be that studies of the hormonal regulation of the ATP-dependent [^{45}Ca] uptake were performed using inside-out oriented vesicles, and were hampered by the difficulty in obtaining large quantities of cardiac sarcolemmal vesicles, freed from contamination by the sarcoplasmic reticulum (SR) membranes. In contrast, Ca^{2+} imaging, combined with the use of specific inhibitors, provided a unique tool to evaluate the activity of Ca^{2+} transport systems in situ.

The effect of mini-glucagon was examined on the sarcolemmal (SL) Ca^{2+} pump activity measured in situ, in single embryonic chick heart ventricular

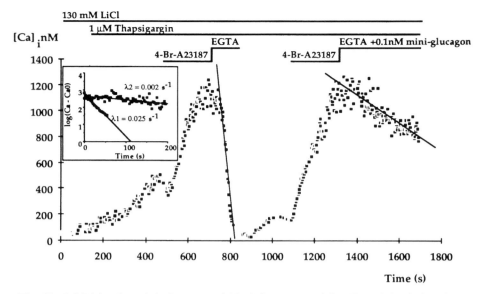

Fig. 10. Inhibition by mini-glucagon of SL Ca^{2+} pump activity. Fura-2-loaded embryonic chick heart cells are bathed in $2\,mM$ Ca^{2+} buffer containing $1\,\mu M$ thapsigargin and in which $130\,mM$ LiCl is substituted for NaCl. $[Ca^{2+}]_i$ pulses are challenged by the addition of $1\,\mu M$ of the Ca^{2+} ionophore 4-bromo-A-23187. The application of an EGTA buffer, free of Ca^{2+} and Ca^{2+} ionophore, containing $130\,mM$ LiCl and $1\,\mu M$ thapsigargin, stops the $[Ca^{2+}]_i$ increase and initiates $[Ca^{2+}]_i$ decay. In these conditions: under null Ca^{2+} gradient conditions, which eliminate passive Ca^{2+} fluxes through the SL membrane, in the presence of LiCl and thapsigargin, which inhibit the Na^+/Ca^{2+} exchange and the SR Ca^{2+} pump, respectively, and with $[Ca^{2+}]_i$ increase of between 0.6 and $1.5\,\mu M$, $[Ca^{2+}]_i$ decay represents the SL Ca^{2+} pump activity. The cell is challenged with two successive $[Ca^{2+}]_i$ pulses: the first pulse is given under basal conditions; in the second pulse, $0.1\,nM$ mini-glucagon is included in the EGTA buffer. *Inset* shows the linear representation of the exponential fits of $[Ca^{2+}]_i$ changes. The kinetic constants are indicated as: $\lambda 1$, first pulse slope in control conditions; $\lambda 2$, second pulse slope in the presence of $0.1\,nM$ mini-glucagon. (From SAUVADET et al. 1995)

cells loaded with Fura-2. Mini-glucagon elicited an immediate, dose-dependent inhibition of the SL Ca^{2+} pump, maximal 80% inhibition being observed with $1\,nM$ mini-glucagon (Fig. 10) (SAUVADET et al. 1995).

The inhibition of the SL Ca^{2+} pump by mini-glucagon was immediate, but without an effect on cytosolic free calcium levels during several minutes. However, after a delay, a $[Ca^{2+}]_i$ oscillatory response was observed. Both effects, immediate inhibition of the SL Ca^{2+} pump and delayed $[Ca^{2+}]_i$ oscillations, were mimicked by vanadate tested at $2\,\mu M$, a concentration at which this compound acted as a specific inhibitor of the SL Ca^{2+} pump (SAUVADET et al. 1995). These observations supported the assumption that $[Ca^{2+}]_i$ spikes and the SL Ca^{2+} pump inhibition were related phenomena, and that inhibition of the SL Ca^{2+} pump might result in an increase in total cellular Ca^{2+}, leading to an overload phenomenon. In this regard, the lethal effect elicited by nanomolar concentrations of mini-glucagon could be explained by such a mechanism (PAVOINE et al. 1991). In fact, it has already been proposed that the development of Ca^{2+} overload and myocardial dysfunction observed in diabetes (HEYLIGER et al. 1987), in genetically linked cardiomyopathy (KUO et al. 1987) and in septic shock (WU and LIU 1992), resulted from the long-term impairment of the SL Ca^{2+} pump.

The SL Ca^{2+} pump cannot be expected to participate, in a quantitatively significant fashion, in a "beat-to-beat" relaxation, due to its high affinity for Ca^{2+} and its low capacity to extrude Ca^{2+} as compared to the Na^+/Ca^{2+} exchanger and the SR Ca^{2+} pump. Nevertheless, the above data highlight the role of the SL Ca^{2+} pump in the regulation of intracellular Ca^{2+} homeostasis.

III. Mini-glucagon Produces Accumulation of Ca^{2+} into the Sarcoplasmic Reticulum Stores

One might have expected an immediate and progressive $[Ca^{2+}]_i$ increase following SL Ca^{2+} pump inhibition by mini-glucagon. In fact, the delay observed in $[Ca^{2+}]_i$ mobilization reflected the temporary ability of the cells to maintain low $[Ca^{2+}]_i$ in the face of a defect in the SL Ca^{2+} pump activity. Mini-glucagon produced a [^{45}Ca] accumulation into intracellular compartments, resistant to digitonin treatment. Those were identified as sarcoplasmic reticulum (SR) stores since they were unloaded upon caffeine application (SAUVADET et al. 1996). When the SR Ca^{2+} storage capacity of the cell was reduced by pretreatment with thapsigargin, a specific inhibitor of the sarcoplasmic Ca^{2+} pump, mini-glucagon produced an immediate rise in $[Ca^{2+}]_i$. These observations illustrated the buffering capacity of the cardiac SR in the face of trans-sarcolemmal Ca^{2+} fluxes and corroborated a previous report by JANCZEWSKI and LAKATTA (1993), who demonstrated a rapid sequestration by the SR of at least 50% of the Ca^{2+} entering the cells during a single post-rest stimulation of guinea pig ventricular myocytes.

Mini-glucagon also evoked delayed $[Ca^{2+}]_i$ oscillations as a result of spontaneous Ca^{2+} release from the SR due to high Ca^{2+} loading (LAKATTA 1992). This phenomenon was interpreted as a mechanism in which the activation of the Ca^{2+}-induced Ca^{2+}-release (CICR) channel did not rely on an increase in resting $[Ca^{2+}]_i$ but, instead, was dependent upon the filling state of the intracellular stores (LIPP and NIGGLI 1994; ENDO 1977; STERN et al. 1988). Thus, the onset of $[Ca^{2+}]_i$ transients elicited by mini-glucagon would account for the time required by the SR compartment to reach the threshold for spontaneous release.

As shown in Fig. 11A,B, mini-glucagon potentiated the $[Ca^{2+}]_i$ transients associated with caffeine contractures (SAUVADET et al. 1996). The application of isradipine, a specific L-type Ca^{2+} channel blocker, did not suppress the effect of mini-glucagon, suggesting that the action of mini-glucagon did not occur

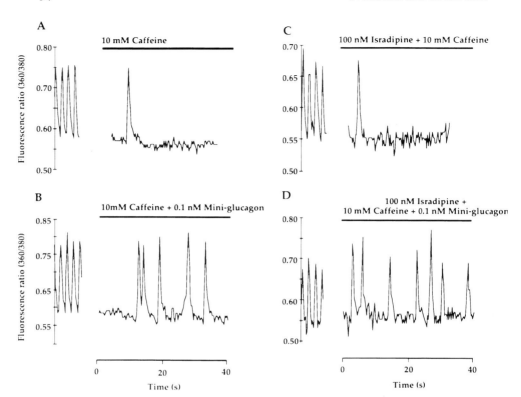

Fig. 11A–D. Mini-glucagon potentiates Ca^{2+} release from the intracellular stores induced by caffeine. Cells are electrically stimulated at 0.5 Hz; stimulation is discontinued a few seconds before the addition of 10 mM caffeine alone (**A**), 10 mM caffeine plus 0.1 nM mini-glucagon (**B**), 10 mM caffeine plus 100 nM isradipine (**C**) or 10 mM caffeine plus 0.1 nM mini-glucagon plus 100 nM isradipine (**D**). **A** Caffeine applied alone evokes a unique $[Ca^{2+}]_i$ transient. **B** In contrast, the application of caffeine together with mini-glucagon results in a train of $[Ca^{2+}]_i$ transients. **C,D** The L-type Ca channel blocker, isradipine, has no effect on those Ca^{2+} responses. (From SAUVADET et al. 1996)

through L-type calcium channel activation (Fig. 11C,D). It is noteworthy that mini-glucagon had no effect on InsP$_3$ production and was unlikely to trigger InsP$_3$-mediated Ca^{2+} mobilization.

Taken together, these results gave further evidence that mini-glucagon action led to Ca^{2+} accumulation in the SR stores.

F. Glucagon and Mini-glucagon Act in Concert

In vivo, since glucagon is converted into mini-glucagon when it interacts with the cardiac cell, one may expect that both peptides act in concert. This proposal was supported by the observation that both peptides in combination reproduced the cardiac-positive inotropic response obtained after intravenous injection of glucagon, while glucagon alone, in conditions in which its conversion to mini-glucagon was limited, was ineffective (PAVOINE et al. 1991). Thus, it was important to evaluate the combined effect of both peptides on $[Ca^{2+}]_i$ metabolism.

The experiments described below were performed using embryonic chick heart cells, electrically prestimulated in order to ensure maximal loading of Ca^{2+} into the SR compartment before the application of the peptide(s). Glucagon (30 nM), perfused alone, produced a small $[Ca^{2+}]_i$ signal, i.e., a single transient over a 180-s period, the amplitude of which was 103% that of control twitch amplitude defined as the mean amplitude of the $[Ca^{2+}]_i$ transients during

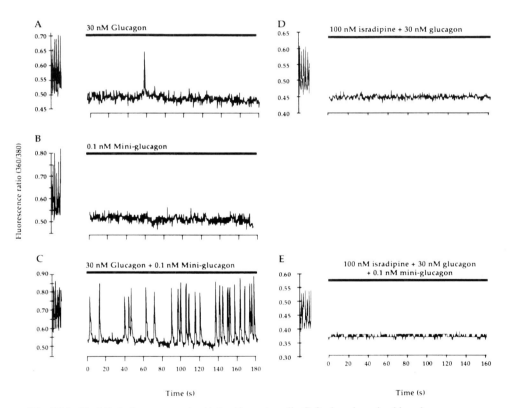

Fig. 12A–E. Mini-glucagon potentiates the cytosolic Ca^{2+} signal evoked by glucagon on cells electrically prestimulated. Cells are stimulated until a steady-state level of the twitches is achieved. Stimulation is then discontinued before the addition of 30nM glucagon (**A**), 0.1 nM mini-glucagon (**B**) or 30 nM glucagon plus 0.1 nM mini-glucagon (**C**). **A** The pattern of Ca^{2+} mobilization elicited by glucagon consists of sporadic Ca^{2+} spikes. **B** Mini-glucagon alone has no effect. **C** Mini-glucagon considerably potentiates Ca^{2+} mobilization elicited by glucagon. **D,E** Isradipine inhibit glucagon action. (From SAUVADET et al. 1996)

electrical stimulation (Fig. 12A). In the same conditions, mini-glucagon alone, at 0.1 nM, did not trigger any $[Ca^{2+}]_i$ signal (Fig. 12B). In contrast, the combination of mini-glucagon with glucagon elicited a train of $[Ca^{2+}]_i$ spikes (23 ± 2 $[Ca^{2+}]_i$ transients over a 180-s period with a mean amplitude of 140% ± 5% of the control twitch amplitude), which represented a considerable amplification of the signal triggered by glucagon alone (Fig. 12C). The potentiation of glucagon action by mini-glucagon was consistent with the statement that CICR is controlled by the filling state of the SR compartment, and more readily triggered under improved SR loading (CHENG et al. 1993). 8-Bromo-cyclic AMP could replace glucagon, and this effect by glucagon was blocked by isradipine, suggesting that it relied on the sequential stimulation of adenylyl cyclase and activation, via phosphorylation, of L-type Ca^{2+} channels (Fig. 12D,E).

Taken together these data gave evidence that the action of glucagon in heart relied not only on a cyclic AMP pathway but also on a cyclic AMP-independent Ca^{2+} loading of the SR. On the same line of evidence, it is well established that β_1- and β_2-adrenergic agonists stimulate adenylyl cyclase activity. Nevertheless, while stimulation of contraction evoked by β_1-agonists paralleled the increase in cAMP, it appeared that the effects of β_2-agonists on contraction were completely dissociated from the production of cAMP (MILANO et al. 1994; XIAO et al. 1994). From these studies it was concluded that changes in contraction produced by the β_2-agonists required the activation of signal transduction pathway(s) other than cAMP. Also stimulation of prostaglandin E_1 receptors in heart cells induced an increase in cyclic AMP but had no effect on contraction (HAYES et al. 1980).

These results suggest the existence of cyclic AMP pools not linked to Ca^{2+} and contractile regulation (BUXTON and BRUNTON 1983), and would prove that cAMP is necessary but may be not sufficient by itself to produce heart cell contraction. One critical observation is that glucagon, through its dual mechanism, has proven more efficient than β-adrenergic agonists in reversing profound myocardial depressions, in particular those induced by Ca^{2+} channel blocker toxicity (ZARITSKY et al. 1988).

G. Conclusion and Perspectives

Data from the past few years have shed new light on the physiological pathways of glucagon action. It now seems necessary to discriminate between the effects of glucagon itself, "the mother" molecule, from those elicited by mini-glucagon, its "daughter" metabolite. Furthermore, while an increase in cAMP still remains the classical message expected following glucagon receptor activation, a new concept concerns the diversity of transduction pathways implicated. In fact, an increase in cAMP may be due to either the activation of adenylyl cyclase via Gs, or to the inhibition of a cAMP phosphodiesterase via Gi.

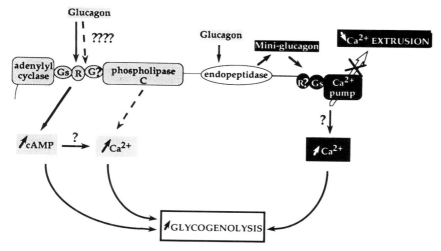

Fig. 13. Pathways of glucagon action in liver. Glucagon increases cellular cAMP via adenylyl cyclase activation, and mobilizes Ca^{2+}. This results in the activation of glycogenolysis. Whether Ca^{2+} mobilization is due to direct or indirect activation of phospholipase C is still a matter of debate. The hormone is also processed into mini-glucagon which inhibits the plasma membrane Ca^{2+} pump and increases glycogenolysis

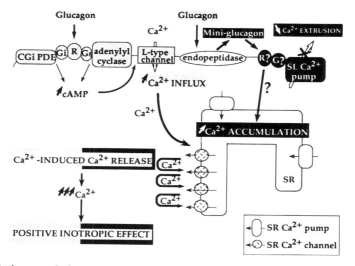

Fig. 14. Pathways of glucagon action in heart. Glucagon action in heart is due to the combination of: (a) the action of glucagon, the "mother molecule" mediated through its receptor and (b) the effect of mini-glucagon, the "daughter" metabolite, released after the processing of glucagon by an endopeptidase located on the outside of the plasma membrane. Glucagon increases cellular cAMP via either adenylyl cyclase activation or cAMP-phosphodiesterase inhibition. This leads to an influx of Ca^{2+} through the L-type Ca^{2+} channel which triggers mobilization of the Ca^{2+} accumulated in the SR stores. Mini–glucagon inhibits the Ca^{2+} pump in the sarcolemmal membrane and induces Ca^{2+} accumulation into the SR stores. The combined actions of both peptides trigger a CICR phenomenon which is likely to support the positive inotropic effect of glucagon, in vivo

An important controversy still exists concerning the unity of the glucagon second messenger in liver. The role of $InsP_3$ as the mediator of glucagon actions, such as glucagon-induced $[Ca^{2+}]_i$ mobilization, is still a matter of active debate. Also, effects of glucagon on both $InsP_3$ and $[Ca^{2+}]_i$ have not been convincingly demonstrated to constitute cAMP-independent features.

Finally, evidence has been presented demonstrating that glucagon may also act as a prohormone. Thus, its interaction with its target tissues leads to the production of an active metabolite, mini-glucagon. Mini-glucagon possesses its own biological activity, independent of cAMP and $InsP_3$ pathways. This peptide, acting on intracellular Ca^{2+}, potentiates glucagon action. In liver, mini-glucagon increases glycogenolysis, thus mimicking glucagon action (Fig. 13). In heart, both molecules act "in concert," to mobilize Ca^{2+} from the intracellular stores (Fig. 14). Their actions may be summarized as the ability of mini-glucagon to accumulate Ca^{2+} in sarcoplasmic reticulum stores (SR) and that of glucagon to induce CICR from the same stores. In this last case, the combined actions of glucagon and mini-glucagon are necessary to support the positive inotropic effect triggered by intravenous injection of glucagon on heart contraction.

Acknowledgments. We gratefully acknowledge the important contributions of colleagues and collaborators in the work reviewed here. We also thank S. Lotersztajn, T. Rohn and B. Van Tine for helpful discussion, and J. Hanoune for his permanent support. This work was supported by the Institut National de la Santé et de la Recherche Médicale, the French Ministère de la Recherche et de l'Enseignement, l'Unité de Formation et de Recherche de Médecine, Créteil, Paris-Val de Marne, the Association Française contre les Myopathies, and the North Atlantic Treaty Organization.

References

Abou-Samra AB, Jüppner H, Force T, Freeman MW, Kong XF, Schipani E, Urena P, Richards J, Bonventre JV, Potts JT Jr, Kronenberg HM, Segre GV (1992) Expression cloning of a common receptor for parathyroid hormone and parathyroid hormone-related peptide from rat osteoblast-like cells: a single receptor stimulates intracellular accumulation of both cAMP and inositol triphosphates and increases free calcium. Proc Natl Acad Sci USA 89:2732–2736

Altin JG, Bygrave FL (1986) Synergistic stimulation of Ca^{2+} uptake by glucagon and Ca^{2+}-mobilizing hormones in the perfused rat liver. A role for mitochondria in long-term Ca^{2+} homeostasis. Biochem J 238:653–661

Beavo JA, Reifsnyder DH (1990) Primary sequence of cyclic nucleotide phosphodiesterase isozymes and the design of selective inhibitors. Trends Pharmacol Sci 11:150–155

Benedetti A, Graf P, Fulceri R, Romani A, Sies H (1989) Ca^{2+} mobilization by vasopressin and glucagon in perfused livers. Effect of prior intoxication with bromotrichloromethane. Biochem Pharmacol 38:1799–1805

Blache P, Kervran A, Martinez J, Bataille D (1988a) Development of an oxyntomodulin/glicentin C-terminal radioimmunoassay using a "thiol-maleoyl" coupling method for preparing the immunogen. Anal Biochem 173:171–179

Blache P, Kervran A, Bataille D (1988b) Oxyntomodulin and glicentin: brain-gut peptides in the rat. Endocrinology 123:2782–2787

Blache P, Kervran A, Le-Nguyen D, Laur J, Cohen-Solal A, Devilliers G, Mangeat P, Martinez J, Bataille D (1988c) The glucagon-containing peptides and their fragments in the rat gastro-entero-pancreatic and central nervous systems. Biomed Res 9 [Suppl 3]:19–28

Blache P, Kervran A, Dufour M, Martinez J, Le-Nguyen D, Lotersztajn S, Pavoine C, Pecker F, Bataille D (1990) Glucagon (19-29), a Ca^{2+} pump inhibitory peptide, is processed from glucagon in the rat liver plasma membrane by a thiol endopeptidase. J Biol Chem 265:21514–21519

Blache P, Kervran A, Le-Nguyen D, Dufour M, Cohen-Solal A, Duckworth W, Bataille D (1993) Endopeptidase from rat liver membranes which generates miniglucagon from glucagon. J Biol Chem 268:21748–21753

Blackmore PF, Exton JH (1986) Studies on the hepatic calcium-mobilizing activity of alúminium fluoride and glucagon. Modulation by cAMP and phorbol myristate acetate. J Biol Chem 261:11056–11063

Bocckino SB, Blackmore PF, Exton JH (1985) Stimulation of 1,2-diacylglycerol accumulation in hepatocytes by vasopressin, epinephrine, and angiotensin II. J Biol Chem 260:14201–14207

Brechler V, Pavoine C, Hanf R, Garbarz E, Fischmeister R, Pecker F (1992) Inhibition by glucagon of the cGMP-inhibited low-K_m cAMP phosphodiesterase in heart is mediated by a pertussis toxin-sensitive G-protein. J Biol Chem 267:15496–15501

Burgess GM, Dooley RK, McKinney JS, Nånberg E, Putney JW Jr (1986) Further studies on the interactions between the calcium mobilization and cyclic AMP pathways in guinea pig hepatocytes. Mol Pharmacol 30:315–320

Burgess GM, Bird G St J, Obie JF Putney JW Jr (1991) The mechanism for synergism between phospholipase C- and adenylylcyclase-linked hormones in liver. Cyclic AMP-dependent kinase augments inositol triphosphate-mediated Ca^{2+} mobilization without increasing the cellular levels of inositol polyphosphates. J Biol Chem 266:4772–4781

Buxton ILO, Brunton LL (1983) Compartments of cyclic AMP and protein kinase in mammalian cardiomyocytes. J Biol Chem 258:10233–10239

Bygrave FL, Benedetti A (1993) Calcium: its modulation in liver by cross-talk between the actions of glucagon and calcium-mobilizing agonists. Biochem J 296:1–14

Bygrave FL, Gamberucci A, Fulceri R, Benedetti A (1993) Evidence that stimulation of plasma-membrane Ca^{2+} inflow is an early action of glucagon and dibutyryl cyclic AMP in rat hepatocytes. Biochem J 292:19–22

Capiod T, Noel J, Combettes L, Claret M (1991) Cyclic AMP-evoked oscillations of intracellular [Ca^{2+}] in guinea-pig hepatocytés. Biochem J 275:277–280

Carafoli E (1994) Biogenesis: plasma membrane calcium ATPase: 15 years of work on the purified enzyme. FASEB J 8:993–1002

Chabre O, Conklin BR, Lin HY, Lodish HF, Wilson E, Ives HE, Catanzariti L, Hemmings BA, Bourne HR (1992) A recombinant calcitonin receptor independently stimulates 3′,5′-cyclic adenosine monophosphate and Ca^{2+}/inositol phosphate signaling pathways. Mol Endocrinol 6:551–556

Charest R, Blackmore PF, Berthon B, Exton JH (1983) Changes in free cytosolic Ca^{2+} in hepatocytes following alpha-1-adrenergic stimulation. Studies on Quin-2-loaded hepatocytes. J Biol Chem 258:8769–8773

Chatelain P, Robberecht P, Waelbroeck M, De Neef P, Camus JC, Huu AN, Roba J, Christophe J (1983) Topographical distribution of the secretin- and VIP-stimulated adenylate cyclase system in the heart of five animal species. Pflugers Arch 397:100–105

Cheng H, Lederer WJ, Cannell MB (1993) Calcium sparks: elementary events underlying excitation-contraction coupling in heart muscle. Science 262:740–744

Clark CM Jr, Waller D, Kohalmi D, Gardner R, Clark J, Levey GS, Wildenthal K, Allen D (1976) Evidence that cyclic AMP is not involved in the chronotropic action of glucagon in the adult mouse heart. Endocrinology 99:23–29

Combettes L, Berthon B, Binet A, Claret M (1986) Glucagon and vasopressin interactions on Ca^{2+} movements in isolated hepatocytes. Biochem J 237:675–683

Connelly PA, Parker Botelho LH, Sisk RB, Garrison JC (1987) A study of the mechanism of glucagon-induced protein phosphorylation in isolated rat hepatocytes using (S_p)-cAMPS and (R_p)-cAMPS, the stimulatory and inhibitory diastereomers of adenosine cyclic 3',5'-phosphorothioate. J Biol Chem 262:4324–4332

Conti M, Jin CSL, Monaco L, Repaske DR, Swinnen JV (1991) Hormonal regulation of cyclic nucleotide phosphodiesterases. Endocr Rev 12:218–234

Corvera S, Huerta-Bahena J, Pelton JT, Hruby VJ, Trivedi D, Garcia-Sainz JA (1984) Metabolic effects and cyclic AMP levels produced by glucagon, (1-N alpha-trinitrophenylhistidine, 12-homoarginine) glucagon and forskolin in isolated rat hepatocytes. Biochim Biophys Acta 804:434–441

Desbuquois B, Krug F, Cuatrecasas P (1974) Inhibitors of glucagon inactivation. Effect on glucagon-receptor interactions and glucagon-stimulated adenylate cyclase activity in liver cell membranes. Biochim Biophys Acta 343:101–120

Endo M (1977) Calcium release from the sarcoplasmic reticulum. Physiol Rev 57:71–108

Exton JH, Robison GA, Sutherland EW, Park CR (1971) Studies on the role of adenosine 3',5'-monophosphate in the hepatic actions of glucagon and catecholamines. J Biol Chem 246:6166–6177

Farah AE (1983) Glucagon and the heart. In: Lefèbvre PJ (ed) Glucagon II. Springer, Berlin Heidelbery New York, pp 552–609 (Handbook of experimental pharmacology, vol 66/II)

Farah AE, Tuttle R (1960) Studies of the pharmacology of glucagon. J Pharmacol Exp Ther 129:49–55

Fischmeister R, Hartzell HC (1987) Cyclic guanosine 3',5'-monophosphate regulates the calcium current in single cells from frog ventricle. J Physiol (Lond) 387:453–472

Fischmeister R, Hartzell HC (1990) Regulation of calcium current by low-Km cyclic AMP phosphodiesterases in cardiac cells. Mol Pharmacol 38:426–433

Force T, Bonventre JV, Flannery MR, Gorn AH, Yamin M, Goldring SR (1992) A cloned porcine renal calcitonin receptor couples to adenylate cyclase and phospholipase C. Am J Physiol 262:F1110–F1115

Fuller RS, Brake AJ, Thorner J (1989) Intracellular targeting and structural conservation of a prohormone-processing endoprotease. Science 246:482–486

Garcia-Sainz JA, Sánchez-Sevilla L, Pelton JT, Trivedi D, Hruby VJ (1986) Effects of [1-N$^\alpha$-trinitrophenylhistidine, 12-homoarginine] glucagon on cyclic AMP levels and free acid release in isolated rat adipocytes. Biochim Biophys Acta 886:310–315

Garcia-Sainz JA, Marcias-Silva M, Hernández-Sotomayor SMT, Torres-Márquez MaE, Trivedi D, Hruby VJ (1990) Modulation of glucagon actions by phorbol myristate acetate in isolated hepatocytes. Effect of hypothyroidism. Cell Signal 2:235–243

Gettys TW, Blackmore PF, Corbin JD (1988) An assessment of phosphodiesterase activity in situ after treatment of hepatocytes with hormones. Am J Physiol 254:E449–E453

Gluschankof P, Cohen P (1987) Proteolytic enzymes in the post-translational processing of polypeptide hormone precursors. Neurochem Res 12:951–958

Grover AK, Khan I (1992) Calcium pump isoforms: diversity, selectivity and plasticity. Cell Calcium 13:9–17

Hall-Boyer K, Zaloga GP, Chernow B (1984) Glucagon: hormone or therapeutic agent? Crit Care Med 12:584–589

Hartzell HC, Fischmeister R (1986) Opposite effects of cyclic GMP and cyclic AMP on Ca^{2+} current in single heart cells. Nature 323:273–275

Hayes JS, Brunton LL, Mayer SE (1980) Selective activation of particulate cAMP-dependent protein kinase by isoproterenol and prostaglandin E1. J Biol Chem 255:5113–5119

Heyliger CE, Prakash A, McNeill JH (1987) Alterations in cardiac sarcolemmal Ca^{2+} pump activity during diabetes mellitus. Am J Physiol 252:540–544

Heyworth CM, Wallace AV, Houslay MD (1983) Insulin and glucagon regulate the activation of two distinct membrane-bound cyclic AMP phosphodiesterases in hepatocytes. Biochem J 214:99–110

Horne WC, Shyu JF, Chakraborty M, Baron R (1994) Signal transduction by calcitonin. Multiple ligands, receptors, and signaling pathways. Trends Endocrinol Metab 5:395–401

Iyengar R, Rich KA, Herberg JT, Premont RT, Codina J (1988) Glucagon receptor-mediated activation of Gs is accompanied by subunit dissociation. J Biol Chem 263:15348–15353

Janczewski AM, Lakatta EG (1993) Thapsigargin inhibits Ca^{2+} uptake, and depletes Ca^{2+} sarcoplasmic reticulum in intact cardiac myocytes. Am J Physiol 265:H517–H522

Jelinek LJ, Lok S, Rosenberg GB, Smith RA, Grant FJ, Biggs S, Bensch PA, Kuijper JL, Sheppard PO, Sprecher CA, O'Hara PJ, Foster D, Walker KM, Chen LHJ, McKernan PA, Kindsvogel W (1993) Expression cloning and signaling properties of the rat glucagon receptor. Science 259:1614–1616

Jouneaux C, Audigier Y, Goldsmith P, Pecker F, Lotersztajn S (1993) Gs mediates hormonal inhibition of the calcium pump in liver plasma membranes. J Biol Chem 268:2368–2372

Kass GEN, Llopis J, Chow SC, Duddy SK, Orrenius S (1990) Receptor-operated calcium influx in rat hepatocytes. Identification and characterization using manganese. J Biol Chem 265:17486–17492

Kiss Z, Tkachuk VA (1984) Guanine-nucleotide-dependent inhibition of adenylate cyclase of rabbit heart by glucagon. Eur J Biochem 142:323–328

Kones RJ, Phillips JH (1971) Glucagon: present status in cardiovascular disease. Clin Pharmacol Ther 12:427–444

Kraus-Friedmann N (1995) Hormonal regulation of cytosolic calcium levels in the liver. Braz J Med Biol Res 28:275–284

Kuo TH, Tsang W, Wiener J (1987) Defective Ca^{2+} pumping ATPase of heart sarcolemma from cardiomyopathic hamster. Biochim Biophys Acta 900:10–16

Lakatta EG (1992) Functional implications of spontaneous sarcoplasmic reticulum Ca^{2+} release in the heart. Cardiovasc Res 26:193–214

Lipp P, Niggli E (1994) Modulation of Ca^{2+} release in cultured neonatal rat cardiac myocytes. Insight from subcellular release patterns revealed by confocal microscopy. Circ Res 74:979–990

Loten EG, Assimacopoulos-Jeannet FD, Exton JH, Park CR (1978) Stimulation of a low Km phosphodiesterase from liver by insulin and glucagon. J Biol Chem 253:746–757

Lotersztajn S, Pecker F (1982) A membrane-bound protein inhibitor of the high affinity Ca ATPase in rat liver plasma membranes. J Biol Chem 257:6638–6641

Lotersztajn S, Hanoune J, Pecker F (1981) A high affinity calcium-stimulated magnesium-dependent ATPase in rat liver plasma membranes. Dependence on an endogenous protein activator distinct from calmodulin. J Biol Chem 256:11209–11215

Lotersztajn S, Epand RM, Mallat A, Pecker F (1984) Inhibition by glucagon of the calcium pump in liver plasma membranes. J Biol Chem 259:8195–8201

Lotersztajn S, Pavoine C, Brechler V, Roche B, Dufour M, Le-Nguyen D, Bataille D, Pecker F (1990) Glucagon (19-29) exerts a biphasic action on the liver plasma membrane Ca^{2+} pump which is mediated by G proteins. J Biol Chem 265:9876–9880

Lotersztajn S, Pavoine C, Deterre P, Capeau J, Mallat A, Le-Nguyen D, Dufour M, Rouot B, Bataille D, Pecker F (1992) Role of G protein $\beta\gamma$ subunits in the regulation of the plasma membrane Ca^{2+} pump. J Biol Chem 267:2375–2379

MacNeil DJ, Occi JL, Hey PJ, Strader CD, Graziano MP (1994) Cloning and expression of a human glucagon receptor. Biochem Biophys Res Commun 198:328–334

Mallat A, Pavoine C, Lotersztajn S, Pecker F (1985) Inhibition of the Ca pump in liver plasma membranes by glucagon is due to a metabolite of the hormone. Fed Proc 44:1392

Mallat A, Pavoine C, Dufour M, Lotersztajn S, Bataille D, Pecker F (1987) A glucagon fragment is responsible for the inhibition of the liver Ca^{2+} pump by glucagon. Nature 325:620–622

Martinez J, Potier P (1986) Peptide hormones as prohormone. Trends Pharmacol Sci 139–147

Mauger JP, Claret M (1986) Mobilization of intracellular calcium by glucagon and cyclic AMP analogues in isolated rat hepatocytes. FEBS Lett 195:106–110

Mauger JP, Poggioli J, Claret M (1985) Synergistic stimulation of the Ca^{2+} influx in rat hepatocytes by glucagon and the Ca^{2+}-linked hormones vasopressin and angiotensin II. J Biol Chem 260:11635–11642

Méry PF, Brechler V, Pavoine C, Pecker F, Fischmeister R (1990) Glucagon stimulates the cardiac Ca^{2+} current by activation of adenylyl cyclase and inhibition of phosphodiesterase. Nature 345:158–161

Méry PF, Pavoine C, Belhassen L, Pecker F, Fischmeister R (1993) Nitric oxide regulates cardiac Ca^{2+} current. Involvement of cGMP-inhibited and cGMP-stimulated phosphodiesterases through guanylyl cyclase activation. J Biol Chem 268:26286–26295

Méry PF, Pavoine C, Pecker F, Fischmeister R (1995) EHNA (erythro-9-[2-hydroxy-3-nonyl] adenine) inhibits cGMP-stimulated phosphodiesterase in isolated cardiac myocytes. Mol Pharmacol 48:121–130

Milano CA, Allen LF, Rockman HA, Dolber PC, McMinn TR, Chien KR, Johnson TD, Bond RA, Lefkowitz RJ (1994) Enhanced myocardial function in transgenic mice overexpressing the β_2-adrenergic receptor. Science 264:582–586

Mine T, Kojima I, Ogata E (1988) Evidence of cyclic AMP-independent action of glucagon on calcium mobilization in rat hepatocytes. Biochim Biophys Acta 970:166–171

Muller B, Lugnier C, Stoclet JC (1990) Implication of cyclic AMP in the positive inotropic effects of cyclic GMP-inhibited cyclic AMP phosphodiesterase inhibitors on guinea pig isolated left atria. J Cardiovasc Pharmacol 15:444–451

Muller B, Komas N, Keravis T, Lugnier C (1993) Les phosphodiestérases des nucléotides cycliques. Med Sci 9:1335–1341

Pavoine C, Lotersztajn S, Mallat A, Pecker F (1987) The high affinity (Ca^{2+}-Mg^{2+}) ATPase in liver plasma membranes is a Ca^{2+} pump. Reconstitution of the purified enzyme into phospholipid vesicles. J Biol Chem 262:5113–5117

Pavoine C, Brechler V, Kervran A, Blache P, Le-Nguyen D, Laurent S, Bataille D, Pecker F (1991) Miniglucagon [glucagon (19-29)] is a component of the positive inotropic effect of glucagon. Am J Physiol 260:C993–C999

Pittner RA, Fain JN (1989) Exposure of cultured hepatocytes to cyclic AMP enhances the vasopressin-mediated stimulation of inositol phosphate production. Biochem J 257:455–460

Pittner RA, Fain JN (1991) Activation of membrane protein kinase C by glucagon and Ca^{2+}-mobilizing hormones in cultured rat hepatocytes. Role of phosphatidylinositol and phosphatidylcholine hydrolysis. Biochem J 277:371–378

Poggioli J, Mauger JP, Claret M (1986) Effect of cyclic AMP-dependent hormones and Ca^{2+}-mobilizing hormones on the Ca^{2+} influx and polyphosphoinositide metabolism in isolated rat hepatocytes. Biochem J 235:663–669

Pohl SL, Birnbaumer L, Rodbell M (1971) The glucagon sensitive adenylyl cyclase system in plasma membranes of rat liver. I. Properties. J Biol Chem 246:1849–1856

Robberecht P, Damien C, Moroder L, Coy DH, Wünsch E, Christophe J (1988) Receptor occupancy and adenylate cyclase activation in rat liver and heart membranes by 10 glucagon analogs modified in position 2,3,4,25,27 and/or 29. Regul Pept 21:117–128

Robison GA, Exton JH, Park CR, Sutherland EW (1967) Effect of glucagon and epinephrine on cyclic AMP levels in rat liver. Fed Proc 26:257

Robson RH (1980) Glucagon for beta-blocker poisoning. Lancet I:1357–1358

Rose K, Savoy LA, Muir AV, Davies JG, Offord RE, Turcatti G (1988) Insulin proteinase liberates from glucagon a fragment known to have enhanced activity against Ca^{2+} + Mg^{2+}-dependent ATPase. Biochem J 256:847–851

Sauvadet A, Pecker F, Pavoine C (1995) Inhibition of the sarcolemmal Ca^{2+} pump in embryonic chick heart cells by mini-glucagon. Cell Calcium 18:76–85

Sauvadet A, Rohn T, Pecker F, Pavoine C (1996) Synergistic actions of glucagon and mini-glucagon on calcium mobilization in cardiac cells. Circ Res 78:102–109

Segre GV, Goldring SR (1993) Receptors for secretin, calcitonin, parathyroid hormone (PTH)/PTH-related peptide, vasoactive intestinal peptide, glucagonlike peptide 1, growth hormone-releasing hormone, and glucagon belong to a newly discovered G-protein-linked receptor family. Trends Endocrinol Metab 4:309–314

Seidah NG, Day R, Marcinkiewicz M, Chrétien M (1993) Mammalian paired basic amino acid convertases of prohormones and proproteins. Ann N Y Acad Sci 680:135–146

Sener EK, Gabe S, Henry JA (1995) Response to glucagon in imipramine overdose. J Toxicol Clin Toxicol 33:51–53

Sheetz MJ, Tager HS (1988) Characterization of a glucagon receptor-linked protease from canine hepatic plasma membranes. Partial purification, kinetic analysis, and determination of sites for hormone processing. J Biol Chem 263:19210–19217

Simmons MA, Hartzell HC (1988) Role of phosphodiesterase in regulation of calcium current in isolated cardiac myocytes [published erratum appears (1988) in Mol Pharmacol 34:604]. Mol Pharmacol 33:664–671

Sistare FD, Picking RA, Haynes RC Jr (1985) Sensitivity of the response of cytosolic calcium in Quin-2-loaded rat hepatocytes to glucagon, adenine nucleosides, and adenine nucleotides. J Biol Chem 260:12744–12747

Smeekens SP, Steiner DF (1990) Identification of a human insulinoma cDNA encoding a novel mammalian protein structurally related to the yeast dibasic processing protease Kex2. J Biol Chem 265:2997–3000

Somogyi R, Zhao M, Stucki JW (1992) Modulation of cytosolic-$[Ca^{2+}]$ oscillations in hepatocytes results from cross-talk among second messengers. Biochem J 286:869–877

Spengler D, Waeber C, Pantaloni C, Holsboer F, Bockaert J, Seeburg SH, Journot L (1993) Differential signal transduction by five splice variants of the PACAP receptor. Nature 365:170–175

Staddon JM, Hansford RG (1986) 4β-Phorbol 12-myristate 13-acetate attenuates the glucagon-induced increase in cytoplasmic free Ca^{2+} concentration in isolated rat hepatocytes. Biochem J 238:737–743

Staddon JM, Hansford RG (1989) Evidence indicating that the glucagon-induced increase in cytoplasmic free Ca^{2+} concentration in hepatocytes is mediated by an increase in cyclic AMP concentration. Eur J Biochem 179:47–52

Stalmans W (1983) Glucagon and liver glycogen metabolism. In: Lefèbvre PJ (ed) Glucagon I. Berlin Heidelberg New York, pp 291–314 (Handbook of experimental pharmacology, vol 66/I)

Stern MD, Capogrossi MC, Lakatta EG (1988) Spontaneous calcium release from the sarcoplasmic reticulum in myocardial cells: mechanisms and consequences. Cell Calcium 9:247–256

Trimble ER, Bruzzone R, Biden TJ, Meehan CJ, Andreu D, Merrifield RB (1987) Secretin stimulates cyclic AMP and inositol triphosphate production in rat pancreatic acinar tissue by two fully independent mechanisms. Proc Natl Acad Sci USA 84:3146–3150

Unger RH, Orci L (1990) Glucagon. In: Rifkin H, Porte D (eds) Diabetes mellitus. Elsevier, New York, pp 104–120

Wakelam MJO, Murphy GJ, Hruby VJ, Houslay MD (1986) Activation or two signal-transduction systems in hepatocytes by glucagon. Nature 323:68–71

Whipps DE, Armston AE, Pryor HJ, Halestrap AP (1987) Effects of glucagon and Ca^{2+} on the metabolism of phosphatidylinositol 4-phosphate and phosphatidylinositol 4,5-bisphosphate in isolated rat hepatocytes and plasma membranes. Biochem J 241:835–845

Wildenthal K, Allen DO, Karlsson J, Wakeland JR, Clark CM Jr (1976) Responsiveness to glucagon in fetal hearts. Species variability and apparent disparities between changes in beating, adenylate cyclase activation and cyclic AMP concentration. J Clin Invest 57:551–558

Wu LL, Liu MS (1992) Heart sarcolemmal Ca^{2+} transport in endotoxin shock: I. Impairment of ATP-dependent Ca^{2+} transport. Mol Cell Biochem 112:125–133

Xiao RP, Hohl C, Altschuld R, Jones L, Livingston B, Ziman B, Tantini B, Lakatta EG (1994) Beta 2-adrenergic receptor-stimulated increase in cAMP in rat heart cells is not coupled to changes in Ca^{2+} dynamics, contractility, or phospholamban phosphorylation. J Biol Chem 269:19151–19156

Zaloga GP, Delacey W, Holmboe E, Chernow B (1986) Glucagon reversal of hypotension in a case of anaphylactoid shock. Ann Intern Med 105:65–66

Zaritsky AL, Horowitz M, Chernow B (1988) Glucagon antagonism of calcium channel blocker-induced myocardial dysfunction. Crit Care Med 16:246–251

CHAPTER 6
Pulsatility of Glucagon

P. LEFÈBVRE, G. PAOLISSO, and A.J. SCHEEN

A. Introduction

As reviewed elsewhere (LEFÈBVRE et al. 1987), various observations made in vivo and in vitro have shown that glucagon secretion is not a continuous, but rather a pulsatile, process. Furthermore, administration of a given amount of glucagon either in vitro or intravenously in vivo is more efficient, especially in stimulating liver glucose production, when done in pulses instead of continuously. These observations will be summarized in the present chapter.

B. Oscillations in Glucagon Plasma Levels

I. Animal Studies

GOODNER et al. (1977) were the first to report, in overnight fasted rhesus monkeys, the occurrence of synchronous, regular oscillations in the peripheral plasma concentrations of glucose, insulin and glucagon. The oscillations displayed a mean period of 9 min, and the amplitudes for insulin and glucagon were ten- and fivefold greater than those for glucose. Insulin cycled in, and glucagon out, of phase with glucose. Subsequent measurements further showed that plasma C-peptide levels were oscillating in synchrony with glucose (KOERKER et al. 1978). Direct sampling in the portal vein of two baboons confirmed that regular insulin cycles (period 9–10 min) were consistently followed by glucagon pulses 4–5 min later (JASPAN et al. 1986). These latter experiments also revealed that peaks in the profile of pancreatic polypeptide (PP) occurred in synchrony with either insulin or glucagon, whereas, in these animals, the fluctuations of somatostatin were erratic, without significant periodicity. In the study reported in the dog by JASPAN et al. (1986), frequent sampling from the portal vein revealed basal, spontaneous oscillations of glucagon, insulin, somatostatin and pancreatic polypeptide. Moreover, the pulsatility of C-peptide levels was remarkably parallel to that of insulin, whereas glucose fluctuations were inconsistent. Insulin and glucagon oscillations were the most prominent, were in phase and had a period of 10–14 min. Pulses of pancreatic polypeptide were less frequent, though always associated with insulin pulses. Somatostatin pulses were less consistently associated with those of other peptides. Interestingly, in one of nine dogs, pulsatility of all four hormones included components of both shorter and longer periods. In longer experiments (up to 12 h) carried out in fasted dogs by SIREK et al. (1985), less frequent blood sampling (every 7.5–15 min) evidenced slow fluctuations in portal and peripheral concentrations of insulin, glucagon and somatostatin, whereas glucose levels did not significantly fluctuate. The period of these oscillations ranged between 32 and 107 min. On the other hand, in three pancreatectomized dogs, extrapancreatic glucagon and somatostatin appeared to be secreted in nonperiodic, randomly occurring pulses (SIREK et al. 1985).

II. Human Studies

The existence of plasma glucagon cycles in the fasted man has been established by LANG et al. (1982). In peripheral venous blood, glucagon cycles had a mean period of 13.7 min and a mean amplitude of 5.5 ng/l. In these subjects, plasma insulin cycles were also demonstrated with a mean period of 10.7 min and a mean amplitude of 1.1 mU/l. There was a significant correlation between the amplitude of simultaneous plasma insulin and glucagon cycles and crosscorrelation showed that the changes in plasma glucagon levels lagged only 2 min behind the changes in plasma insulin levels. In another study, performed in fasted volunteers by HANSEN et al. (1982), plasma levels of insulin were found to oscillate with a sustained periodicity of 11–13 min. The periodicity was similar (10–13 min) for plasma glucose levels, but ranged from 7 to 26 min for plasma glucagon levels. In that study, no consistent relationship could be found by cross-correlation analysis between the periodic fluctuations in insulin, glucose and glucagon. Taken together these observations make it unlikely that fluctuations in glucose levels are driving pulsatile secretion by A and B cells of the islets of Langerhans.

C. Pulsatile Glucagon Secretion In Vitro

Sustained oscillations in the release of glucagon from the isolated canine pancreas perfused at a constant concentration of glucose were convincingly demonstrated by STAGNER et al. (1980). Insulin and somatostatin cycles were also identified. Insulin and somatostatin cycles were in phase, but glucagon cycles were less regular, shorter and not consistently 90° out of phase with insulin cycles. Factors affecting these cycles have been reviewed by STAGNER (1991). More recently, LECLERCQ-MEYER and MALAISSE (1994) confirmed the existence of a pulsatile pattern of glucagon, insulin and somatostatin release in the arginine-stimulated isolated perfused rat pancreas. In this system, glucagon pulses had a periodicity of 5.6 ± 0.4 min. Furthermore, BERTS et al. (1995) have demonstrated that mice A cells have intrinsic abilities to generate oscillatory Ca^{2+} signals, a finding that indicates that the pulsatile release of glucagon does not require functional coupling to nearby B cells. BODE et al. (1994) have also described spontaneous calcium oscillations in clonal endocrine pancreatic glucagon-secreting cells.

D. Pulsatile Glucagon Delivery In Vitro

GOODNER (1994) has recently reviewed the in vitro systems available for modelling the target tissue responses to secretory pulses of glucagon. Indeed, WEIGLE et al. (1984) were the first to show that administering glucagon as a series of brief pulses to perifused rat hepatocytes resulted in the production of a greater total amount of glucose than was obtained when the same amount of glucagon was administered as a continuous infusion. The response augmentation by a pulsatile glucagon administration was interpreted as a delayed relaxation in hepatocyte glucose production after termination of each hormone pulse. Using a model based on the waveform of the hepatocyte response to a transient glucagon stimulus, WEIGLE et al. (1985) demonstrated that the time constant for response decay was an important determinant of the relative efficacy of continuous and intermittent hormone delivery. In further studies using the same in vitro system, WEIGLE and GOODNER (1986) also reported that the enhancement of hepatic glucose production by glucagon pulses is a frequency-dependent phenomenon and that hepatic glucose production is optimized for interpulse intervals of 10–20 min, a period close to the physiological secretory period of 10 min observed in non-human primates (GOODNER et al. 1977; JASPAN et al. 1986) and of 14–20 min reported in humans (LANG et al. 1982; HANSEN et al. 1982). KOMJATI et al. (1986) investigated the effect of pulsatile versus continuous glucagon exposure on glucose production from the isolated perfused rat liver. They observed that continuous exposure to glucagon (35 pmol/l) induced a twofold increase in hepatic glucose production, while intermittent exposure (3 min on/off intervals; total dose 50%) to the same glucagon concentration elicited an almost identical increase in hepatic glucose output. Therefore, all these in vitro studies concur to demonstrate that pulsatile delivery of glucagon is more efficient than continuous exposure to stimulate hepatic glucose production.

E. Pulsatile Glucagon Delivery In Vivo

I. Pulsatile Glucagon Administration in Normal Man

As already reviewed (LEFÈBVRE et al. 1994), the first study investigating the respective effects of continuous and intermittent glucagon administration in man was negative in the sense that no greater effect of pulsatile glucagon administration was identified (PAOLISSO et al. 1987). In that study, however, a relatively high plasma glucagon concentration was achieved: 189 ± 38 ng/l in the continuous infusions and oscillations between 95 and 501 ng/l in the intermittent mode of glucagon administration. It is likely that such high plasma levels were out of the "window" of glucagon concentrations for which the glucagon enhancement effect on liver glucose output had been observed in vitro (WEIGLE et al. 1984, 1985; WEIGLE and GOODNER 1986). Indeed, in a subsequent study (PAOLISSO et al. 1989), lower glucagon doses were administered (under pancreatic hormone secretion inhibition by somatostatin and continuous insulin infusion to replace basal insulin levels); in this setting, pulsatile glucagon administration, resulting in plasma glucagon levels oscillating between 115 and 254 ng/l, induced a greater stimulation of hepatic glucose output than an identical dose of the hormone infused continuously for 325 min. Finally, in the presence of somatostatin-induced insulin deficiency (with no insulin replacement), pulsatile glucagon in young healthy volunteers induced greater rises in blood glucose, plasma nonesterified fatty acid, glycerol, and β-hydroxybutyrate levels than did its continuous delivery (PAOLISSO et al. 1990a). Interestingly, in the elderly (69.4 ± 2.0 years), the lipolytic and ketogenic, but not the hyperglycemic, responses to pulsatile glucagon were significantly reduced when compared to the effects observed in the healthy young volunteers (24.2 ± 1.2 years), an observation that may be relevant for the understanding of the pathophysiology of hyperosmolar nonketotic coma in elderly people (PAOLISSO et al. 1990a). Finally, ATTVALL et al. (1992) investigated the insulin-antagonistic effects of pulsatile (3-min pulses every 20 min) and continuous glucagon over 4 h with the euglycemic clamp technique in healthy subjects. Glucose production and utilization were evaluated with D-[3-^{3}H]glucose while, again, somatostatin was used in all studies to inhibit endogenous insulin and glucagon release. In this setting, the amount of glucagon given during the pulsatile infusion (27% of that during continuous infusion) was adjusted so that the peak glucagon levels (371 ± 22 ng/l) were similar to those used during the continuous infusion (365 ± 20 ng/l). The insulin-antagonistic effects of pulsatile and continuous glucagon infusions on glucose production were similar during the 1st h and impaired the insulin effect by 44% ± 8% and 47% ± 6%, respectively. However, the effect of glucagon when infused continuously declined rapidly, whereas the effect of the pulsatile infusion decreased more slowly and was evident for 3 h. Raising the glucagon level four-fold restored the insulin-antagonistic effect again, suggesting that the liver had become desensitized.

II. Combined Pulsatile Administration of Glucagon and Insulin in Normal Man

PAOLISSO et al. (1989) investigated the respective effects of continuous intravenous delivery of both insulin and glucagon compared with those of pulsatile insulin (and continuous glucagon), pulsatile glucagon (and continuous insulin), and both hormones administered in a pulsatile manner (but out of phase) on various parameters of glucose turnover. The study was performed on six healthy male volunteers submitted to a 325-min glucose-controlled glucose intravenous infusion using the Biostator. The endogenous secretion of pancreatic hormones was inhibited by somatostatin ($2\,\mu g/min$). The four combinations of continuous and pulsatile infusions of insulin and glucagon were performed on different days and in random order. The amounts of hormone infused were identical in all instances. In the case of pulsatile administration of both hormones, the pulses of insulin and glucagon were given out of phase with a 6-min interval (Figs. 1, 2). Blood glucose levels and glucose infusion rate were monitored continuously by the Biostator, and classic methodology using a D-[3-^{3}H]glucose infusion allowed glucose turnover to be studied (Fig. 3). When compared with pulsatile insulin and continuous glucagon, pulsatile glucagon and continuous insulin were characterized by significantly higher endogenous (hepatic) glucose production. When both insulin and glucagon were delivered in a pulsatile manner, the effect of pulsatile glucagon was predominant, maintaining a higher endogenous glucose production than when both hormones were infused in a continuous manner. Under no circumstance was an effect on glucose utilization or clearance detected. This study demonstrated that pulsatile delivery of insulin or glucagon in humans has a greater effect on modulating endogenous glucose production than continuous infusion of the hormones. Furthermore, when both insulin and glucagon were delivered intermittently and out of phase, the stimulatory effect of glucagon on endogenous glucose production clearly prevailed over the inhibitory effect exerted by insulin.

III. Pulsatile Administration of Glucagon in Diabetic Patients

To our knowledge, only one study has investigated the respective effects of continuous and pulsatile intravenous glucagon delivery in type 1 diabetic patients (PAOLISSO et al. 1990b). The study was performed in seven insulin-dependent (type 1) diabetic subjects proved to have no residual insulin secretion. In random order and on different days, each subject was submitted to glucagon delivery given continuously for 325 min (58 ng/min) or in a pulsatile manner (same total dose but given in 2-min pulses separated by 11-min periods during which no glucagon was infused). Endogenous pancreatic hormone secretion was inhibited by somatostatin. Insulin was infused overnight and continued until 120 min before the experiment. In the continuous glucagon infusion protocol, plasma glucagon averaged 109 ± 13 ng/l. In the intermittent glucagon administration, plasma glucagon levels oscillated between 28 and 197 ng/l. Pulsatile glucagon delivery resulted in a higher plasma glucose, glycerol, β-hydroxybutyrate, and triglyceride levels than did continuous delivery.

IV. Pulsatile Administration of Glucagon in Dogs

In contrast to all in vitro (WEIGLE et al. 1984, 1985; WEIGLE and GOODNER 1986; KOMJATI et al. 1986) and most in vivo human investigations (PAOLISSO et al. 1989, 1990a,b), DOBBINS et al. (1994) reported that, in the conscious dog, the

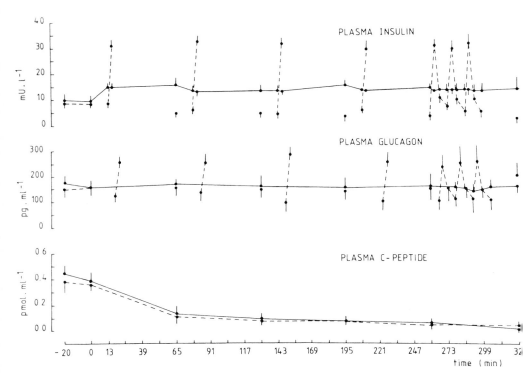

Fig. 1. Plasma insulin, glucagon, and C-peptide levels during continuous (—) or pulsatile (---) insulin and glucagon infusion. Multiple collections of blood permitting demonstration of oscillations in plasma insulin and glucagon were performed from 260 to 305 min only. Results are expressed as means ± SE ($n = 6$). (From Paolisso et al. 1989)

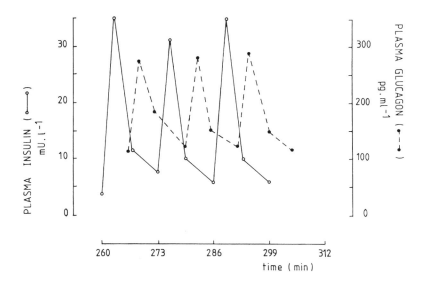

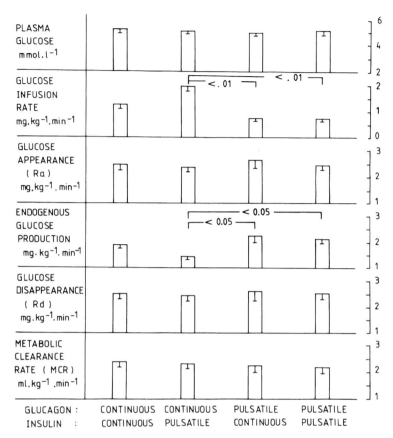

Fig. 3. Comparison of various parameters of glucose metabolism during the last 65 min of each test under four experimental conditions: continuous glucagon-continuous insulin, continuous glucagon-pulsatile insulin, pulsatile glucagon-continuous insulin, and pulsatile glucagon-pulsatile insulin infusions. For each subject, mean values of five measurements carried out during last 65 min of study were calculated. Results are expressed as means ± SE ($n = 6$). (From PAOLISSO et al. 1989)

response of liver glucose production to pulsatile or continuous glucagon infusion was not different. In that study, endogenous insulin and glucagon secretion were inhibited by somatostatin infusion. Insulin was replaced at an infusion rate of 960–980 μmol/kg per minute, resulting in circulating plasma insulin levels between 30 and 45 pmol/l. With continuous infusion, plasma glucagon concentrations increased from 56 ± 7 to 194 ± 27 ng/l. With pulsatile infusion, plasma glucagon concentrations started at 53±ng/l and then oscil-

◀ **Fig. 2.** Oscillations in plasma insulin and glucagon levels during pulsatile infusion of both hormones. Insulin was infused during the first 2 min and glucagon from 6 to 8 min of each cycle of 13 min, leading to a 6-min phase displacement between insulin and glucagon pulses. Results are expressed as means ± SE ($n = 6$). (From PAOLISSO et al. 1989)

lated between 157 ± 15 and 253 ± 28 ng/l and the glucagon pulse period was about 9 min. Whatever the mode of glucagon administration, the effects on liver glucose production were similar. The reasons for the discrepancy between the results of this dog study and those of previous in vitro (WEIGLE et al. 1984, 1985; WEIGLE and GOODNER 1986; KOMJATI et al. 1986) and human (PAOLISSO et al. 1989, 1990a,b) investigations are presently unclear.

F. Conclusions

Numerous studies performed in vivo and in vitro have shown that glucagon is secreted in a pulsatile manner, with a periodicity of 10–15 min (probably less in vitro). The mechanisms controlling these oscillations and their relationship with the secretory pattern of the other pancreatic hormones are still poorly understood. Most in vitro and in vivo studies, with the exception of the conscious dog, have shown that glucagon delivered in a pulsatile manner exerts a greater effect on liver glucose production than the same amount of the hormone delivered continuously. It seems, therefore, that the physiological pulsatile secretory pattern of glucagon is important for obtaining an optimal response of its main target cell, namely the hepatocyte.

Acknowledgments. Our research has been supported by grants of the Fonds National de la Recherche Scientifique, the Fonds de la Recherche Scientifique Médicale and the Fonds de la Recherche de la Faculté de Médecine de l'Université de Liège. I am indebted to E. Vaessen-Petit for expert secretarial assistance.

References

Attvall S, Fowelin J, Von Schenck H, Smith U, Lager I (1992) Insulin antagonistic effects of pulsatile and continuous glucagon infusions in man – a comparison with the effect of adrenaline. J Clin Endocrinol Metab 74:1110–1115

Berts A, Gylfe E, Hellman B (1995) Ca^{2+} oscillations in pancreatic islet cells secreting glucagon and somatostatin. Biochem Biophys Res Commun 208:644–649

Bode H-P, Yule DI, Fehmann H-C, Göke B, Williams JA (1994) Spontaneous calcium oscillations in clonal endocrine pancreatic glucagon-secreting cells. Biochem Biophys Res Commun 205:435–440

Dobbins RL, Davis SN, Neal DW, Cobelli C, Cherrington AD (1994) Pulsatility does not alter the response to a physiological increment in glucagon in the conscious dog. Am J Physiol 266 (Endocrinol Metab 29):E467–E478

Goodner CJ (1994) In vitro systems for modeling target tissue responses to secretory pulses of the islet hormones: glucagon and insulin. In: Levine JE (ed) Methods in neurosciences, vol 20. Pulsatility in neuroendocrine systems. Academic Press, San Diego, p 203

Goodner CJ, Walike BC, Koerker DJ, Ensinck JE, Brown AC, Chideckel EW, Palmer J, Kalnasy L (1977) Insulin, glucagon and glucose exhibit synchronous sustained oscillations in fasting monkeys. Science 195:177–179

Hansen BC, Jen K-L, Belbez-Pek S, Wolfe RA (1982) Rapid oscillations in plasma insulin, glucagon and glucose in obese and normal weight humans. J Clin Endocrinol Metab 54:785–792

Jaspan JB, Lever E, Polonsky KS, Van Cauter E (1986) In vivo pulsatility of pancreatic islet peptides. Am J Physiol 251:E215–E226

Koerker DJ, Goodner CJ, Hansen BW, Brown AC, Rubenstein AH (1978) Synchronous sustained oscillations of C-peptide and insulin in the plasma of fasting monkeys. Endocrinology 102:1649–1652

Komjati M, Bratusch-Marrain P, Waldhäusl W (1986) Superior efficacy of pulsatile versus continuous hormone exposure on hepatic glucose production in vitro. Endocrinology 118:312–319

Lang DA, Matthews DR, Burnett M, Ward GM, Turner RC (1982) Pulsatile, synchronous basal insulin and glucagon secretion in man. Diabetes 31:22–26

Leclercq-Meyer V, Malaisse WJ (1994) Pulsatility of arginine-stimulated insulin, glucagon and somatostatin release. Med Sci Res 22:151–154

Lefèbvre PJ, Paolisso G, Scheen AJ, Henquin JC (1987) Pulsatility of insulin and glucagon release: physiological significance and pharmacological implications. Diabetologia 30:443–452

Lefèbvre PJ, Paolisso G, Scheen AJ (1994) Pulsatile administration of insulin and glucagon in man. In: Levine JE (ed) Methods in neurosciences, vol 20. Pulsatility in neuroendocrine systems. Academic Press, San Diego, p 488

Paolisso G, Scheen AJ, Luyckx AS, Lefèbvre PJ (1987) Pulsatile hyperglucagonemia fails to increase hepatic glucose production in normal man. Am J Physiol 251:E1–E7

Paolisso G, Scheen AJ, Albert A, Lefèbvre PJ (1989) Effects of pulsatile delivery of insulin and glucagon in humans. Am J Physiol 257 (Endocrinol Metab 20) E686–E696

Paolisso G, Buonocore S, Gentile S, Sgambato S, Varricchio M, Scheen A, D'Onofrio F, Lefèbvre PJ (1990a) Pulsatile glucagon has greater hyperglycaemic, lipolytic and ketogenic effects than continuous hormone delivery in man: effect of age. Diabetologia 33:272–277

Paolisso G, Sgambato S, Giunta R, Varricchio M, D'Onofrio F (1990b) Pulsatile rather than continuous glucagon infusion leads to greater metabolic derangements in insulin-dependent diabetic subjects. Diabete Metab (Paris) 16:42–47

Sirek A, Vaitkus P, Norwich KH, Sirek OV, Unger RH, Harris V (1985) Secretory patterns of glucoregulatory hormones in prehepatic circulation of dogs. Am J Physiol 249:E34–E42

Stagner JI (1991) Pulsatile secretion from the endocrine pancreas: metabolic, hormonal and neural modulation. In: Samols E (ed) The endocrine pancreas. Raven, New York, p 283

Stagner JL, Samols E, Weir GC (1980) Sustained oscillations of insulin, glucagon, and somatostatin from the isolated canine pancreas during exposure to a constant glucose concentration. J Clin Invest 65:939–942

Weigle DS, Goodner CJ (1986) Evidence that the physiological pulse frequency of glucagon secretion optimizes glucose production by perifused rat hepatocytes. Endocrinology 118:1606–1613

Weigle DS, Koerker DJ, Goodner CJ (1984) Pulsatile glucagon delivery enhances glucose production by perifused rat hepatocytes. Am J Physiol 247:E564–E568

Weigle DS, Koerker DJ, Goodner CJ (1985) A model for augmentation of hepatocyte response to pulsatile glucagon stimuli. Am J Physiol 248 (Endocrinol Metab 11):E681–E686

CHAPTER 7
Glucagon and Diabetes

P.J. Lefèbvre

A. Introduction

Since the review on glucagon and diabetes by Unger and Orci (1983) in volume 66/II of this series (Lefèbvre 1983a), several reviews on the role of glucagon in both insulin-dependent (IDDM) and non-insulin-dependent (NIDDM) diabetes mellitus have appeared (Gerich 1989; Lefèbvre et al. 1991; Unger and Orci 1995; Lefèbvre 1991, 1995). In this chapter, we will consider: (1) the diabetogenic effects of glucagon; (2) the morphology of the A cells in the islets of Langerhans of diabetic patients; (3) the data available on glucagon plasma levels in the various forms of diabetes mellitus; (4) the glucagon dysfunction that is present in diabetes; (5) the role of glucagon excess in the metabolic abnormalities of diabetes; and (6), briefly, the attempts to improve diabetes control by reducing excessive glucagon secretion and/or action. Specific characteristics of the diabetic state associated with the glucagonoma syndrome are reviewed in more detail in Chap. 15 of this volume while the search for new glucagon antagonists is considered in Chap. 8.

B. Diabetogenic Effects of Glucagon

If one accepts the view, for operational purposes, that diabetes is a disease characterized by hyperglycemia associated, in the more severe cases, with accelerated lipolysis and excessive ketogenesis, one must recognize that glucagon, when investigated *in isolated tissues and organs*, has precisely the property of increasing the mobilization of glucose, free fatty acids (FFAs) and ketone bodies, metabolites which are found in excessive concentrations in the blood of diabetic patients (Lefèbvre and Luyckx 1979). It is now accepted that glucagon acts mainly through binding to specific receptors located at the target cell plasma membrane. A major effect of glucagon is to activate adenylate cyclase and to increase the intracellular production of cyclic AMP. As reviewed in Chap. 5, this volume, the intimate mechanism of action of glucagon is much more complex than previously thought. The hepatocyte is a major target cell of glucagon. The main effect of glucagon on the liver is to increase glucose output, an effect which results from inhibition of glycogen synthesis (Roden et al. 1995) and stimulation of both liver glycogenolysis and gluconeogenesis (see reviews in Stalmans 1983 and Claus et al. 1983). There is ample evidence that most of these effects are mediated by cyclic AMP, but the possibility has been raised that part of the glycogenolytic effect of glucagon may occur by a cAMP-independent mechanism (see Chap. 5, this volume). Another major effect of glucagon on the liver is to stimulate ketogenesis. The elegant studies of McGarry and Foster (1983) have convincingly shown that liver ketogenesis depends on both the flux of FFAs into the liver and the enzymatic setting of this organ, which is influenced in a crucial manner by the glucagon/insulin ratio in the blood perfusing the liver. The studies of these authors have shown that a high glucagon/insulin ratio increases the intracellular level of cAMP, reduces glycogenolysis and acetyl coenzyme (CoA) carboxylase activity, and reduces the intracellular concentration of malonyl-CoA. This fall in malonyl-CoA brings fatty acid synthesis to a halt and causes derepression of the enzyme carnitine acyltransferase such that incoming fatty acids (made abundant through stimulation of lipolysis) are efficiently converted into the ketone bodies, acetoacetate and β-hydroxybutyrate. The effects of glucagon on the adipocyte depend markedly upon the species considered. While glucagon is a potent lipolytic hormone in birds and in rodents, its effects on the human adipose cell have long been disputed (review in Lefèbvre 1983b). More recent investigations have shown that indeed glucagon is strongly lipolytic in the human adipocyte in vitro but that this effect is difficult to evidence using incubation of adipose cells or adipose tissue pieces because glucagon is rapidly destroyed by a proteolytic activity associated with those cells (Koràny 1983). When appropriate techniques are used, the lipolytic effect of glucagon on human adipocytes can easily be demonstrated (Koranyi 1983; Richter et al. 1989; Perea et al. 1994, 1995). Thus, in isolated systems, glucagon induces metabolic changes similar to those that are observed in diabetes. However, experiments based on infusion of glucagon *in normal man* do not help much in our understanding of the diabetogenic properties of glucagon. In fact, such infusions simultaneously stimulate the release of insulin, which potently antagonizes the effects of glucagon on liver glucose output, adipose tissue lipolysis and liver ketogenesis. Critical experiments have been performed by infusing glucagon in subjects whose insulin secretion was inhibited by a simultaneous infusion of somatostatin or in insulin-deprived diabetic patients. In these conditions (Gerich et al. 1976), modest increases in glucagon levels cause a significant rise in plasma glucose, glycerol, FFA and 3-hydroxybutyrate, all changes which are reversed by the infusion of insulin. Such experiments have clearly demonstrated that exogenous glucagon administration in insulin-deficient subjects aggravates the metabolic abnormalities typical of diabetes mellitus. Important studies on the role of glucagon in maintaining or increasing liver glucose output have been performed in dogs by the group of Cherrington (Stevenson et al. 1987; Hendrik et al. 1992). Stevenson et al. (1987) demonstrated that gluconeogenesis and glycogenolysis were similarly sensitive to stimulation by glucagon in vivo, and that the dose-response curves were markedly parallel. Hendrik et al. (1992) emphasized the role of basal glucagon in maintaining hepatic glucose production during a prolonged fast in conscious dogs and demonstrated the major role of glycogenolysis in this maintenance. More recently, MAGNUSSON et al. (1995), using ^{13}C-labeled nuclear magnetic resonance spectroscopy in man, showed that during the initial response to a physiological increment in plasma glucagon (from 136 ± 18 to 304 ± 57 ng/l): (1) net hepatic glycogenolysis accounted for virtually all of the increase in hepatic glucose production and (2) the glucagon's evanescent effect on hepatic glucose production was not caused by depletion of hepatic glycogen stores. Using the same technique, RODEN et al. (1995) have shown that basal glucagon plays a crucial role in inhibiting liver glycogen synthesis. Concerning the effect of glucagon on lipid metabolism, CARLSON et al. (1993) have convincingly shown, in healthy volunteers, that moderate hyperglucagonemia undoubtedly stimulates the rate of appearance in the plasma of both glycerol and FFAs.

C. The A Cell in Diabetes

Quantitative morphological data on the A cells in the islets of Langerhans of the pancreas of human diabetic patients are not abundant and, to some extent, contradictory (RAHIER 1988). *In type 1 diabetes of long duration*, B cells have almost completely disappeared, resulting in a relative increase in the non-B cells that make up the endocrine pancreas. Studies by ORCI et al. (1983) have shown that, in addition, there seems to be an absolute increase in the number of A cells, which make up about 30% of the nondiabetic islet and 70% of the diabetic islet. The total volume of A cells per pancreas has been estimated to be about doubled. Obviously, the microanatomy of the islets of Langerhans is completely disrupted with the functional consequences that have been reviewed by SAMOLS et al. (1983) and SAMOLS and STAGNER (1991). Attempts to quantitate the number of A cells in type 1 diabetes have been performed on pancreatic tissue from two patients by STEFAN et al. (1982) and four patients by RAHIER et al. (1983). On these six samples, although the volume density of A cells was increased, their total mass was not increased, and in one of the studies it was slightly *decreased* (RAHIER et al. 1983).

In *type 2 diabetes*, with the exception of one study (SAITO et al. 1979), the ratio of B to A cells in the islets of Langerhans has been found lower than in the pancreas from normal subjects (STEFAN et al. 1982; RAHIER et al. 1983). According to RAHIER et al. (1983), this modification is due to an absolute increase in the mass of A cells.

D. Circulating Glucagon Levels in Diabetes

ASSAN et al. (1969) were the first to demonstrate that plasma glucagon levels are usually grossly elevated in diabetic ketoacidosis, a finding confirmed by MÜLLER et al. (1973). Elevated glucagon levels have also been reported in hyperosmolar nonketotic coma (LINDSEY et al. 1974) and in poorly controlled diabetes mellitus (GERICH 1977; UNGER 1978; UNGER and ORCI 1981). In mild or moderate diabetes, plasma glucagon levels are usually slightly elevated (REAVEN et al. 1987): these values, however, are usually higher than those found in normal subjects made similarly hyperglycemic by glucose ingestion or infusion. This situation has been termed "relative hyperglucagonemia" (GERICH 1977); it was recognized almost 30 years ago and has been regularly confirmed since. Further data have shown that the hyperglucagonemia of type 1 diabetes is totally normalized (KAWAMORI et al. 1980) or markedly reduced (OHNEDA et al. 1975) by perfect metabolic control using the so-called artificial pancreas or artificial B cell, a finding suggesting that it is likely a consequence of relative or absolute insulin deficiency (LEFÈBVRE and LUYCKX 1979). The mechanisms involved in the hyperglucagonemia of type 2 or non-insulin-dependent diabetes mellitus will be discussed below.

E. Glucagon Dysfunction in Diabetes

I. Hyperglucagonemia of Diabetes

As we have recently reviewed (LEFÈBVRE 1995), understanding the mechanisms leading to hyperglucagonemia in diabetes requires analysis of the intraislet insulin-glucagon relationship. In most species, including humans, the glucagon-secreting A cells are located at the periphery of the islets of Langerhans (ORCI et al. 1983), and fine morphological studies have shown that, in a given islet, the microcirculation goes from the core to the mantle (BONNER-WEIR and ORCI 1982). This suggests that, in vivo, peripheral A cells are exposed to high concentrations of insulin released from the more centrally located B cells, a suggestion supported by several experimental observations (GREENBAUM et al. 1991). The proposal has been made that, under normal conditions, insulin inhibits glucagon secretion by the A cell (SAMOLS et al. 1983, 1988; MARUYAMA et al. 1984), a mechanism that is supported by our observations using the isolated perfused dog stomach model (LEFÈBVRE and LUYCKX 1978), in which factors influencing glucagon secretion can be investigated without the confounding influence of nearby-located B cells (LEFÈBVRE and LUYCKX 1977). Such a concept that intrainsular insulin may control glucagon secretion by the peripherally located A cells would imply classifying the A cell in the group of cells in the body that are sensitive to insulin and therefore express insulin receptors at their plasma membrane. The study performed by VAN SCHRAVENDIJK et al. (1985) on isolated rat A cells concluded that, if A cells possess high-affinity insulin receptors, their number should be much lower than in classic target cells such as the hepatocyte or the adipocyte. More recently, KISANUKI et al. (1995) have reported their results on cloned A cells from a hamster insulinoma and on A cells subcloned from a transgenic mouse adenoma. In both types of cells, glucagon secretion was decreased by the addition of insulin, insulin receptors were present, the mRNA specific for insulin receptors was expressed, positive immunostainings for insulin receptors were observed, and immunoprecipitation using anti-insulin receptor antibody and [^{35}S]methionine labeling showed a correct size for insulin receptor protein. Finally, the insulin receptors expressed in these cells underwent autophosphorylation by insulin stimulation. It is well known that IDDM results from the progressive disappearance, by an autoimmune mechanism, of the insulin-secreting B cells in the islets of Langerhans (review in PIPELEERS et al. 1988). One is permitted to suggest that the hyperglucagonemia of IDDM results from the loss of a paracrine control exerted by intra-islet insulin on glucagon release. This hypothesis is supported by the fact that perfect insulin replacement in IDDM patients results in complete normalization of basal and stimulated glucagon release (KAWAMORI et al. 1980). However, STARKE et al. (1985) have shown that correction of hyperglycemia with phloridzin-induced glycosuria restores the glucagon inhibitory response to glucose in insulin-deficient dogs. Thus, the hyperglucagonemia of IDDM results from both insulin deficiency and chronic hyperglycemia. The mechanism by which hyperglycemia leads to glucose "blindness" of the A cells or glucose desensitization remains to be elucidated (UNGER and ORCI 1995).

As we have discussed elsewhere (LEFÈBVRE et al. 1991), the mechanisms leading to hyperglucagonemia in NIDDM are less clear. GERICH (1989) and HAMAGUCHI et al. (1991) have provided some evidence that the A cell participates in the insulin resistance that is characteristic of NIDDM; suppression of glucagon secretion by insulin infusion in NIDDM patients was not as complete as in nondiabetic subjects. An alternative hypothesis, also advanced by GERICH (1989), is that prolonged hyperglycemia in NIDDM might in some way desensitize the A cell to the suppressive effect of glucose on glucagon secretion in a manner similar to the desensitization reported at the B-cell level and likely to be involved in the reduced insulin response to glucose. With our colleagues, Paolisso and Scheen, we have proposed a third hypothesis to explain the hyperglucagonemia of NIDDM (LEFÈBVRE et al. 1991). As reviewed in detail in Chap. 6, this volume, it is now well recognized that insulin and glucagon are secreted in a pulsatile manner, with some indication that these two hormones are secreted out of phase (see LEFÈBVRE et al. 1987). In NIDDM, the short-term oscillations of plasma insulin are more rapid and generally less regular than in normal subjects (LANG et al. 1981). We have suggested that disturbances of the normal oscillatory secretory pattern of insulin contribute to the hyperglucagonemia of NIDDM described above. Such a hypothesis is supported by the facts that (1) pulsatile insulin delivery has advantageous metabolic effects compared with continuous hormone administration in patients with NIDDM (PAOLISSO et al. 1988a) and (2) pulsatile insulin is also more efficient than continuous delivery in inhibiting the excessive glucagon response to intravenous arginine seen in diabetic patients (PAOLISSO et al. 1988b). This hypothesis agrees with the concept expressed above that a major mechanism controlling glucagon release is paracrine and insulin mediated (SAMOLS et al. 1983, 1988; SAMOLS and STAGNER 1991; MARUYAMA et al. 1984).

The intimate mechanism by which insulin inhibits glucagon release is still unclear. However, recent data indicate that insulin may negatively regulate the glucagon gene (see Chap. 2, this volume). Whatever the mechanism involved, if we assume that (1) glucagon circulating levels are higher in NIDDM patients than in healthy subjects (see above), (2) glucagon is a major determinant of hepatic glucose output (see above), (3) liver glucose output is increased in NIDDM (BEST et al. 1982; CAMPBELL et al. 1988), and (4) a high liver glucose output is a major factor contributing to fasting (CONSOLI et al. 1987) and probably postprandial (KELLY et al. 1988) hyperglycemia in NIDDM, the conclusion is that glucagon is likely to be involved in the hyperglycemia of NIDDM, a concept that we have termed "the glucagon logic" (LEFÈBVRE et al. 1991).

A recent study by LARSSON et al. (1995) suggests that A cell dysfunction may occur early during the development of NIDDM; indeed the acute glucagon response to arginine is increased and its inhibition by circulating glucose decreased in women with impaired glucose tolerance compared to women with normal glucose tolerance.

II. Defective Glucose Counterregulation

Glucose counterregulation has been defined as "the sum of processes that protect against development of hypoglycemia and that restore euglycemia if hypoglycemia should occur" (GERICH 1983). Numerous studies, reviewed by GERICH (1983) and Cryer in Chap. 9, this volume, have shown that, in order of importance, the key counterregulatory factors are glucagon, epinephrine, growth hormone, cortisol, and hepatic autoregulation (MITRAKOU et al. 1991). Consequently, adequate glucagon response has been termed "the first line of defense against hypoglycemia." We have just seen that diabetes is characterized, almost universally, by excessive plasma glucagon levels. Paradoxically, and as stated by UNGER and ORCI (1995), ironically, a defective glucagon response to insulin-induced hypoglycemia is one of the features of long-duration diabetes. The impaired glucagon response to acute hypoglycemia is almost universally present after 5 years of diabetes; it is selective because the glucagon secretory responses to other stimuli are either normal or exaggerated (GERICH et al. 1973). Impaired glucagon response is sometimes partially compensated by an increase in norepinephrine secretion (CRYER and GERICH 1985), but evidence that this compensatory mechanism is efficacious is lacking (DEFEO et al. 1991). This glucagon hyporesponsiveness to hypoglycemia is usually more severe in IDDM than in NIDDM (HELLER et al. 1987). It does not seem to be systematically correlated to autonomic diabetic neuropathy (KENNEDY et al. 1987) but is more frequent when autonomic neuropathy is present. It is not restored by long-term optimal insulin therapy delivered either conventionally or by continuous subcutaneous insulin infusion (BERGENSTAL et al. 1983). Loss of adequate counterregulation is associated with a major increase in the incidence of severe hypoglycemic episodes (WHITE et al. 1983). Recent studies have also shown that chronic hyperinsulinemia, either endogenous as in insulinoma (MITRAKOU et al. 1993) or exogenous as in intensive insulin therapy (VENEMAN et al. 1993), can induce defective counterregulation of hypoglycemia, including defective glucagon release (LIU et al. 1991). Removal of the insulinoma (MITRAKOU et al. 1993) or reduction of chronic overinsulinization (FANELLI et al. 1993) restores counterregulatory mechanisms, including a normal glucagon response to insulin-induced hypoglycemia. Again, negative regulation of the glucagon gene by insulin (see Chap. 2, this volume) might explain these phenomena.

F. Role of Glucagon Excess in the Metabolic Abnormalities of Diabetes

I. Postpancreatectomy Diabetes

As reviewed by LEFÈBVRE and LUYCKX (1983), pancreatectomized patients have no or only extremely small circulating levels of "true" glucagon in their plasma (BRINGER et al. 1981). On the basis of this finding, a crucial experiment has been reported by BARNES et al. (1977) in which the importance of glucagon in the development of diabetic hyperglycemia and ketoacidosis has been evaluated by withholding insulin from six patients with type 1 diabetes and four totally pancreatectomized subjects. In type 1 diabetic patients, insulin withholding was associated with an abrupt rise in plasma glucagon and a sharp increase in blood glucose and plasma 3-hydroxybutyrate; in contrast, in the pancreatectomized patients, no significant rise in glucagon was observed, while the rises in blood glucose and plasma 3-hydroxybutyrate were much smaller than those observed in the type 1 diabetic patients. This experiment demonstrated clearly that, in the presence of glucagon, both hyperglycemia and ketoacidosis were enhanced and accelerated, as expected from the known effects of the hormone recalled p. 115.

As recently reviewed by UNGER and ORCI (1995), a possible role of minimal glucagonemia from extrapancreatic sources in pancreatectomized human subjects remains possible. Because insulin markedly suppresses the secretion of glucagon, particularly by the exquisitely sensitive extrapancreatic A cells, a substantial period of insulin deprivation may be necessary to determine, in man, the maximal level of extrapancreatic glucagon production.

II. Insulin-Dependent Diabetes

Critical studies using insulin withholding, with or without somatostatin infusion, in IDDM patients have clearly demonstrated the role of glucagon in the metabolic abnormalities of diabetes (GERICH et al. 1975; KRZENTOWSKI et al. 1983). The most impressive study, a classic in metabolic medicine, was that by

GERICH et al. (1975). In short, a series of insulin-dependent, ketosis-prone IDDM patients were made normoglycemic by an overnight intravenous infusion of insulin. Terminating insulin infusion (under intravenous saline administration) prompted a rapid rise in blood glucose, plasma FFA, and 3-hydroxybutyrate levels, which were paralleled by a rise in plasma glucagon. In contrast, if the glucagon rise was prevented by an intravenous somatostatin infusion, the rises in blood glucose, plasma FFA, and 3-hydroxybutyrate were markedly reduced (but not completely prevented). Similar studies were performed subsequently by our group, which investigated in depth the metabolic consequences of a nocturnal interruption of continuous subcutaneous insulin infusion or CSII (KRZENTOWSKI et al. 1983). Here again the metabolic deterioration is impressive and occurs after a delay of 1–2 h due to the resorption of the subcutaneous insulin depot; it is markedly attenuated either by an intravenous infusion of somatostatin (SCHEEN et al. 1983) or by the subcutaneous injection of a somatostatin analogue (SCHEEN et al. 1989). In both instances, glucagon suppression by somatostatin or its analogue seems to be a crucial parameter in the attenuation of the metabolic deterioration due to the interruption of CSII (SCHEEN et al. 1989).

III. Non-Insulin-Dependent Diabetes

As already pointed out, studies by REAVEN et al. (1987) have clearly documented significant hyperglucagonemia throughout 24 h in nonobese and obese patients with NIDDM. We have seen that one of the major effects of glucagon is on the liver, where this hormone enhances hepatic glucose output by stimulating liver glycogenolysis and gluconeogenesis and by inhibiting glycogen synthesis (see p. 115). It is now accepted that hepatic glucose output is significantly increased in patients with type 2 diabetes. The role of glucagon in this process has been evaluated by BARON et al. (1987) using tracer turnover studies and somatostatin infusions. The authors concluded that: (1) elevated glucagon levels contribute significantly to the elevated rates of basal hepatic glucose output in NIDDM patients; (2) basal glucagon sustains close to 60% of basal hepatic glucose output in these patients, and (3) the excess glucagon effect is largely responsible for the apparent insulin resistance seen in NIDDM patients. It should be pointed out that chronic overexposure of the hepatocytes to glucagon may lead to some desensitization to the action of the hormone. Indeed, ARNER et al. (1987) have observed marked alterations in the adenylate cyclase activity of liver plasma membranes of NIDDM patients chronically exposed to hyperglucagonemia. This enzymatic activity was reduced by 35%–50% when compared to the activity found in control nondiabetic subjects. It has been suggested that this loss of response to glucagon may be an adaptive compensation for the hyperglucagonemia in these patients. In maturity-onset diabetes mellitus of the youth (MODY), both glucagon excess and insulin deficiency have been evidenced (AVRUSKIN et al. 1994).

G. Therapeutic Implications

Since it is obvious that excessive plasma glucagon levels contribute to the metabolic abnormalities of diabetes, it naturally follows that glucagon suppression or the use of glucagon antagonists should be considered as potential new approaches to the treatment of diabetes. The search for glucagon antagonists has really exploded over the last 10 years; the approach is discussed in detail in Chap. 8, this volume. We will briefly review here the attempts that have been made to improve metabolic control in diabetes by reducing glucagon *secretion*.

I. Insulin-Dependent Diabetes

Glucagon suppression might be considered as a useful adjunct to insulin therapy since excessive plasma glucagon levels are usually found in patients receiving *conventional* insulin therapy. In contrast, in patients under *intensive* insulin therapy, either by multiple injection or by protable pump, hyperglucagonemia is fully (KAWAMORI et al. 1980) or markedly (OHNEDA et al. 1975) corrected. It follows that perfect substitution of insulin deficiency is probably the best means to normalize the hypersecretion of glucagon. When this is not feasible, other pharmacological means to reduce hyperglucagonemia may be considered. The pharmacological compounds reducing glucagon secretion have been reviewed by LUYCKX (1983). They include atropine, β-receptor-blocking agents, verapamil, diphenylhydantoin, diazepam, prostaglandin synthesis inhibitors, somatostatin, and somatostatin analogues. Only the latter of these compounds has been tested clinically.

The prostaglandin synthesis inhibitors, acetylsalicylic acid and indomethacin, are potent inhibitors of glucagon release in vitro (review in LUYCKX and LEFÈBVRE 1983). However, clinical trials in diabetic patients did not demonstrate any effect on basal or arginine-induced glucagon release (LUYCKX et al. 1981). Numerous studies have shown reduction of the hyperglycemia of diabetes by somatostatin or various of its analogues (review in DAVIES et al. 1989). Part of the improvement is likely to be due to glucagon suppression; other mechanisms, however, are possibly involved including delay in nutrient absorption and inhibition of growth hormone release. At the date of writing, the use of somatostatin or its analogues is not part of the treatment of type 1 diabetes mellitus. Recent data of WILLMS et al. (1995) indicate that glucagon-like peptide 1 (GLP-1) lowers but does not normalize blood glucose levels in IDDM patients, mainly by inhibition of glucagon secretion.

II. Non-Insulin-Dependent Diabetes

Treatment of non-insulin-dependent diabetes includes diet, physical activity, and administration of sulfonylureas, biguanides, and α-glucosidase inhibitors (reviews in LEFÈBVRE and SCHEEN 1992; SCHEEN and LEFÈBVRE 1993, 1995). SAVAGE et al. (1979) reported a case of severe ketosis-resistant, non-insulin-dependent (type 2) diabetes mellitus associated with massive obesity. In this patient, *prolonged diet therapy* induced a 42-kg weight loss, a normalization of fasting blood glucose and oral glucose tolerance test and a complete normalization of the excessive glucagon response to an intravenous arginine challenge. This observation suggests that aggresive diet therapy might normalize glucagon secretion in obese NIDDM patients. Whether this effect is due to improvement of insulin release, normalization of blood glucose, or both is still unknown. The effects of *sulfonylureas* on glucagon release are still a matter of discussion. Indeed glucagon release has been reported to be unaffected (AGUILAR-PARADA et al. 1969; PEK et al. 1972; KAJINUMA et al. 1974), stimulated (LOUBATIÈRES et al. 1974; SAMOLS and HARRISON 1976), inhibited (SAMOLS et al. 1969; LAUBE et al. 1971; OHNEDA et al. 1974; TAKANASHI et al. 1994), or dually stimulated-inhibited (SAMOLS and HARRISON 1976; GRODSKY et al. 1977) by this group of compounds. Our group has observed that one of these compounds, glipizide, had no effect on plasma glucagon levels and that it preserved the glucagon response to hypoglycemia, a fact considered as an advantage (LECOMTE et al. 1977). As mentioned earlier, we have introduced the possibility that the loss of the intra-islet pulsatility of insulin release may be a factor contributing to the hyperglucagonemia of NIDDM (LEFÈBVRE et al. 1991). If this hypothesis is correct, a compound restoring a normal intra-islet pattern of insulin release is likely to reduce the excessive glucagon release seen in NIDDM. *Biguanides* seem to have little or no effect of their own on pancreatic A cell function. We have conducted a study in which we investigated the possible influence of metformin (500mg, three times a day) on arginine-induced glucagon response in 11 NIDDM patients. The clear-cut reduction in blood glucose and plasma FFA levels was not accompanied by any significant change in insulin or glucagon plasma concentrations during the test (CARPENTIER et al. 1975). Similarly in a group of ten obese patients with mild-to-moderate glucose intolerance, a 14-day administration of butylbiguanide did not modify the plasma glucagon concentrations after an overnight fast and during an oral glucose load (LEFÈBVRE et al. 1978). In NIDDM, *somatostatin and its analogues* aggravate hyperglycemia by their simultaneous inhibitory effect on insulin secretion (TAMBORLANE et al. 1977); these compounds can therefore not be considered as useful therapeutic adjuncts for the management of this type of patient. Here, the ideal pharmacological suppressor would be a compound which (1) does not inhibit insulin secretion (in contrast to somatostatin); (2) reduces the hyperresponsiveness to stimuli such as amino acids; but (3) preserves the response of glucagon to hypoglycemia, a useful safeguard if hypoglycemic agents acting by other mechanisms, such as the sulfonylureas, are given simultaneously. Such an ideal pharmacological glucagon-suppressor has not yet been identified.

As discussed in detail in Chaps. 17 and 18, glucagon-like peptide 1 (GLP-1) probably has great potential in the treatment of diabetes. An interesting feature of this peptide is that, in addition to stimulating insulin secretion, it seems also to be a potent *inhibitor* of glucagon secretion (NAUCK et al. 1993; WILLMS et al. 1994; RITZEL et al. 1995; DUPRÉ et al. 1995).

H. Conclusions

Circulating glucagon levels are usually elevated in diabetes, the highest levels being found in the absence of insulin. There is strong evidence that excessive glucagon levels contribute to the major metabolic abnormalities of diabetes, which are increased liver glucose output, accelerated lipolysis, and excessive ketogenesis but, as in complete pancreatectomy, diabetes can occur in the almost complete absence of glucagon. The glucagon dysfunction of diabetes is characterized by (1) hyperglucagonemia and a hyperresponsiveness of glucagon secretion in response to various stimuli and (2), in some patients with diabetes of long duration, a reduction of the capacity of the A cell of the islets of Langerhans to adequately secrete glucagon in response to insulin-induced hypoglycemia. This latter defect contributes to the so-called defective counterregulation to hypoglycemia observed in some diabetic patients. In *insulin-dependent diabetes*, there is therefore strong evidence that hyperglucagonemia is the consequence of the insulin deficiency and that the absence of high intra-islet levels of insulin may explain the persistence of abnormally high plasma concentrations of glucagon in patients receiving conventional, or even intensive, insulin therapy. In *non-insulin-dependent diabetes*, the cause of the hyperglucagonemia is less clear. Among the possible mechanisms involved are a loss of the inhibitory effect of insulin on glucagon release (may be due to the loss of the normal oscillatory pattern of insulin secretion) or a "resistance" to insulin of the pancreatic A cell (may be as a consequence of chronic hyperglycemia) or an intrinsic defect affecting both A and B cells.

A search for compounds selectively inhibiting glucagon release or, as discussed in Chap. 8, this volume, for glucagon analogues with appropriate antagonistic properties represents an attractive new way to improve diabetes management.

Acknowledgements. Our research has been supported by grants of the Fonds National de la Recherche Scientifique, the Fonds de la Recherche Scientifique Médicale, and the Fonds de la Recherche de la Faculté de Médecine de l'Université de Liège. I am indebted to E. Vaessen-Petit for expert secretarial assistance.

References

Aguilar-Parada E, Eisentraut AM, Unger RH (1969) Effect of HB-419 induced hypoglycemia on pancreatic glucagon secretion. Horm Metab Res 1 [Suppl]:48–50.
Arner P, Einarsson K, Ewerth S, Livingston JN (1987) Altered action of glucagon on human liver in type 2 (non-insulin-dependent) diabetes mellitus. Diabetologia 30:323–326

Assan R, Hautecouverture G, Guillemant S, Dauchy F, Protin P, Derot M (1969) Evolution de paramètres hormonaux (glucagon, cortisol et hormone somatotrope) et énergétiques (glucose, acides gras libres, glycérol) dans dix acidocétoses diabétiques graves traitées. Pathol Biol (Paris) 17:1095–1105

AvRuskin TW, Obilessetty V, Jabbar M, Prasad V, Greenfield E, Greig F, Juan CS (1994) Both glucagon excess and insulin deficiency characterize maturity-onset diabetes mellitus of youth (MODY). J Pediatr Endocrinol 7:335–341

Barnes AJ, Bloom SR, Mashiter K, Alberti KGMM, Smythe P, Turnell D (1977) Persistent metabolic abnormalities in diabetes in the absence of glucagon. Diabetologia 13:71–75

Baron AD, Schaeffer L, Shragg P, Kolterman OG (1987) Role of hyperglucagonemia in maintenance of increased rates of hepatic glucose output in type II diabetics. Diabetes 36:274–283

Bergenstal R, Polonsky K, Pons G, Jaspan J, Rubenstein A (1983) Lack of glucagon response to hypoglycemia in type 1 diabetes after long-term optimal therapy with a continued subcutaneous insulin infusion pump. Diabetes 32:398–402

Best J, Judzewitsch RG, Pfeifer MA, Beard JC, Halter JB, Porte D Jr (1982) The effect of chronic sulfonylurea therapy on hepatic glucose production in non-insulin-dependent diabetes. Diabetes 31:333–338

Bonner-Weir S, Orci L (1982) New perspectives on the microvasculature of the islets of Langerhans in the rat. Diabetes 31:883–889

Bringer J, Mirouze J, Marchal G, Pham TC, Luyckx A, Lefèbvre P, Orsetti A (1981) Glucagon immunoreactivity and antidiabetic action of somatostatin in the totally duodeno-pancreatectomized and gastrectomized human. Diabetes 30:851–856

Campbell PJ, Mandarino LJ, Gerich JE (1988) Quantification of the relative impairment in actions of insulin on hepatic glucose production and peripheral glucose uptake in non-insulin-dependent diabetes mellitus. Metabolism 37:15–21

Carlson MG, Snead WL, Campbell PJ (1993) Regulation of free fatty acid metabolism by glucagon. J Clin Endocrinol Metab 77:11–15

Carpentier JL, Luyckx AS, Lefèbvre PJ (1975) Influence of metformin on arginine-induced glucagon secretion in human diabetes. Diabete Metab 1:23–28

Claus TH, Park CR, Pilkis SJ (1983) Glucagon and gluconeogenesis. In: Lefèbvre PJ (ed) Glucagon I. Springer, Berlin Heidelberg New York, p 315 (Handbook of experimental pharmacology, vol 66/I)

Consoli A, Nurjhan N, Kennedy F, Gerich J (1987) Accelerated gluconeogenesis accounts for all of the increase in basal hepatic glucose output (HGO) of noninsulin-dependent diabetes mellitus (NIDDM) (abstract). Diabetes 36 [Suppl 1]:4A

Cryer P, Gerich J (1985) Glucose counterregulation, hypoglycemia, and intensive insulin therapy in diabetes mellitus. N Engl J Med 313:232–241

Davies RR, Turner SJ, Alberti KGMM, Johnston DG (1989) Somatostatin analogues in diabetes mellitus. Diabetic Med 6:103–111

De Feo P, Perriello G, Torlone E, Fanelli C, Ventura MM, Santeusanio F, Brunetti P, Gerich JE, Bolli GB (1991) Evidence against important catecholamine compensation for absent glucagon counterregulation. Am J Physiol 260 (Endocrinol Metab 23):E203–E212

Dupré J, Behme MT, Hramiak IM, McFarlane P, Williamson MP, Zabel P, McDonald TJ (1995) Glucagon-like peptide I reduces postprandial glycemic excursions in IDDM. Diabetes 44:626–630

Fanelli CG, Epifano L, Rambotti AM, Pampanelli S, Di Vincenzo A, Modarelli F, Lepore M, Annibale B, Ciofetta M, Bottini P, Porcellati F, Scionti L, Santeusanio F, Brunetti P, Bolli GB (1993) Meticulous prevention of hypoglycemia normalizes the glycemic thresholds and magnitude of most of neuroendocrine responses to, symptoms of, and cognitive function during hypoglycemia in intensively treated patients with short-term IDDM. Diabetes 42:1683–1689

Gerich J (1977) On the causes and consequences of abnormal glucagon secretion in human diabetes mellitus. In: Foà PP, Bajaj JS, Foà NL (eds) Glucagon: its role in physiology and clinical medicine. Springer, Berlin Heidelberg New York, p 617

Gerich JE (1983) Glucagon as a counterregulatory hormone. In: Lefèbvre PJ (ed) Glucagon II. Springer, Berlin Heidelberg New York, p 275 (Handbook of experimental pharmacology, vol 66/II)

Gerich JE (1989) Abnormal glucagon secretion in type-2 (non insulin-dependent) diabetes mellitus: causes and consequences. In: Creutzfeldt W, Lefèbvre PJ (eds) Diabetes mellitus: pathophysiology and therapy. Springer, Berlin Heidelberg New York, p 127

Gerich J, Langlois M, Noacco C, Karam J, Forsham P (1973) Lack of glucagon response to hypoglycemia in diabetes: evidence for an intrinsic pancreatic alpha-cell defect. Science 182:171–173

Gerich JE, Lorenzi M, Bier DM, Schneider V, Tsalikian E, Karam JH, Forsham PH (1975) Prevention of human diabetic ketoacidosis by somatostatin: evidence for an essential role of glucagon. N Engl J Med 292:985–989

Gerich JE, Lorenzi M, Bier DM, Tsalikian E, Schneider V, Karam JH, Forsham PH (1976) Effect of physiologic levels of glucagon and growth hormone on human carbohydrate and lipid metabolism. Studies involving administration of exogenous hormone during suppression of endogenous hormone secretion with somatostatin. J Clin Invest 57:875–884

Greenbaum CJ, Havel PJ, Taborsky GJ Jr, Klaff LJ (1991) Intra-islet insulin permits glucose to directly suppress pancreatic A cell function. J Clin Invest 88:767–773

Grodsky GM, Epstein GH, Franska R, Karam JH (1977) Pancreatic action of the sulfonylureas. Fed Proc 36:2714–2719

Hamaguchi T, Fukushima H, Uehara M, Wada S, Shirotani T, Kishikawa H, Ichinose K, Yamaguchi K, Shichiri M (1991) Abnormal glucagon response to arginine and its normalization in obese hyperinsulinaemic patients with glucose intolerance: importance of insulin action on pancreatic alpha cells. Diabetologia 34:801–806

Heller SR, Macdonald IA, Tattersall RB (1987) Counterregulation in type 2 (non-insulin-dependent) diabetes mellitus. Normal endocrine and glycaemic responses, up to ten years after diagnosis. Diabetologia 30:924–929

Hendrick GK, Wasserman DH, Frizzell RT, Williams PE, Lacy DB, Jaspan JB, Cherrington AD (1992) Importance of basal glucagon in maintaining hepatic glucose production during a prolonged fast in conscious dogs. Am J Physiol 263 (Endocrinol Metab 26):E541–E549

Kajinuma H, Kuzuya T, Ide T (1974) Effects of hypoglycemic sulfonamides on glucagon and insulin secretion in ducks and dogs. Diabetes 23:412–417

Kawamori R, Schichiri M, Kikuchi M, Yamasaki Y, Abe H (1980) Perfect normalization of excessive glucagon responses to intravenous arginine in human diabetes mellitus with the artificial beta-cell. Diabetes 29:762–765

Kelley D, Mitrakou A, Marsh H, Schwenk F, Benn J, Sonnenberg G, Archangeli M, Aoki T, Sorenson T, Berger M, Sonksen P, Gerich J (1988) Skeletal muscle glycolysis, oxidation and storage of an oral glucose load. J Clin Invest 81:1563–1571

Kennedy F, Bolli G, Go V, Cryer P, Gerich J (1987) The significance of impaired pancreatic polypeptide and epinephrine responses to hypoglycemia in patients with insulin-dependent diabetes mellitus. J Clin Endocrinol Metab 64:602–608

Kisanuki K, Kishikawa H, Araki E, Shirotani T, Uehara M, Isami S, Ura S, Jinnouchi H, Miyamura N, Shichiri M (1995) Expression of insulin receptor on clonal pancreatic alpha cells and its possible role for insulin-stimulated negative regulation of glucagon secretion. Diabetologia 38:422–429

Korànyi L (1983) Lipolytic effect of glucagon on perifused isolated human fat cells (abstract). Diabetologia 25:172

Krzentowski G, Scheen A, Castillo M, Luyckx AS, Lefèbvre PJ (1983) A 6-hour nocturnal interruption of a continuous subcutaneous insulin infusion: I. Metabolic and hormonal consequences and scheme for a prompt return to adequate control. Diabetologia 24:314-318

Lang DA, Matthews DR, Barnett M, Turner RC (1981) Brief irregular oscillations of basal plasma insulin and glucose concentrations in diabetic man. Diabetes 30:435-439

Larsson H, Berglund G, Ahren B (1995) Glucose modulation of insulin and glucagon secretion is altered in impaired glucose tolerance. J Clin Endocrinol Metab 80:1778:1782

Laube H, Fussganger R, Goberna R, Schroder K, Straub K, Sussman K, Pfeiffer EF (1971) Effects of tolbutamide on insulin and glucagon secretion of the isolated perfused rat pancreas. Horm Metab Res 3:238-242

Lecomte MJ, Luyckx AS, Lefèbvre PJ (1977) Plasma glucagon and diabetes control in maturity-onset type diabetics. Respective effects of diet, placebo and glipizide. Diabete Metab 3:2714-2719

Lefèbvre PJ (1983a) Glucagon II. Springer, Berlin Heidelberg New York (Handbook of experimental pharmacology, vol 66/II)

Lefèbvre PJ (1983b) Glucagon and adipose tissue lipolysis. In: Lefèbvre PJ (ed) Glucagon I. Springer, Berlin Heidelberg New York, p 418 (Handbook of experimental pharmacology, vol 66/I)

Lefèbvre PJ (1991) Abnormal secretion of glucagon. In: Samols E (ed) The endocrine pancreas. Raven, New York, p 191

Lefèbvre PJ (1995) Glucagon and its family revisited. Diabetes Care 18:715-730

Lefèbvre PJ, Luyckx AS (1977) Factors controlling gastric-glucagon release. J Clin Invest 59:716-722

Lefèbvre PJ, Luyckx AS (1978) Glucose and insulin in the regulation of glucagon release from the isolated perfused dog stomach. Endocrinology 103:1579-1582

Lefèbvre PJ, Luyckx AS (1979) Glucagon and diabetes: a reappraisal. Diabetologia 16:347-354

Lefèbvre PJ, Luyckx AS (1983) Extrapancreatic glucagon and its regulation. In: Lefèbvre PJ (ed) Glucagon II Springer, Berlin Heidelberg New York, p 205 (Handbook of experimental pharmacology, vol 66/II)

Lefèbvre PJ, Scheen AJ (1992) Update on the treatment of NIDDM. In: Lefèbvre PJ, Standl E (eds) New aspects in diabetes. De Gruyter, Berlin, p 71

Lefèbvre PJ, Luyckx AS, Mosora F, Lacroix M, Pirnay F (1978) Oxidation of an exogenous glucose load using naturally labelled ^{13}C-glucose. Effect of butylbiguanide therapy in obese mildly diabetic subjects. Diabetologia 14:39-45

Lefèbvre PJ, Paolisso G, Scheen AJ, Henquin JC (1987) Pulsatility of insulin and glucagon release: physiological significance and pharmacological implications. Diabetologia 30:443-452

Lefèbvre PJ, Paolisso G, Scheen A (1991) The role of glucagon in non-insulin-dependent (type 2) diabetes mellitus. In: Sakamoto N, Angel A, Hotta H (eds) New directions in research and clinical works for obesity and diabetes mellitus. Elsevier Science, Amsterdam, p 25

Lindsey C, Faloona G, Unger RH (1974) Plasma glucagon in non-ketotic hyperosmolar coma. JAMA 229:1771-1773

Liu D, Moberg E, Kollind M, Lins P-E, Adamson U (1991) A high concentration of circulating insulin suppresses the glucagon response to hypoglycemia in normal man. J Clin Endocrinol Metab 73:1123-1128

Loubatières AL, Loubatières-Mariani MM, Alric R, Ribes G (1974) Tolbutamide and glucagon secretion. Diabetologia 10:271-276

Luyckx AS (1983) Pharmacological compounds affecting glucagon secretion. In: Lefèbvre PJ (ed) Glucagon II. Springer, Berlin Heidelberg New York, p 175 (Handbook of experimental pharmacology, vol 66/II)

Luyckx AS, Lefèbvre PJ (1983) Prostaglandins and glucagon secretion. In: Lefèbvre PJ (ed) Glucagon II. Springer, Berlin Heidelberg New York, p 83 (Handbook of experimental pharmacology, vol 66/II)

Luyckx AS, Mendoza E, Lefèbvre PJ (1981) Failure of indomethacin to affect arginine-induced C-peptide and glucagon release in insulin-treated diabetics. Major role of residual B-cell function in conditioning the magnitude of the blood glucose rise after intravenous arginine. Diabetologia 21:376–382

Magnusson I, Rothman DL, Gerard DP, Katz LD, Shulman GI (1995) Contribution of hepatic glycogenolysis to glucose production in humans in response to a physiological increase in plasma glucagon concentration. Diabetes 44:185–189

Maruyama H, Hisatomi A, Orci L, Grodsky GM, Unger RH (1984) Insulin within islets is a physiologic glucagon release inhibitor. J Clin Invest 74:2296–2299

McGarry JD, Foster DW (1983) Glucagon and ketogenesis. In: Lefèbvre PJ (ed) Glucagon I. Springer, Berlin Heidelberg New York, p 383 (Handbook of experimental pharmacology, vol 66/I)

Mitrakou A, Ryan C, Veneman T, Mokan M, Jenssen T, Kiss I, Durrant J, Cryer P, Gerich J (1991) Hierarchy of glycemic thresholds for counterregulatory hormone secretion, symptoms, and cerebral dysfunction. Am J Physiol 260 (Endocrinol Metab 23):E67–E74

Mitrakou A, Fanelli C, Veneman T, Perriello G, Calderone S, Platanisiotis D, Rambotti A, Raptis S, Brunetti P, Cryer P, Gerich J, Bolli G (1993) Reversibility of unawareness of hypoglycemia in patients with insulinomas. N Engl J Med 329:834–839

Müller W, Faloona G, Unger RH (1973) Hyperglucagonemia in diabetic ketoacidosis: Its prevalence and significance. Am J Med 54:52–57

Nauck MA, Kleine N, Orskov C, Holst JJ, Willms B, Creutzfeldt W (1993) Normalization of fasting hyperglycaemia by exogenous glucagon-like peptide 1 (7–36 amide) in type 2 (non-insulin-dependent) diabetic patients. Diabetologia 36:741–744

Ohneda A, Sato M, Matsuda K, Itabashi H, Horigome K, Chiba M, Yamagata S (1974) Suppression of pancreatic glucagon secretion by tolbutamide in dogs. Horm Metab Res 6:478–483

Ohneda A, Ishii S, Horigome K, Yamagata S (1975) Glucagon response to arginine after treatment of diabetes mellitus. Diabetes 24:811–819

Orci L, Bordi C, Unger RH, Perrelet A (1983) Glucagon- and glicentin-producing cells. In: Lefèbvre PJ (ed) Glucagon I. Springer, Berlin Heidelberg New York, p 57 (Handbook of experimental pharmacology, vol 66/I)

Paolisso G, Sgambato S, Gentile S, Memoli P, Varricchio M, D'Onofrio F (1988a) Advantageous metabolic effects of pulsatile insulin delivery in non-insulin-dependent diabetic subjects. J Clin Endocr Metab 67:1005–1010

Paolisso G, Sgambato S, Torella R, Varricchio M, Scheen A, D'Onofrio F, Lefèbvre PJ (1988b) Pulsatile insulin delivery is more efficient than continuous infusion in modulating islet cell function in normal subjects and patients with Type 1 diabetes. J Clin Endocrinol Metab 66:1220–1226

Pek S, Fajans SS, Floyd JC, Knopf RF, Conn JW (1972) Failure of sulfonylureas to suppress plasma glucagon in man. Diabetes 21:216–222

Perea A, Clemente F, Martinell J, Villanueva-Peñacarrillo, Valverde I (1994) Lipolytic effect of glucagon in human isolated adipocytes (abstract). Diabetologia 37 [Suppl 1]:A129

Perea A, Clemente F, Martinell J, Nillanueva-Peñacarrillo, Valverde I (1995) Physiological effect of glucagon in human isolated adipocytes. Horm Metab Res 27:372–375

Pipeleers DG, In't Veld PA, Van De Winkel (1988) Death of the pancreatic B cell. In: Lefèbvre PJ, Pipeleers DG (eds) The pathology of the endocrine pancreas in diabetes. Springer, Berlin Heidelberg New York, p 106

Rahier J (1988) The diabetic pancreas: a pathologist's view. In: Lefèbvre PJ, Pipeleers DG (eds) The pathology of the endocrine pancreas in diabetes. Springer, Berlin Heidelberg New York, p 17

Rahier J, Goebbels RM, Henquin JC (1983) Cellular composition of the human diabetic pancreas. Diabetologia 24:366–371

Reaven GM, Chen Y-DI, Golay A, Swilocki ALM, Jaspan JB (1987) Documentation of hyperglucagonemia throughout the day in nonobese and obese patients with noninsulin-dependent diabetes mellitus. J Clin Endocrinol Metab 64:106–110

Richter WO, Robl W, Schwandt P (1989) Human glucagon and vasoactive intestinal polypeptide (VIP) stimulate free fatty acid release from human adipose tissue in vitro. Peptides 10:333

Ritzel R, Ørskov C, Holst JJ, Nauck MA (1995) Pharmacokinetic, insulinotropic, and glucagonostatic properties of GLP-1 [7–36 amide] after subcutaneous injection in healthy volunteers. Dose-response relationships. Diabetologia 38:720–725

Roden M, Perseghin G, Hwang J-H, Petersen KF, Cline G, Rothman DL, Shulman GI (1995) Important role for glucagon in the regulation of hepatic glycogen synthesis and turnover (abstract). Diabetologia 38 [Suppl 1]:A66

Saito K, Yaginuma N, Takahashi T (1979) Differential volumetry of A, B and D cells in the pancreatic islets of diabetic and non diabetic subjects. Tohoku J Exp Med 129:273–283

Samols E, Harrison J (1976) Intraislet negative insulin-glucagon feedback. Metabolism 25 [Suppl 1]:1443–1447

Samols E, Stagner J (1991) Intraislet and islet-acinar portal systems and their significance. In: Samols E (ed) The endocrine pancreas. Raven, New York, p 93

Samols E, Tyler J, Miahle (1969) Suppression of pancreatic glucagon release by the hypoglycemic sulfonylureas. Lancet I:174–176

Samols E, Weir GC, Bonner-Weir S (1983) Intraislet insulin-glucagon-somatostatin relationship. In: Lefèbvre PJ (ed) Glucagon II. (Springer, Berlin Heidelberg New York (Handbook of experimental pharmacology, vol 66/I)

Samols E, Stagner J, Ewart RBL, Marks V (1988) The order of islet perfusion is B-A-D in the perfused rat pancreas. J Clin Invest 82:350–354

Savage PJ, Bennion LJ, Bennett PH (1979) Normalization of insulin and glucagon secretion in ketosis-resistant diabetes mellitus with prolonged diet therapy. J Clin Endocrinol Metab 49:830–833

Scheen AJ, Lefèbvre PJ (1993) Pharmacological treatment of the obese diabetic patient. Diabete Matab 19:547–559

Scheen AJ, Lefèbvre PJ (1995) Antihyperglycaemic agents. Drug interactions of clinical importance. Drugs Safety 12:32–45

Scheen AJ, Krzentowski G, Castillo M, Lefèbvre PJ, Luyckx AS (1983) A 6-hour nocturnal interruption of a continuous subcutaneous insulin infusion: 2. Marked attenuation of the metabolic deterioration by somatostatin. Diabetologia 24:319–325

Scheen AJ, Gillet J, Rosenthaler J, Guiot J, Henrivaux Ph, Jandrain B, Lefèbvre PJ (1989) Sandostatin, a new analogue of somatostatin, reduces the metabolic changes induced by the nocturnal interruption of continuous subcutaneous insulin infusion in Type 1 (insulin-dependent) diabetic patients. Diabetologia 32:801–809

Stalmans W (1983) Glucagon and liver glycogen metabolism. In: Lefèbvre PJ (ed) Glucagon I. Springer, Berlin Heidelberg New York, p 291 (Handbook of experimental pharmacology, vol 66/I)

Starke A, Grundy S, McGarry JD, Unger RH (1985) Correction of hyperglycemia with phloridizin restores the glucagon response to glucose in insulin-deficient dogs: implications for human diabetes. Proc Natl Acad Sci USA 82:1544–1546

Stefan Y, Orci L, Malaisse-Lagae F, Perrelet A, Patel Y, Unger RH (1982) Quantitation of endocrine cell content in the pancreas of nondiabetic and diabetic humans. Diabetes 31:694–700

Stevenson RW, Steiner KE, Davis MA, Hendrick GK, Williams PE, Lacy WW, Brown L, Donahue P, Lacy DB, Cherrington AD (1987) Similar dose responsiveness of hepatic glycogenolysis and gluconeogenesis to glucagon in vivo. Diabetes 36:382–389

Takahashi K, Yamatani K, Hara M, Sasaki H (1994) Gliclazide directly suppresses arginine-induced glucagon secretion. Diab Res Clin Pract 24:143–151

Tamborlane WV, Sherwin RS, Hendler R, Felig P (1977) Metabolic effects of somatostatin in maturity-onset diabetes. N Engl J Med 297:181–183
Under RH (1978) Role of glucagon in the pathogenesis of diabetes: the status of the controversy. Metabolism 27:1691–1709
Unger RH, Orci L (1981) Glucagon and the A cell. Physiology and pathophysiology. N Engl J Med 304:1518–1524, 1575–1580
Unger RH, Orci L (1983) Glucagon in diabetes mellitus. In: Lefèbvre PJ (ed) Glucagon II. Springer, Berlin Heidelberg New York (Handbook of experimental pharmacology, vol 66/II)
Unger RH, Orci L (1995) Glucagon secretion, alpha cell metabolism, and glucagon action. In: DeGroodt LJ (ed) Endocrinology, vol 2, 3rd edn. Saunders, Philadelphia, 1337
Van Schravendijk CFH, Foriers A, Hoghe-Peters EL, Rogiers B, De Meyts P, Sodoyez JC, Pipeleers DG (1985) Pancreatic hormone receptors on islet cells. Endocrinology 117:841–848
Veneman T, Mitrakou A, Mokan M, Cryer P, Gerich J (1993) Induction of hypoglycemia unawareness by asymptomatic nocturnal hypoglycemia. Diabetes 42:1233–1237
White NH, Skor DA, Cryer PE, Levandorsky LA, Bier DM (1983) Identification of type-1 diabetic patients at increased risk for hypoglycemia during intensive therapy. N Engl J Med 308:485–491
Willms B, Werner J, Creutzfeldt W, Ørskov C, Holst JJ, Nauck M (1994) Inhibition of gastric emptying by glucagon-like peptide-1 (7–36 amide) in patients with type-2 diabetes mellitus (abstract). Diabetologia 37 [Suppl 1]:A118
Willms B, Kleine N, Creutzfeldt W, Ørskov C, Holst J, Nauck M (1995) Glucagon-like peptide 1 (7–36 amide) lowers blood glucose also in type-1-diabetic patients (abstract). Diabetologia 38 [Suppl 1]:A40

CHAPTER 8

The Search for Glucagon Antagonists

J.M. AMATRUDA and J.N. LIVINGSTON

A. Glucagon as a Drug Target

The concept of diabetes as a bihormonal disease was championed by UNGER (UNGER and ORCI 1975; UNGER 1976). This concept was based on data showing that glucagon levels are absolutely or relatively elevated in all forms of hyperglycemia (REAVEN et al. 1987) and that somatostatin reduces plasma glucose in both normal and diabetic subjects.

For many reasons, glucagon is an attractive drug target (LEFÈBVRE and LUYCKX 1979; LEFÈBVRE et al. 1991). Fasting plasma glucose and hepatic glucose output are closely related (DEFRONZO 1988) and glucagon is a major hormonal regulator of gluconeogenesis and glycogenolysis. In man, glucagon has been demonstrated to regulate hepatic glucose output in both the fasted and postprandial states (BARON et al. 1987; DINNEEN et al. 1995). For example, isolated glucagon suppression with serum glucose maintained at basal levels leads to suppression of hepatic glucose output of 71% in normal control subjects and 58% in patients with non-insulin-dependent diabetes mellitus (NIDDM) (BARON et al. 1987). Also, since carbohydrate ingestion leads to glucagon suppression in nondiabetic but not in diabetic individuals, DINNEEN et al. studied patients with insulin-dependent diabetes mellitus (IDDM) during glucose ingestion with and without suppression of glucagon. The lack of postprandial suppression of glucagon resulted in a 40 mg% increase in postprandial glucose levels due to higher rates of hepatic postprandial glucose release. Thus, the lack of postprandial suppression of glucagon in patients with IDDM contributes to hyperglycemia.

An important role for hepatic glucose output and glucagon in impaired glucose tolerance (IGT) has also been suggested (MITRAKOU et al. 1992). Suppression of hepatic glucose production during an oral glucose load was approximately 28% in patients with IGT, but 48% in subjects with normal glucose tolerance. In all subjects studied, glucose appearance correlated with peak postprandial plasma glucose ($r = 0.72$) and the ratio of insulin to glucagon levels correlated inversely ($r = -0.62$) with the rates of systemic glucose appearance. Patients with IGT had smaller reductions in plasma glucagon during the oral glucose test.

Glucagon has also been demonstrated in animals (MYERS et al. 1991) and man (DEL PRATO et al. 1987) to affect hepatic sensitivity to insulin. There is a close inverse correlation between the ability of insulin to suppress hepatic glucose output and plasma glucagon levels (MYERS et al. 1991) and, in man, chronic glucagon infusion leads to hepatic insulin resistance (DEL PRATO et al. 1987). In dogs, 70% of the fall in hepatic glucose production during intraportal insulin infusion is due to the insulin-induced decrease in plasma glucagon levels (STEVENSON et al. 1987).

The role of glucagon is critical in determining the ketogenic set of the liver. Suppression of plasma glucagon prevents the development of ketoacidosis in man while altering the relationship between plasma alanine and glucose (GERICH et al. 1975).

Glucagon action is also relatively specific for the liver and a glucagon antagonist would be expected to have few adverse effects stemming from its effects on other organs. Hypoglycemia is prevented by redundant systems including catecholamines and the lowering of insulin levels (GERICH et al. 1979; RIZZA et al. 1979). Thus, there are several cogent reasons to consider glucagon antagonism as a viable drug target for patients with both NIDDM and IDDM. In addition to its use alone as monotherapy, a glucagon antagonist might also be useful in combination with any of the current pharmacologic agents available.

While the preponderance of evidence in animals and man suggests that glucagon is important in the control of fasting and postprandial plasma glucose, there are several unknown and some negative data. First, no small molecular weight compounds have been available to test the short- and long-term effects of antagonizing glucagon action. Thus, there is no proof of principle in man. Furthermore, glucagon antibodies have been shown to have no effect on fasting plasma glucose in streptozotocin diabetic rats with very high fasting glucose levels (ALMDAL et al. 1992). Monoclonal antibodies to glucagon affect postprandial glucose only in mildly diabetic animals (BRAND et al. 1994). In severely diabetic animals, glucose is lowered by a glucagon monoclonal antibody only if animals are simultaneously treated with insulin (BRAND et al. 1994). These results are not surprising, however, since antagonizing glucagon action should not have a major effect on fasting plasma glucose under situations where hepatic glucose output is driven by substrate availability and absolute insulin deficiency. Finally, patients with IDDM have been shown to have a reduced hepatic glucose output response to glucagon (ORSKOV et al. 1991).

Despite the animal studies cited, the bulk of the evidence suggests that, in the presence of insulin, glucagon is important in the control of hepatic glucose output and hepatic insulin resistance and that an antagonist to glucagon will be an important addition to the treatment of NIDDM. Such an antagonist should reduce both fasting and postprandial glucose as monotherapy and, in combination with insulin and with agents that increase insulin secretion (sulfonylureas), delay glucose absorption (α-glucosidase inhibitors) and enhance insulin action (insulin sensitizers). A glucagon antagonist may also be useful for the prevention and treatment of diabetic ketoacidosis.

B. Search for a Glucagon Antagonist

I. Other Effects of Glucagon

The discussion of the role of glucagon in the pathophysiology of IDDM and NIDDM (see Chap. 8, this volume) argues for the utility of blocking glucagon action. However, before discussing the specific approaches for blocking glucagon action, a consideration is in order of the potential negative effects that might occur when the actions of this hormone are disrupted.

A large number of studies have documented the role of glucagon in carbohydrate metabolism (reviewed in CHRISTOPHE 1995). These findings help form the basis for inhibiting the actions of glucagon in the diabetic patient. In addition, there are glucagon effects on processes other than hepatic glycogenolysis and gluconeogenesis (Table 1). For example, effects of glucagon on amino acid transport and metabolism have been described (CARIAPPA and KILBERG 1992; HEINDORFF et al. 1993). Changes in renal glomerular filtration rate and reabsorption of urea and water have been reported (AHLOULAY et al. 1992; LANG et al. 1992). The effects of glucagon on kidney function are supported by the recent finding of significant amounts of glucagon receptor mRNA in this tissue (YOO-WARREN et al. 1994).

An action of glucagon on fat cell lipolysis has been clearly established in rat (LIVINGSTON et al. 1974) and in human (see Chap. 7, this volume) adipose tissue. Gastric acid secretion may be inhibited by glucagon (HOLST et al. 1993) and the hormone can reduce the motility of the gastrointestinal tract (SANTAMARIA et al. 1991). Finally, the possibility that glucagon

Table 1. Main effects of glucagon

1. Increases in glucose metabolism
    a) Glycogenolysis
    b) Gluconeogenesis
    c) Glycolysis
    d) Glucose cycling
2. Increases in ketogenesis
3. Increases in protein metabolism
    a) Increased proteolysis
    b) Inhibition of protein synthesis
    c) Increased amino acid transport
    d) Increased urea production
4. Increases in lipolysis
5. Satiety signal
6. Enhanced insulin secretion
7. Positive inotropic effects on heart
8. Decrease in intestinal motility

influences appetite has been reported (GEARY et al. 1993; see Chap. 14, this volume).

II. Glucagon-Receptor Knockout Mice

These and other findings raise the possibility that blocking glucagon action may lead to effects in body physiology beyond those changes desired, i.e., reducing the production of glucose by diabetic liver. In order to evaluate the possibility of unwanted effects from treatment with a glucagon antagonist, scientists at Bayer have generated a strain of mice that lack a glucagon receptor. Research on these animals is not complete but it is clear that lack of a glucagon receptor and thus lack of glucagon action is not seriously detrimental. The characteristics of glucagon-receptor-deficient mice are summarized in Table 2.

Table 2. Characteristics of glucagon-receptor-deficient mice

1. Normal life span
2. Normal weight
3. Fertile
4. Reduction in fed and fasting blood glucose levels
5. Normal liver glycogen levels
6. Extreme elevation in circulating immunoreactive glucagon levels
7. Normal plasma insulin levels
8. No gross abnormalities in various organ systems

Such mice appear normal, reach normal body weight, and have no obvious signs of impairments in organ development.

Endocrine and metabolic parameters are also shown in Table 2. As expected, blood glucose levels are reduced both in the fasted and fed states, findings that support the usefulness of a glucagon antagonist to reduce the hyperglycemia of diabetes. In the receptor-depleted animals, fasting plasma insulin levels are normal but plasma glucagon levels are greatly elevated. The reduction in blood glucose supports the usefulness of a glucagon antagonist for the treatment of elevated glucose generated by an elevated plasma glucagon/insulin ratio. The very interesting marked increase in circulating immunoreactive glucagon suggests that secretion or clearance of glucagon or both are altered by the lack of a glucagon receptor or by the lack of glucagon action. Since precise studies of glucagon clearance or secretion have not been carried out, the cause of elevated plasma glucagon is unclear. If the glucagon receptor is required for efficient removal of glucagon as some reports suggest (BARAZZONE et al. 1980; SHEETZ and TAGER 1988), then the elevation could be due to reduced clearance. On the other hand, the chronic decrease in blood glucose may provide a signal that over time results in very high glucagon levels. These and other possibilities are being examined.

Regardless of the reason(s), the elevation in plasma glucagon in mice that lack a glucagon receptor raises an important issue regarding the development of a glucagon antagonist. If the antagonist is competitive with native glucagon and if this phenomenon operates in human subjects, then the levels of the hormone may rise to concentrations high enough to overcome the antagonist. Although a concern based on these initial findings, this analysis may be overly simplistic. For example, in diabetes the blood glucose level will usually be elevated rather than decreased as is the case in knockout mice, and this chronic stimulation of A cells for glucagon release will be absent. Also, insulin levels should remain normal or increased in patients with NIDDM as blood glucose falls with the blockade of glucagon action. Theoretically, glucagon levels should not rise until glucose concentrations reach hypoglycemic levels and then the levels may rise high enough to overcome the competitive blockade if hypoglycemia does occur. Another important caveat is that the degradation of glucagon may differ between man and mouse. Clearance of glucagon is reported to be mainly by the kidney in man and by the liver in rodents (HOLST 1991).

Overall the findings with the glucagon-receptor-deficient mice argue that a highly specific glucagon antagonist should be a safe drug. Moreover, the findings of lowered blood glucose support the utility of an antagonist in treating diabetes.

III. Targets for a Glucagon Antagonist

Blocking glucagon action can be accomplished either by blocking the synthesis and secretion of glucagon, or by blocking any one of the numerous processes involved in producing an action of glucagon on target cells. The utility of each site for developing a glucagon antagonist will be discussed.

1. Synthesis, Processing and Secretion of Glucagon

It is theoretically possible to inhibit the expression of the glucagon gene and by this means block glucagon action. This approach is more problematic than the other approaches because of the limited information on the control of the glucagon gene (see Chap. 11, this volume) and, importantly, on the availability of cell lines that can mimic the A cell in the controlled secretion of glucagon. Since regions in the 5' flanking region of the gene have been shown to confer regulation (KNEPEL et al. 1990; EFRAT et al. 1988), it is possible to construct a reporter gene to use in finding compounds that block expression of the glucagon gene. Thus, a 1.3-kb portion of the 5'-flanking region has shown appropriate activity for gene regulation in transgenic animals (EFRAT et al. 1988). However, a major difficulty is finding the appropriate cell line in which to express the reporter gene. This is critical since the cell line must demonstrate appropriate regulation of the reporter construct in order that compounds which reduce glucagon gene transcription can be identified during the screening of large chemical libraries. If the cell environment does not contain the appropriate transcription factors which are responsive to extracellular stimuli, the search for a chemical that blocks gene transcription will be fruitless. At present, limitations are evident in some of the immortalized cell lines currently used in studies of glucagon gene regulation (PHILIPPE et al. 1988; DRUCKER et al. 1988). Until a relatively easily maintained cell line is available that regulates glucagon gene transcription in a physiologic manner, this approach will encounter a host of difficulties in finding drugs that regulate the transcription of the glucagon gene.

Blocking the appropriate processing of proglucagon to glucagon is another site for drug targeting. Again the difficulties involve understanding the steps in enough molecular detail and having the appropriate cell model that closely mimics the processing of proglucagon in the A cell. Post-translational processing of proglucagon has been examined in several different cell lines and aberrant processing for the cellular phenotype has been noted in some lines (DRUCKER et al. 1988; DRUCKER et al. 1994). These findings do not mean that this line of work lacks feasibility. One cell line, mouse a-TC-1 islet cells, appears to process proglucagon in an islet-specific manner (POWERS et al. 1990). Therefore it may be possible to develop drugs that block the synthesis of glucagon from proglucagon and not alter the synthesis of GLP-1 from gut proglucagon.

An area that offers a more direct approach is the targeting of glucagon secretion. In this case, the prototype inhibitor is somatostatin, a 14-amino-acid peptide produced by D cells in the pancreatic islet. This peptide acts on A cells to block glucagon secretion and thus eliminates the systemic effects of glucagon (SCHUSDZIARRA et al. 1978; GERICH 1981).

Unfortunately, there are a number of problems in developing a drug that operates on this principle. First is the issue of specificity: somatostatin blocks the secretion of insulin and growth hormone as well as glucagon (GERICH 1981; GOLDBERG et al. 1979; MENEILLY et al. 1988). Some attempts have been made to engineer modifications into the peptide that would enable it to associate specifically with the somatostatin receptor on pancreatic A cells (EFENDIC et al. 1975; MEYERS et al. 1977; ROSSOWSKI and COY 1994). Five subtypes of the somatostatin receptor have been reported of which the A cell has a form that differs from that of the B cell (ROSSOWSKI and COY 1994; ROSSOWSKI et al. 1994). This difference supports the possibility that a somatostatin analogue specific for the A cell could be developed.

To this end, a series of peptide analogues have been synthesized and their effects on glucagon and insulin release compared (ROSSOWSKI and COY 1994). Peptides were selected for having different selectivity for binding to somatostatin receptor subtypes 2, 3 or 5. The results show that peptide analogues can be found which act preferentially on glucagon rather than insulin release. For example, an EC_{50} value of $18 nM$ was found for the inhibition of glucagon release by an analogue that also had an IC_{50} of $56 nM$ for inhibiting insulin release. Also, this analogue was more potent than somatostatin for inhibiting glucagon release.

Although the results are encouraging, a number of difficulties remain in developing peptides for use in the chronic treatment of diabetes. One example is that the data from the analogue series argue that glucagon release by A cells is controlled by type 2 receptors, which may also control growth hormone and gastric acid secretion (ROSSOWSKI et al. 1994). Thus the selectivity obtained between glucagon and insulin secretion may not be obtained for glucagon secretion and the secretion of some other substances.

Other drawbacks are those typically associated with the use of peptides as therapeutic agents. Usually they are not orally active and their half-life in the circulation is short. Moreover, peptides are costly to manufacture in comparison with small organic compounds. These are significant problems, especially for a drug that would be administered over a long period of time.

2. Inhibition of the Actions of the Glucagon Receptor

This approach is one of the most attractive for the development of a glucagon antagonist. Its attractiveness stems from the conceptual simplicity of blocking the action of one specific protein, the hepatic glucagon receptor. Moreover, the cloning of the glucagon receptor cDNA (YOO-WARREN et al. 1994; JELINEK et al. 1993; SVOBODA et al. 1993a; SVOBODA et al. 1993b) and the subsequent demonstration that there is only one gene for the receptor (CHRISTOPHE 1995) have contributed to the simplicity of the approach. A single gene coupled with the lack of any differentially spliced forms of the receptor (CHRISTOPHE 1995) make for a well-defined target, i.e., a protein with no family variants. In addition, there is a rapid accumulation of information on specific aspects of the interaction of the receptor with glucagon (BUGGY et al. 1995) and more information should be available soon on the association of the Gs protein with the receptor. This new wealth of knowledge coupled with the well-understood features of the signaling pathway used by the receptor increase its utility as a target for drug development.

A straightforward approach of blocking glucagon action is to prevent the receptor from binding glucagon. It is convenient to divide the types of binding inhibitors into peptide analogues of glucagon and nonpeptide agents that prevent receptor/glucagon association. Each will be separately discussed.

a) Peptide Analogues of Glucagon

A large amount of research has been conducted with peptide analogues of glucagon to evaluate glucagon's binding and activation properties. Also this research has attempted to find peptide inhibitors of glucagon action that can be used in the treatment of diabetes.

Over 100 glucagon analogues have been synthesized (SMITH et al. 1993), of which a number have been tested for their ability to bind to the glucagon receptor and/or to activate signal transduction, i.e., increase the production of cAMP. From this work have come general ideas of the regions of glucagon involved in the binding to the glucagon receptor and the regions that induce activation of the receptor. For example, the results argue that the N-terminus portion of glucagon is mainly involved in the activation of the receptor and only partially contributes to receptor binding (CHRISTOPHE 1995; KRSTENANSKY et al. 1986; HRUBY et al. 1986). Furthermore, the COOH-terminus portion is felt to be responsible mainly for binding. A third region composed of the amino acid residues 9–14 forms a hinge that allows the two regions to communicate (DHARANIPRAGADA et al. 1993). Overall, the research demonstrates that the entire sequence of glucagon is needed for high-affinity binding by the receptor and for initiating the proper signal transduction.

For treating diabetes, our interest is in the antagonist action of the analogues. Of course some analogues are agonists instead of antagonists, e.g., removal of the C-terminal negative charge by amidation increases the binding affinity as does the replacement of selected amino acids in the C-terminal region (UNSON et al. 1987; UNSON et al. 1991). In contrast to these changes, the replacement of amino acids 15, 17 or 18 with Glu, Lys or Lys, respectively, to increase the helical content of glucagon reduces signal transduction more than binding to the receptor (HRUBY et al. 1993). This difference, which allows binding to predominate over activation of the receptor, is an essential requirement of a peptide antagonist.

An example of a weak agnoist which can be used as a partial antagonist is des-His-1 glucagon. This analogue has been described as the first glucagon antagonist because of its ability to reduce binding of endogenous glucagon while generating a much less potent activation of the receptor (LIN et al. 1975). However, only a few peptide analogues of glucagon are potent full antagonists, i.e., they have no agonist activity. This is understandable since the peptide analogue must compete effectively with glucagon for binding to the receptor, which requires that the analogue's binding affinity remain high. In turn the high-affinity binding of the analogue must be accomplished in the absence of receptor activation.

One analogue with good inhibition activity is des-His [Glu-9] glucagon amide (UNSON et al. 1987; UNSON et al. 1989). This analogue retains about 40% of the binding affinity of native glucagon while not generating any significant activation of signal transduction by the receptor. By its ability at elevated concentrations to block glucagon from interacting with the receptor, des-His [Glu-9] glucagon amide was able to suppress glycogenolysis in rabbits and reduce the hyperglycemia in diabetic rats (UNSON et al. 1989).

Another analogue, 1-N-trinitrophenylhistidine, 12 homoarginine-glucagon, also is a full antagonist (JOHNSON et al. 1982). This analogue has been infused in diabetic rats and shown to cause a 50% decrease in blood glucose levels. Finally other derivatives that are interesting full antagonists are cyclic glucagon amide analogues (DHARANIPRAGADA et al. 1993). These analogues are from the des His-1 [Glu-9] series and they include analogues cycled through Glu-9 to lys-12 or lys-12 to Asp-15. Both analogues demonstrate biphasic binding curves to the glucagon receptor in which the high-affinity K_d is 1.5nM for the Glu-Lys analogue and 4nM for the Lys-Asp derivative.

Neither analogue produced an effect on adenylate cyclase activity in liver membranes.

A large number of glucagon analogues were prepared by random mutagenesis of glucagon cDNA and this library was expressed in yeast (SMITH et al. 1993). A screening assay was used to test over 3500 transformant yeast clones for their ability to block glucagon-stimulated adenylate cyclase activity. More than 20 different glucagon analogues with antagonist activity were identified by this route. A general finding was that an antagonist required a change in one of the first four N-terminal amino acids. Also some synergism was obtained by having more than one change in this region as well as by incorporating changes in a number of other positions, i.e., amino acids 5, 7, 9, 11, 13, 21 and 29. Although none of the peptides identified in the random library were better than the des His-1 [Glu-9] analogue, the information helps identify the sites at which mutations may be engineered to generate a more effective peptide antagonist of glucagon.

The work on peptide antagonists has helped to identify the regions in glucagon important for binding to the receptor and for activation of signal transduction. These results considered with our findings with glucagon/GLP-1 receptor chimeras (BUGGY et al. 1995) argue that a relatively large portion of glucagon is needed to associate with a relatively large area of the receptor to obtain high-affinity binding. The practical implication of these considerations is that a small peptide analogue of glucagon with high affinity for the receptor will be difficult to design. Thus, the development of a peptide antagonist for use in the chronic treatment of diabetic subjects will continue to pose a challenge. Of course there are a number of other obstacles as well, including the cost of peptides and their usual short biological half-life, all of which must be overcome. As an example of the biological half-life problem, the peptide antagonist [1-N α-trinitrophenylhistidine, 12-homoarginine]-glucagon was effective in lowering blood glucose of diabetic rats for only a 10-min period after an i.v. bolus injection of 1 mg/kg (JOHNSON et al. 1982).

b) Small Molecular Weight Synthetic Compounds

It may seem counterintuitive to argue that small organic compounds are potential therapeutic candidates for inhibiting glucagon action in diabetic subjects if useful peptide antagonists will be quite difficult to develop. However, the difficulties faced by peptide antagonists of systemic delivery, short half-life and the cost of production are more easily overcome by a non-peptide drug.

These advantages of a small synthetic organic antagonist make it the optimal choice for pharmaceutical companies. The major question is simply one of practicality, i.e., can a nontoxic, orally active, small synthetic compound be developed that blocks glucagon action? We believe that the answer is yes.

One reason is the success that others have had in developing compounds which block the action of peptide hormones that act on G-protein-linked receptors. These compounds include ones that are effective against the neurokinin-1 receptor (SNIDER et al. 1991; SACHAIS et al. 1993; FONG et al. 1993), the cholecystokinin-B receptor (EVANS et al. 1986; BEINBORN et al. 1993) and the neuropeptide Y Y1 receptor (GAL et al. 1995; JACQUES et al. 1995). For example, the activity of the small organic, orally active Y1 antagonist, SR 120819A, demonstrates that a small molecule can inhibit the receptor binding of a relatively large ligand, in this instance a 36-amino-acid peptide hormone (GAL et al. 1995). SR 120819A shows a highly selective and competitive affinity for the Y1 receptor (K_i of 15 nM). Moreover, this compound was able to block the pressor response of i.v. administered Leu-31, Pro-34, NPY in the guinea pig, an effect that lasted for 4 h following an oral dose of 5 mg/kg body wt. (GAL et al. 1995).

Such examples argue that an effective glucagon antagonist can indeed be developed. However, a compound ready for clinical development has not been reported although a styryl quinoxaline, CP-99,711, has been shown to inhibit glucagon-binding activity with an IC_{50} value of $4 \mu M$ (COLLINS et al. 1992). This compound inhibited glucagon-generated cAMP production by liver membranes with an IC_{50} of $7 \mu M$. The compound alone did not cause an increase in basal cAMP production, indicating the lack of agonist activity. No information on in vivo action of the compound has been provided.

Although these data are encouraging, the findings are tempered by the relatively low effectiveness of the compound and, more importantly, by results that show a lack of specificity for the glucagon receptor. The styryl quinoxalines were stated to inhibit the binding of several different ligands to their respective G-protein-linked receptors (COLLINS et al. 1992). In the absence of specificity or at least high selectivity for the glucagon receptor, it is not feasible to develop a glucagon antagonist from this family of compounds.

The latter consideration highlights the most difficult aspect of developing drugs against G-protein-linked receptors. The common structural elements that involve the seven transmembrane domains and the possible similar packing of these domains in the plane of the membrane may provide common sites for drug interactions.

Although compounds may cross-react with a number of different G-protein-linked receptors, it is possible to find compounds that are highly specific for a particular receptor. In fact, small organic compounds in some cases may be more specific for a particular receptor than the much larger peptide ligand. Two examples of this type of specificity involve the receptors for neurokinin-1 and for CCK-B. The synthetic antagonist CP 96,345 has a 90-fold increase in affinity for the human neurokinin-1 receptor than for the rat neurokinin-1 receptor (SACHAIS et al. 1993). A 20-fold part of the overall 90-fold difference is explained by the presence of isoleucine at position 290 in the human receptor rather than the serine present at this position in the rat receptor.

The second example is the ability of the nonpeptide benzodiazepine-based antagonist, L364718, to inhibit the binding of CCK-8 to the CCK-B receptor (BEINBORN et al. 1993). This antagonist is 20-fold more potent in blocking the CCK-8-binding activity of the canine CCK-B receptor than in blocking the binding activity of the human CCK-B receptor. The relationship between the receptors from the two species can be reversed by mutating 319-Val to Leu in the human receptor and the corresponding 355-Leu to Val in the dog receptor (BEINBORN et al. 1993).

These species differences highlight one of the important principles of the action of small synthetics on G-protein-linked receptors, i.e., the domain involved in interacting with the compound is small. From the above examples it is clear that a change in the side group of one amino acid can markedly alter the potency of a small synthetic antagonist. Thus, the human receptor should be used for screening chemical libraries to detect compounds that block the binding of ligands to G-protein-linked receptors. Although the use of animal receptors may be more convenient, a compound developed with nonprimate receptors may not work efficiently in man.

IV. Humanized Mice

An important advance in drug development is the increasing availability of human receptors that can be used as the basis for a drug discovery program. Although this is certainly an advantage, the approach can generate problems as well, especially if the compound discovered fails to act on any species other than primates. Progress with the development of the compound can be hampered by the lack of a convenient animal model in which to conduct efficacy and toxicity experiments. At the extreme, such a drug development program may require testing in very expensive nonhuman primates.

A solution for this situation that we have undertaken is to "humanize" mice. In this approach homologous recombination technology is used to replace the endogenous mouse gene with either the cDNA for the human receptor or with the entire human gene if it is relatively small. We have carried out both approaches with the glucagon receptor. The human cDNA for the receptor has been used to target and disrupt the mouse glucagon receptor gene. With this approach Bayer scientists were able to disrupt the mouse gene and to elicit production of the human glucagon receptor. Since the controls for the gene were not disrupted by the knockout procedure and human replacement, the expression pattern of the human cDNA was identical to that of the endogenous mouse gene (data not shown, manuscript in preparation). One difference, however, was that the level of human receptor expressed in liver was about 50% that of mouse glucagon receptors in normal animals. Experiments are underway on animals in which the mouse gene-coding body has been replaced by the entire human gene-coding body for the glucagon receptor, a maneuver that should normalize the expression level of the human receptor. Overall, these studies show the feasibility of humanization of mice.

Whether or not they will be important in drug development will depend on the species specificity of the glucagon antagonists under development.

References

Ahloulay M, Bouby N, Machet F, Kubrusly M, Coutaud C, Bankir L (1992) Effects of glucagon on glomerular filtration rate and urea and water excretion. Am J Physiol 263:F24–F36

Almdal TP, Holst JJ, Heindorff H, Vilstrup H (1992) Glucagon immunoneutralization in diabetic rats normalizes urea synthesis and decreases nitrogen wasting. Diabetes 41:12–16

Barazzone P, Gorden P, Carpentier JL, Orci L, Freychet P, Canivet B (1980) Binding, internalization, and lysosomal association of ^{125}I-glucagon in isolated rat hepatocytes. J Clin Invest 66:1081–1093

Baron AD, Schaeffer L, Shragg P, Kolterman OG (1987) Role of hyperglucagonemia in maintenance of increased rates of hepatic glucose output in type II diabetics. Diabetes 36:274–283

Beinborn M, Lee Y-M, McBride E, Quinn S, Kopin A (1993) A single amino acid of the cholecystokinin-B/gastrin receptor determines specificity for non-peptide antagonists. Nature 362:348–350

Brand CL, Rolin B, Jorgensen PN, Svendsen I, Kirstensen JS, Holst JJ (1994) Immunoneutralization of endogenous glucagon with monoclonal glucagon antibody normalizes hyperglycaemia in moderately streptozotocin-diabetic rats. Diabetologia 37:985–993

Buggy JJ, Livingston JN, Rabin DU, Yoo-Warren H (1995) Glucagon-like peptide 1 receptor chimeras reveal domains that determine specificity of glucagon binding. J Biol Chem 270:7474–7478

Cariappa R, Kilberg MS (1992) Plasma membrane domain localization and transcytosis of the glucagon-induced hepatic system A carrier. Am J Physiol 263:E1021–E1028

Christophe J (1995) Glucagon receptors: from genetic structure and expression to effector coupling and biological responses. Biochim Biophys Acta 1241:45–57

Collins JL, Dambek PJ, Goldstein SW, Faraci WS (1992) CP-99,7711: a non-peptide glucagon receptor antagonist. Bioorg Med Chem Let 2:915–918

DeFronzo RA (1988) The triumvirate: β-Cell, muscle, liver: a collusion responsible for NIDDM. Diabetes 37:667–687

Del Prato S, Castellino P, Simonson DC, DeFronzo RA (1987) Hyperglucagonemia and insulin-mediated glucose metabolism. J Clin Invest 79:547–556

Dharanipragada R, Trivedi D, Bannister A, Siegel M, Tourwe D, Mollova N, Schram K, Hruby V (1993) Synthetic linear and cyclic glucagon antagonists. Int J Pept Protein Res 42:68–77

Dinneen S, Alzaid A, Turk D, Rizza R (1995) Failure of glucagon suppression contributes to postprandial hyperglycaemia in IDDM. Diabetologia 38:337–343

Drucker DJ, Philippe J, Mojsov S (1988) Proglucagon gene expression and posttranslational processing in a hamster islet cell line. Endocrinology 123:1861–1867

Drucker DJ, Jin T, Asa SL, Young TA, Brubaker PL (1994) Activation of proglucagon gene transcription by protein kinase-A in a novel mouse enteroendocrine cell line. Mol Endocrinol 8:1646–1655

Efendic S, Luft R, Sievertsson H (1975) Relative effects of somatostatin and two somatostatin analogues on the release of insulin, glucagon and growth hormone. FEBS Lett 58:302–305

Efrat S, Teitelman G, Anwar M, Ruggiero D, Hanahan D (1988) Glucagon gene regulatory region directs oncoprotein expression in neurons and pancreatic alpha cells. Neuron 1:605–613

Evans B, Bock M, Rittle E, DiPardo R, Whitter W, Veber D, Anderson P, Freidinger R (1986) Design of potent, orally effective, nonpeptidal antagonists of the peptide hormone cholecystokinin. Proc Natl Acad Sci USA 83:4918–4922

Fong T, Cascieri M, Yu H, Bansai A, Swain C, Strader C (1993) Amino-aromatic interaction between histidine 197 of the neurokinin-1 receptor and CP96345. Nature 362:350–353

Gal C, Valette G, Rouby P-E, Pellet A, Oury-Donat F, Brossard G, Lespy L, Marty E, Neliat G, deCointet P, Maffrand J-P, LeFur G (1995) SR120819A, an orally-active and selective neuropeptide YY1 receptor antagonist. FEBS Lett 362:192–196

Geary N, Le-Sauter J, Noh U (1993) Glucagon acts in liver to control spontaneous meal size in rats. Am J Physiol 264:R116–R122

Gerich JE (1981) Somatostatin. In: Brownlee M (ed) Handbook of diabetes mellitus. Garland, New York, pp 297–354

Gerich JE, Lorenzi M, Bier DM, Schneider V, Tsalikian E, Karam JH, Forsham PH (1975) Prevention of human diabetic ketoacidosis by somatostatin. Evidence for an essential role of glucagon. N Engl J Med 292:985–989

Gerich J, Davis J, Lorenzi M, Rizza R, Bohannon N, Karam J, Lewis S, Kaplan S, Schultz T, Cryer P (1979) Hormonal mechanisms of recovery from insulin-induced hypoglycemia in man. Am J Physiol 236:E380–E385

Goldberg DJ, Walesky M, Sherwin RS (1979) Effects of somatostatin on the plasma amino acid response to ingested protein in man. Metabolism 28:866–872

Heindorff H, Holst JJ, Almdal T, Vilstrup H (1993) Effect of glucagon immunoneutralization on the increase in urea synthesis after hysterectomy in rats. Eur J Clin Invest 23:166–170

Holst JJ (1991) Degradation of glucagons. In: Henriksen JH (ed) Degradation of bioactive substances: physiology and pathophysiology. CRC, Boca Raton, p 167

Holst JJ, Rasmussen TN, Harling H, Schmidt P (1993) Effect of intestinal inhibitory peptides on vagally induced secretion from isolated perfused porcine pancreas. Pancreas 8:80–87

Hruby VJ, Gysin B, Trivedi D, Johnson DG (1993) New glucagon analogues with conformational restrictions and altered amphiphilicity: effects on binding, adenylate cyclase and glycogenolytic activities. Life Sci 52:845–855

Hruby VJ, Krstenansky J, Gysin B, Pelton JT, Trivedi D, Mckee RL (1986) Conformational considerations in the design of glucagon agonists and antagonists: examination using synthetic analogs. Biopolymers 25:s135–s155

Jacques D, Cadieux A, Dumont Y, Quirion R (1995) Apparent affinity and potency of BIBP3226, a non-peptide neuropeptide Y receptor antagonist, on purported neuropeptide Y, Y_1, Y_2, and Y_3 receptors. Eur J Pharmacol 278:R3–R5

Jelinek LJ, Lok S, Rossenberg GB, Smith RA, Grant FJ, Biggs S, Bensch PA, Kuijper JL, Sheppard PO, Sprecher CA (1993) Expression cloning and signaling properties of the rat glucagon receptor. Science 259:1614–1616

Johnson D, Goebel G, Hruby V, Bregman M, Trivedi D (1982) Hyperglycemia of diabetic rats decreased by a glucagon receptor antagonist. Science 215:1115–1116

Knepel W, Jepal L, Habener J (1990) A pancreatic islet cell-specific enhancer-like element in the glucagon gene contains two domains binding distinct cellular proteins. J Biol Chem 265:8725–8735

Krstenansky J, Trivedi D, Hruby VJ (1986) Importance of the 10–13 region of glucagon for its receptor interactions and activation of adenylate cyclase. Biochemistry 25:3833–3839

Lang F, Häussinger D, Tschernko E, Capasso G, DeSanto, NG (1992) Proteins, the liver and the kidney-hepatic regulation of renal function. Nephron 61:1–4

Lefèbvre PJ, Luyckx AS (1979) Glucagon and diabetes: a reappraisal. Diabetologia 16:347–354

Lefèbvre P, Paolisso G, Scheen A (1991) The role of glucagon in non-insulin-dependent (type 2) diabetes mellitus. In: Sakamoto N, Angel A, Hotta N (eds)

New directions in research and clinical works for obesity and diabetes mellitus. Elsevier, Amsterdam, p 25

Lin MC, Wright DE, Hruby VJ, Rodbell M (1975) Structure-function relationships in glucagon: properties of highly purified des-His-1-, monoiodo, and (des-Asn-28, Thr-29) (homoserine lactone-27)-glucagon. Biochemistry 14:1559–1563

Livingston J, Cuatrecasas P, Lockwood DH (1974) Studies of glucagon resistance in large rat adipocytes: ^{125}I-labeled glucagon binding and lipolytic capacity. J Lipid Res 15:26–32

Meneilly GS, Minaker KL, Elahi D, Rowe JW (1988) Somatostatin infusion enhances hepatic glucose production during hyperglucagonemia. Metabolism 37:252–256

Meyers C, Arimura A, Gordin A, Fernandez-Durango R, Coy DH, Schally AV, Drouin J, Ferland L, Beaulieu M, Labrie F (1977) Somatostatin analogs which inhibit glucagon and growth hormone more than insulin release. Biochem Biophys Res Commun 74:630–636

Mitrakou A, Kelley D, Mokan M, Veneman T, Pangburn T, Reilly J, Gerich J (1992) Role of reduced suppression of glucose production and diminished early insulin release in impaired glucose tolerance. N Engl J Med 326:22–29

Myers SR, Diamond MP, Adkins-Marshall BA, Williams PE, Stinsen R, Cherrington AD (1991) Effects of small changes in glucagon on glucose production during a euglycemic, hyperinsulinemic clamp. Metabolism 40:66–71

Orskov L, Alberti KGMM, Mengel A, Moller N, Pedersen O, Rasmussen O, Seefeldt T, Schmitz O (1991) Decreased hepatic glucagon responses in type 1 (insulin-dependent) diabetes mellitus. Diabetologia 34:521–526

Philippe J, Drucker DJ, Knepel W, Jepeal L, Misulovin Z, Habener JK (1988) Alpha-cell specific expression of the glucagon gene is conferred to the glucagon promoter element by the interactions of DNA-binding proteins. Mol Cell Biol 8:4877–4888

Powers AC, Efrat S, Mojsov S, Spector D, Habener JF, Hanahan D (1990) Proglucagon processing similar to normal islets in pancreatic a-like cell line derived from transgenic mouse tumor. Diabetes 39:406–414

Reaven GM, Chen Y-D I, Golay A, Swislocki ALM, Jaspan JB (1987) Documentation of hyperglucagonemia throughout the day in nonobese and obese patients with noninsulin-dependent diabetes mellitus. J Clin Endocrinol Metab 64:106–110

Rizza RA, Cryer PE, Gerich JE (1979) Role of glucagon epinephrine, and growth hormone in human glucose counterregulation: effects of somatostatin and adrenergic blockage on plasma glucose recovery and glucose flux rates following insulin-induced hypoglycemia. J Clin Invest 64:62–71

Rossowski W, Coy DH (1994) Specific inhibition of rat pancreatic insulin or glucagon release by receptor-selective somatostatin analogs. Biochem Biophys Res Commun 205:341–346

Rossowski W, Gu ZF, Akarca US, Jensen RT, Coy DH (1994) Characterization of somatostatin receptor subtypes controlling rat gastric acid and pancreatic amylase release. Peptides 15:1421–1424

Sachais B, Snider M, Lowe J, Krause J (1993) Molecular basis for the species selectivity of the substance P antagonist CP-96,345. J Biol Chem 268:2319–2323

Santamaria L, de-Miguel E, Codesal J, Ramirez JR, Picazo J (1991) Identification of glucagon binding sites on smooth muscle tissue of dog intestine. Quantification by means of ultrastructural autoradiography. J Anatom Anz 172:149–157

Schusdziarra V, Rivier R, Dobbs RE, Brown M, Vale W, Unger RH (1978) Somatostatin analogs as glucagon suppressants in diabetes. Horm Metab Res 10:563–565

Sheetz MJ, Tager HS (1988) Characterization of a glucagon receptor-linked protease from canine hepatic plasma membranes. J Biol Chem 263:19210–10217

Smith RA, Sisk R, Lockhart P, Matthews S, Gilbert T, Walker K, Piggot J (1993) Isolation of glucagon antagonists by random molecular mutagenesis and screening. Mol Pharmacol 43:741–748

Snider R, Constantine J, Lowe J, Longo K, Lebel W, Woody H, Drozda S, Desai M, Vinick F, Spencer R, Hess H (1991) Potent nonpeptide antagonist of the substance P (NK_1) receptor. Science 251:435–437

Stevenson RW, Williams PE, Cherrington AD (1987) Role of glucagon suppression on gluconeogenesis during insulin treatment of the conscious diabetic dog. Diabetologia 30:782–790

Svoboda M, Ciccarelli E, Tastenoy M, Cauvin A, Stievenart M, Christophe J (1993a) Small introns in a hepatic cDNA encoding a new glucagon-like peptide 1-type receptor. Biochem Biophys Res Commun 191:479–486

Svoboda M, Ciccarelli E, Tastenoy M, Robberecht R, Christophe J (1993b) A cDNA construct allowing the expression of rat hepatic glucagon receptors. Biochem Biophys Res Commun 192:135–142

Unger RH (1976) Diabetes and the alpha cell. Diabetes 25:136–151

Unger RH, Orci L (1975) The essential role of glucagon in the pathogenesis of diabetes mellitus. Lancet 1:14–19

Unson C, MacDonald D, Ray K, Durrah T, Merrifield R (1991) Position-9 replacement analogs of glucagon uncouple biological activity and receptor binding. J Biol Chem 266:2763–2766

Unson C, Gurzenda EM, Merrifield RB (1989) Biological activities of des-His 1 [Glu9] glucagon amide, a glucagon antagonist. Peptides 10:1171–1177

Unson C, Andreu D, Gurzenda E, Merrifield R (1987) Synthetic peptide antagonists of glucagon. Proc Natl Acad Sci USA 84:4083–4087

Yoo-Warren H, Willse AG, Hancock N, Hull J, McCaleb M, Livingston JN (1994) Regulation of rat glucagon receptor expression. Biochem Biophys Res Commun 205:347–353

CHAPTER 9
Glucagon and Glucose Counterregulation

P.E. CRYER

A. Introduction

Glucose is the predominant metabolic fuel utilized by the brain (MCCALL 1993; SOKOLOFF 1989; HASSELBALCH et al. 1994). Under physiological conditions glucose oxidation accounts for virtually all of the oxygen consumed by the brain, and the brain respiratory quotient approaches 1.0. The brain can utilize additional substrates, such as ketone bodies, but only when the circulating levels of these are elevated well above normal postabsorptive levels, as during prolonged fasting (HASSELBALCH et al. 1994). Because the brain cannot synthesize glucose or store more than a few minutes' supply as glycogen, it requires a continuous supply of glucose from the circulation for its survival and, therefore, for survival of the individual. At normal (or elevated) plasma glucose concentrations the rate of blood-to-brain glucose transport exceeds the rate of brain glucose metabolism. However, as the plasma glucose concentration falls below the physiological range blood-to-brain glucose transport becomes rate limiting to brain glucose metabolism. Given the survival value of maintenance of the plasma glucose concentration, it is not surprising that physiological mechanisms that very effectively prevent or correct hypoglycemia evolved.

The integrated physiology of systemic glucose balance, specifically the mechanisms of glucose counterregulation (the prevention or correction of hypoglycemia) and the role of glucagon in that physiology in humans, is the focus of this chapter. The physiology of glucose counterregulation (CRYER 1993a, 1993b) and its relevant pathophysiology in patients with insulin-dependent diabetes mellitus (CRYER et al. 1994; CRYER 1994) have been reviewed. The principles of glucose counterregulation are three: First, the prevention or correction of hypoglycemia involves both decrements in insulin and activation of glucose-raising (glucose counterregulatory) systems. These are not due solely to dissipation of insulin. Second, whereas insulin is the dominant glucose-lowering factor, there are redundant glucose counterregulatory factors. Thus, there is fail-safe system that prevents failure of the counterregulatory process when one or more of its components fails. Third, there is a hierarchy among the glucose counterregulatory factors. Some are more important than others. Glucagon stands high in that hierarchy. Indeed, the prevention or correction of hypoglycemia is the most well-established biological role of the hormone.

B. Glycemic Action of Glucagon

Glucagon, secreted from pancreatic islet A cells, potently and rapidly stimulates hepatic glucose production (FRIZELL et al. 1988; UNGER and ORCI 1995). It opposes insulin in this among other actions. Glucagon does so primarily through intrahepatic actions on glucogenic enzymes although the hormone also increases hepatic extraction of glucogenic amino acids including alanine (FRIZELL et al. 1988). It does not mobilize peripheral glucogenic precursors, nor is it known to exert physiologically important effects on lipid or protein metabolism.

Binding of glucagon to its hepatocyte plasma membrane receptor activates adenylate cyclase and, thus, increases intracellular cyclic AMP levels and protein kinase A activity. Through a series of regulatory enzyme phosphorylations, the latter plausibly explains the action of glucagon to increase glucose formation. It does so by stimulating both hepatic glycogenolysis (phosphorylase activity is increased, glycogen synthase activity decreased) and hepatic gluconeogenesis (fructose-1,6-bisphosphatase activity is increased).

The glucagon-stimulated increase in hepatic gluconeogenesis is sustained, but that in glycogenolysis and, in large part, that in hepatic glucose release are transient (CHERRINGTON et al. 1981, 1982; FERRANINNI et al. 1982) although the hormone continues to support glucose production (BLOOMGARDEN et al. 1978; RIZZA and GERICH 1979; CHERRINGTON et al. 1981). Thus, the resulting increase in the plasma glucose concentration is transient although it increases again if the glucagon level rises further (RIZZA and GERICH 1979).

Factors in addition to glucagon raise plasma glucose concentrations and participate in the glucose counterregulatory process. Whereas glucagon only stimulates glucose production, the adrenomedullary hormone epinephrine (adrenaline) both stimulates hepatic glucose production and limits glucose utilization by insulin-sensitive tissues such as skeletal muscle (RIZZA et al. 1980; CRYER 1993c). Because the latter action is sustained, the hyperglycemic response to epinephrine, unlike that to glucagon, is sustained. Epinephrine stimulates hepatic glycogenolysis and gluconeogenesis through cyclic AMP-mediated intrahepatic mechanisms similar to those of glucagon. However, it also mobilizes peripheral gluconeogenic precursors (lactate, alanine, glycerol) to the liver (FRIZELL et al. 1988) and modulates glucose production by limiting insulin secretion (BERK et al. 1985; CRYER 1993c) among other indirect actions (FANELLI et al. 1992). Pituitary growth hormone (DE FEO et al. 1989a) and adrenocortical cortisol (DE FEO et al. 1989b) limit glucose utilization but also support glucose production. In addition to hormones, neurotransmitters and metabolic substrates regulate glucose metabolism. Among the neurotransmitters, sympathetic neural norepinephrine raises plasma glucose levels (CRYER 1993c), whereas parasympathetic acetylcholine tends to lower plasma glucose by suppressing hepatic glucose production (BOYLE et al. 1988). Among the effects of substrates, there may be an inverse relationship between glucose production and the ambient plasma glucose concentration independent of hormonal and neural factors [glucose autoregulation (BOLLI et al. 1985)]. Finally, fatty acids both limit glucose utilization and drive glucose production (FANELLI et al. 1992).

C. Glucagon Secretion

I. Regulatory Mechanisms

The regulation of glucagon secretion is complex in its details, involving effects of several metabolic substrates, hormones and neurotransmitters (UNGER and ORCI 1995). However, three factors – the plasma glucose concentration, the insulin level and autonomic nervous system activity – are the major determinants. Glucagon secretion is inversely related to glucose and insulin levels, and directly related to the level of autonomic input into the islets.

A falling plasma glucose concentration is a major stimulus to glucagon secretion. The mechanisms (Fig. 1) include both intraislet and CNS-mediated stimuli to the A cells (UNGER and ORCI 1995). Within the islets a decrease in glucose is sensed directly by A cells (STARKE et al. 1985), resulting in increased glucagon secretion. In addition, that signal is sensed by B cells, resulting in decreased insulin secretion. The latter reduces tonic intraislet A cell inhibition by insulin (MARUYAMA et al. 1984; SAMOLS et al. 1988), thus increasing glucagon secretion. A decrease in the plasma glucose concentration also triggers a CNS-mediated increase in autonomic nervous system activity (GARBER et al. 1976), which acts on A cells through both neural and hormonal signals (HAVEL and TABORSKY 1994). Adrenergic (norepinephrine, or noradrenaline), cholinergic and peptidergic neural inputs to A cells increase glucagon secretion as do hormonal inputs including those of adrenomedullary epinephrine (a component of the autonomic response) as well as growth hormone and cortisol among others.

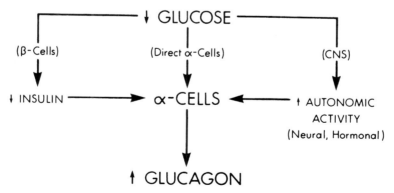

Fig. 1. Schematic diagram of the mechanisms of the glucagon secretory response to falling plasma glucose concentrations

The relative contributions of intraislet and autonomic signals to the glucagon secretory response to hypoglycemia remain a matter of debate (HAVEL and TABORSKY 1994). A substantial body of evidence indicates that autonomic activation plays a major role in experimental animals, particularly in dogs (HAVEL and TABORSKY 1994). However, intraislet mechanisms appear to play a major role in humans. First, in contrast to the findings in experimental animals, combined adrenergic and cholinergic blockade has no effect on the glucagon response to hypoglycemia in humans (HILSTED et al. 1991; TOWLER et al. 1993). Second, although reduced glucagon responses to hypoglycemia have been reported in patients with autonomic failure (LONG et al. 1980; SASAKI et al. 1983), glucose levels were not clamped and tended to be lower in the control subjects compared with the patients. Third, the glucagon response to hypoglycemia is not reduced substantially in adrenalectomized (ENSINCK et al. 1976; BRODOWS et al. 1976), spinal cord sectioned (BRODOWS et al. 1976; PALMER et al. 1976; FRIER et al. 1991) or vagotomized (PALMER et al. 1979) humans. Although the negative autonomic antagonist data (HILSTED et al. 1991; TOWLER et al. 1993) do not exclude peptidergic mediation, the latter surgical/traumatic lesions would be expected to reduce the release of all neurotransmitters including the neuropeptides.

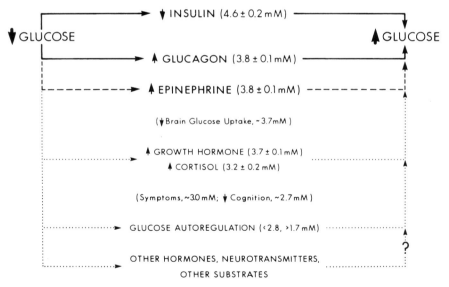

Fig. 2. Diagram of the current construct of the physiology of normal glucose counterregulation in humans. Arterialized venous glycemic thresholds for responses to declining plasma glucose concentrations SCHWARTZ et al. (1987) are also shown. This figure first appeared in CRYER (1993b) and is reproduced with permission of the American Physiological Society

II. Glycemic Thresholds

There is a now well-defined sequence of responses to falling plasma glucose concentrations in humans (SCHWARTZ et al. 1987; MITRAKOU et al. 1991; FANELLI et al. 1994). Glucose decrements within the physiological range signal a decrease in insulin secretion (SCHWARTZ et al. 1987; CRYER 1993b; FANELLI et al. 1993). Those just below the physiological range (e.g., less than 4.0 mmol/l in normal humans) trigger increments in counterregulatory hormone, including glucagon and epinephrine, secretion. Lower glucose levels (about 3.0 mmol/l) are required to produce symptomatic hypoglycemia, even lower levels to produce hypoglycemic cognitive dysfunction and substantially lower levels to ultimately produce seizures, coma and death. This sequence is consistent with the possibility that decrements in insulin and increments in glucagon and epinephrine are involved in the prevention of symptomatic hypoglycemia. As discussed shortly, direct evidence supports that concept. Arterialized venous glycemic thresholds for many of these responses, determined with the hyperinsulinemic, stepped hypoglycemic clamp technique in normal humans (SCHWARTZ et al. 1987; MITRAKOU et al. 1991), are shown in Fig. 2.

D. Role of Glucagon in Glucose Counterregulation

I. Physiology (Fig. 2)

1. Insulin

The first defense against falling plasma glucose concentrations is a decrease in insulin secretion. As noted earlier, decreased insulin secretion occurs with plasma glucose decrements within the physiological range (SCHWARTZ et al. 1987; CRYER 1993b). Glucose recovery from hypoglycemia is inversely related to insulin levels during the recovery phase (HELLER and CRYER 1991). Thus, there is direct evidence that decreased insulin secretion plays a major role in the physiology of glucose counterregulation. However, several lines of evidence indicate that the prevention and correction of hypoglycemia are not due solely to dissipation of insulin. First, in a model of hypoglycemia produced by bolus intravenous insulin injection the changes in glucose kinetics (decreased utilization, increased production) that ultimately restore euglycemia begin while plasma insulin concentrations are still more than tenfold higher than baseline levels (GARBER et al. 1976). Second, during low-dose insulin infusion during the recovery phase, biological glucose recovery from hypoglycemia occurs despite peripheral hyperinsulinemia in the absence of portal hypoinsulinemia (HELLER and CRYER 1991). Third, hypoglycemia develops or progresses when the secretion or actions of certain counterregulatory hormones are suppressed in the absence of hyperinsulinemia as discussed shortly.

Thus, factors in addition to dissipation of insulin, specifically activation of glucose counterregulatory systems, must be involved in the prevention or correction of hypoglycemia.

2. Glucagon

Among the counterregulatory factors, glucagon plays a primary role in both the prevention and the correction of hypoglycemia. This was first demonstrated when it was shown that recovery from hypoglycemia is impaired, by about 40%, during infusion of somatostatin (GERICH et al. 1979; RIZZA et al. 1979). This was an effect of suppression of glucagon rather than growth hormone (which are both suppressed by somatostatin) since it was prevented by glucagon but not by growth hormone replacement. With respect to the prevention of hypoglycemia, suppression of glucagon secretion with partial insulin replacement was shown to lower plasma glucose concentrations after an overnight fast (ROSEN et al. 1984), after a 3-day fast (BOYLE et al. 1989), following an oral glucose load (TSE et al. 1983) and during moderate physical exercise (HIRSCH et al. 1991; MARKER et al. 1991) in humans. These studies demonstrated the primacy of glucagon in the correction or prevention of hypoglycemia. However, they also demonstrated that glucagon is not the exclusive counterregulatory factor. Albeit reduced, recovery from hypoglycemia was approximately 60% of normal in the absence of glucagon secretion (GERICH et al. 1979; RIZZA et al. 1979). Similarly, although glucose levels fell when glucagon secretion was suppressed in the studies of the prevention of hypoglycemia (ROSEN et al. 1984; BOYLE et al. 1989; TSE et al. 1983; HIRSCH et al. 1991; MARKER et al. 1991), they then plateaued and did not progress to hypoglycemic levels. Thus, a factor, or factors, in addition to glucagon must play a role in the prevention and correction of hypoglycemia.

3. Epinephrine

That additional factor is epinephrine. While epinephrine is not normally critical, it becomes critical to the correction (GERICH et al. 1979; RIZZA et al. 1979) and the prevention (ROSEN et al. 1984; BOYLE et al. 1989; TSE et al. 1983; HIRSCH et al. 1991; MARKER et al. 1991) of hypoglycemia when glucagon is deficient. Hypoglycemia develops (ROSEN et al. 1984; BOYLE et al. 1989; TSE et al. 1983; HIRSCH et al. 1991; MARKER et al. 1991) or progresses (GERICH et al. 1979; RIZZA et al. 1979) when both glucagon and epinephrine are deficient and insulin is present despite the counterregulatory actions of other hormones, neurotransmitters and substrate effects (Fig. 2). Thus, insulin, glucagon and epinephrine stand high in the hierarchy of redundant glucoregulatory factors involved in glucose counterregulation.

4. Other Counterregulatory Factors

Nonetheless, additional factors are involved. Growth hormone (DE FEO et al. 1989a) and cortisol (DE FEO et al. 1989b) are involved in defense against prolonged hypoglycemia, but neither is critical to recovery from even prolonged hypoglycemia, or to the prevention of hypoglycemia after an overnight fast at least in adults (BOYLE and CRYER 1991). Glucose autoregulation may be operative in humans but only during very severe hypoglycemia (BOLLI et al. 1985; HANSEN et al. 1986). Finally, additional factors may play roles (CONNOLLY et al. 1992) but these are minor (CRYER 1993b).

II. Pathophysiology

Disease-associated, in contrast to experimentally induced, deficiencies of glucagon and epinephrine provide further support for the construct of the physiology of glucose counterregulation just developed. Patients with insulin-dependent diabetes mellitus commonly develop deficient glucagon responses and, later, reduced epinephrine responses to falling plasma glucose concentrations. Combined deficiencies of the glucagon and epinephrine responses have been found, in prospective studies, to be associated with a 25-fold (WHITE et al. 1983) or greater (BOLLI et al. 1984) increased frequency of severe iatrogenic hypoglycemia. Conversely, patients with histories of recurrent severe iatrogenic hypoglycemia almost invariably exhibit deficient glucagon and epinephrine responses to hypoglycemia (SJÖBOM et al. 1989).

E. Conclusions

Acting in concert with decrements in insulin, increments in glucagon play a primary, albeit not an exclusive, role in the prevention or correction of hypoglycemia. That is the primary biological role of glucagon.

The conventional view is that the postabsorptive plasma glucose concentration is determined by the feedback-regulated interaction of the glucose-lowering action of insulin and the glucose-raising action of glucagon (UNGER and ORCI 1995). However, it may well be that the postabsorptive glucose level is normally regulated primarily by variations in insulin secretion, and that glucagon comes into play only when the plasma glucose falls below the physiological range (e.g., below ~4.0mmol/l). For example, although biological glucose recovery from hypoglycemia occurred despite ~ two fold (infused) peripheral hyperinsulinemia during the recovery phase in the study of HELLER and CRYER (1991), the glucose level rose to only ~4.3mmol/l rather than to the control level of ~5.0mmol/l. The glucose counterregulatory systems, including increased glucagon and epinephrine secretion, disengaged completely as the glucose level rose above the glycemic threshold for their secretion (~3.8mmol/l) and, therefore, did not continue to drive the glucose level up to the control level. Nonetheless, even if the latter interpretation is correct it is clear that glucagon becomes important during relatively small decrements in plasma glucose and at plasma glucose concentrations well above those that produce symptoms of hypoglycemia and hypoglycemic cognitive dysfunction let alone manifestations of severe neuroglycopenia.

Acknowledgments. The author's work cited was supported, in part, by National Institutes of Heath grants 5 R01 DK27085, 1 R01 DK44235, 5 M01 RR00036, 5 P60 DK20579 and 5 T32 DK07120 and by the American Diabetes Association.

The author acknowledges the substantive input of several collaborators and postdoctoral fellows, many of whose names appear in the list of references; the technical assistance of Mr. Suresh D. Shah among others; the assistance of the nursing staff of the Washington University General Clinical Research Center headed by Ms. Carolyn E. Havlin; and the assistance of Ms. Kay Logsdon in the preparation of this manuscript.

References

Berk MA, Clutter WE, Skor D, Shah SD, Gingerich RP, Parvin CA, Cryer PE (1985) Enhanced glycemic responsiveness to epinephrine in insulin dependent diabetes mellitus is the result of the inability to secrete insulin. J Clin Invest 75:1842–1851

Bloomgarden ZT, Liljenquist JE, Cherrington AD, Rabinowitz D (1978) Persistent stimulatory effect of glucagon on glucose production despite downregulation. J Clin Endocrinol Metab 47:1152–1155

Bolli GB, De Feo P, DeCosmo S, Perriello G, Ventura MM, Massi-Benedetti M, Santeusanio F, Gerich JE, Brunetti P (1984) A reliable and reproducible test for adequate glucose counterregulation in type 1 diabetes mellitus. Diabetes 33:732–737

Bolli G, DeFeo P, Perriello G, De Cosmo S, Ventura M, Campbell P, Brunetti P, Gerich JE (1985) Role of hepatic autoregulation in defense against hypoglycemia in humans. J Clin Invest 75:1623–1631

Boyle PJ, Cryer PE (1991) Growth hormone, cortisol, or both are involved in defense against, but are not critical to recovery from, prolonged hypoglycemia in humans. Am J Physiol 260:E395–E402

Boyle PJ, Liggett SB, Shah SD, Cryer PE (1988) Direct muscarinic cholinergic inhibition of hepatic glucose production in humans. J Clin Invest 82:445–449

Boyle PJ, Shah SD, Cryer PE (1989) Insulin, glucagon and catecholamines in the prevention of hypoglycemia during fasting in humans. Am J Physiol 256:E651–E661

Brodows RG, Ensinck JW, Campbell RG (1976) Mechanism of cyclic AMP response to hypoglycemia in man. Metabolism 25:659–663

Cherrington AD, Williams PE, Shulman GI, Lacy WW (1981) Differential time course of glucagon's effect on glycogenolysis and gluconeogenesis in the conscious dog. Diabetes 30:180–187

Cherrington AD, Diamond MP, Green DR, Williams PE (1982) Evidence for an intrahepaic contribution to the waning effect of glucagon on glucose production in the conscious dog. Diabetes 31:917–922

Connolly CC, Adkins-Marshall BA, Neal DW, Pugh W, Jaspan JB, Cherringon AD (1992) Relationship between decrements in glucose level and metabolic response to hypoglycemia in absence of counterregulatory hormones in the conscious dog. Diabetes 41:1308–1319

Cryer PE (1993a) Glucose counterregulation: the physiological mechanisms that prevent or correct hypoglycemia. In: Frier BM, Fisher BM (eds) Hypoglycaemia and diabetes. Arnold, London, pp 34–55

Cryer PE (1993b) Glucose counterregulation: the prevention and correction of hypoglycemia in humans. Am J Physiol 264:E149-55

Cryer PE (1993c) Catecholamines, pheochromocytoma and diabetes. Diabetes Rev 1:309–317

Cryer PE (1994) Hypoglycemia: the limiting factor in the management of IDDM. Diabetes 43:1378–1389

Cryer PE, Fisher JN, Shamoon H (1994) Hypoglycemia. Diabetes Care 17:734–755

De Feo P, Perriello G, Torlone E, Ventura MM, Santeusanio F, Brunetti P, Gerich JE, Bolli GB (1989a) Demonstration of a role of growth hormone in glucose counterregulation. Am J Physiol 256:E835–E843

DeFeo P, Perriello G, Torlone E, Ventura MM, Fanelli C, Santeusanio F, Brunetti P, Gerich JE, Bolli GB (1989b) Contribution of cortisol to glucose counterregulation in humans. Am J Physiol 257:E35–E42

Ensinck JW, Walter RM, Palmer JP, Brodows RG, Campbell RG (1976) Glucagon responses to hypoglycemia in adrenalectomized man. Metabolism 25:227–232

Fanelli C, DeFeo P, Porcellatti F, Perriello G, Torlone E, Santeusanio F, Brunetti P, Bolli GB (1992) Adrenergic mechanisms contribute to the late phase of hypoglycemic glucose counterregulation by stimulating lipolysis. J Clin Invest 89:2005–2013

Fanelli C, Pampanelli S, Epifano L, Rambotti AM, Ciofetta M, Modarelli F, DiVincenzo A, Annibale B, Lepore M, Lalli C, Del Sindaco P, Brunetti P, Bolli GB (1994) Relative roles of insulin and hypoglycaemia on induction of neuroendocrine responses to, symptoms of and deterioration of cognitive function in hypoglycaemia in male and female humans. Diabetologia 37:797–807

Ferraninni E, DeFronzo RA, Sherwin RS (1982) Transient hepatic response to glucagon in man: role of insulin and hyperglycemia. Am J Physiol 242:E73–E81

Frier BM, Corrall RJM, Ratcliffe JG, Ashby JP, McClemont EJW (1991) Autonomic neural control mechanisms of substrate and hormonal responses to acute hypoglycaemia in man. Clin Endocrinol (Oxf) 14:425–433

Frizzell RT, Campbell PJ, Cherrington AD (1988) Gluconeogenesis and hypoglycemia. Diabetes Metab Rev 4:51–70

Garber AJ, Cryer PE, Santiago JV, Haymond MW, Pagliara AS, Kipnis DM (1976) The role of adrenergic mechanisms in the substrate and hormonal response to insulin induced hypoglycemia in man. J Clin Invest 58:7–15

Gerich J, Davis J, Lorenzi M, Rizza R, Bohannon N, Karam J, Lewis S, Kaplan R, Schultz T, Cryer P (1979) Hormonal mechanisms of recovery from insulin-induced hypoglycemia in man. Am J Physiol 236:E380–E385

Hansen I, Firth R, Haymond M, Cryer P, Rizza R (1986) The role of autoregulation of hepatic glucose production in man: response to a physiologic decrement in plasma glucose. Diabetes 35:186–191

Hasselbalch SG, Knudsen GM, Jakobsen J, Hageman LP, Holm S, Paulson OB (1994) Brain metabolism during short-term starvation in humans. J Cereb Blood Flow Metab 14:125–131

Havel PJ, Taborsky GJ Jr (1994) The contribution of the autonomic nervous system to increased glucagon secretion during hypoglycemic stress: update 1994. In: Underwood LE (ed) The endocrine pancreas, insulin action and diabetes. The Endocrine Society, Bethesda, pp 201–204

Heller SR, Cryer PE (1991) Hypoinsulinemia is not critical to glucose recovery from hypoglycemia in humans. Am J Physiol 261:E41–E48

Hilsted J, Frandsen H, Holst JJ, Christensen NJ, Nielsen SL (1991) Plasma glucagon and glucose recovery after hypoglycemia: the effect of total autonomic blockade. Acta Endocrinol (Copenh) 125:466–469

Hirsch IB, Marker JC, Smith LJ, Spina RJ, Parvin CA, Holloszy JO, Cryer PE (1991) Insulin and glucagon in the prevention of hypoglycemia during exercise in humans. Am J Physiol 260:E695–E704

Long RG, Albuquerque RH, Prata A, Barnes AJ, Adrian TE, Christofides ND, Bloom SR (1980) Responses of pancreatic and gastrointestinal hormones and growth hormone to oral and intravenous glucose and insulin hypoglycemia in Chaga's disease. Gut 21:772–777

Marker JC, Hirsch IB, Smith LJ, Parvin CA, Holloszy JO, Cryer PE (1991) Catecholamines in prevention of hypoglycemia during exercise in humans. Am J Physiol 260:E705–E712

Maruyama H, Hisatomi A, Orci L, Grodsky GM, Unger RH (1984) Insulin within islet is a physiologic glucagon release inhibitor. J Clin Invest 74:2296–2299

McCall AL (1993) Effects of glucose deprivation on glucose metabolism in the central nervous system. In: Frier BM, Fisher BM (eds) Hypoglycaemia and diabetes. Arnold, London, pp 56–71

Mitrakou A, Ryan C, Veneman T, Mokan M, Jenssen T, Kiss I, Durrant J, Cryer P, Gerich J (1991) Hierarchy of glycemic thresholds for counterregulatory hormone secretion, symptoms and cerebral dysfunction. Am J Physiol 260:E67–E74

Palmer JP, Henry DP, Benson JW, Johnson DG, Ensinck JW (1976) Glucagon response to hypoglycemia in sympathectomized man. J Clin Invest 57:522–525

Palmer JP, Werner PL, Hollander P, Ensinck JW (1979) Evaluation of the control of glucagon secretion by the parasympathetic nervous system in man. Metabolism 28:549–552

Rizza RA, Gerich JE (1979) Persistent effect of sustained hyperglucagonemia on glucose production in man. J Clin Endocrinol Metab 48:352–355

Rizza RA, Cryer PE, Gerich JE (1979) Role of glucagon, catecholamines and growth hormone in human glucose counterregulation. J Clin Invest 64:62–71

Rizza RA, Cryer PE, Haymond MW, Gerich JE (1980) Adrenergic mechanisms for the effect of epinephrine on glucose production and clearance in man. J Clin Invest 65:682–689

Rosen SG, Clutter WE, Berk MA, Shah SD, Cryer PE (1984) Epinephrine supports the postabsorptive plasma glucose concentration and prevents hypoglycemia when glucagon secretion is deficient in man. J Clin Invest 73:405–411

Samols E, Stagner JI, Ewart RBL, Marks V (1988) The order of islet microvascular cellular perfusion is B → A → D in the perfused rat pancreas. J Clin Invest 82:350–353

Sasaki K, Matsuhashi A, Murabayashi S, Aoyagi K, Baha T, Matsunaga M, Takebe K (1983) Hormonal response to insulin-induced hypoglycemia in patients with Shy-Drager syndrome. Metabolism 32:977–981

Schwartz NS, Clutter WE, Shah SD, Cryer PE (1987) Glycemic thresholds for activation of glucose counterregulatory systems are higher than the threshold for symptoms. J Clin Invest 79:777–781

Sjöbom NC, Adamson U, Lins PE (1989) The prevalence of impaired glucose counterregulation during an insulin infusion test in insulin-treated patients prone to severe hypoglycaemia. Diabetologia 32:818–825

Sokoloff L (1989) Circulation and energy metabolism of the brain. In: Siegel G, Agranoff B, Albers RW, Molinoff P (eds) Basic neurochemistry. Raven, New York, pp 565–590

Starke A, Grundy S, McGarry JD, Unger RH (1985) Correction of hyperglycemia by inducing renal malabsorption of glucose restores the glucagon response to glucose in insulin deficient dogs: implications for human diabetes. Proc Natl Acad Sci USA 82:1544–1546

Towler DA, Havlin CE, Craft S, Cryer PE (1993) Mechanism of awareness of hypoglycemia: perception of neurogenic (predominantly cholinergic) rather than neuroglycopenic symptoms. Diabetes 42:1791–1798

Tse TF, Clutter WE, Shah SD, Cryer PE (1983) Mechanisms of postprandial glucose counterregulation in man. J Clin Invest 72:278–286

Unger RH, Orci L (1995) Glucagon secretion, alpha cell metabolism, and glucagon action. In: De Groot LJ (ed) Endocrinology, 3rd edn. Saunders, Philadelphia, pp 1337–1353

White NH, Skor DA, Cryer PE, Bier DM, Levandoski L, Santiago JV (1983) Identification of type 1 diabetic patients at increased risk for hypoglycemia during intensive therapy. N Engl J Med 308:485–491

CHAPTER 10
Modes of Glucagon Administration

G. SLAMA, D. BENNIS, and P. ZIRNIS

A. Introduction

Although the main use of glucagon lies in the management of severe hypoglycemic episodes in insulin-treated patients (MARKS 1983), the polypeptide is also available for diagnostic procedures (BARTNICKE and BROWN 1993) and, less frequently, for clinical use with other conditions such as iatrogenic cardiac poisoning, or as heart relief agent, on the basis of its positive inotropic and chronotropic effects (HENRY 1994).

Management of severe hypoglycemia by glucagon is, as the prevention of mild hypoglycemic reactions, a vital part of the management of diabetes (MÜLHAUSER et al. 1985b). Diabetic patients and their relatives must therefore be educated to manage hypoglycemia with glucagon in order to limit admittance to emergency units (VUKMIR et al. 1991). This issue will probably gain in importance in the near future since intensive insulin therapy is rapidly becoming the standard reference (DCCT Research Group 1991). In addition, in most countries, glucagon is commercially available as powder requiring extemporaneous solubilization. The need to first bring glucagon to a solution form before injecting it under stressful and emergency conditions remains a real problem, especially for relatives who have not received specialized training. The current management of severe hypoglycemia includes parenteral injections of either glucose or glucagon and sometimes emergency hospitalization, which involves additional costs and increased patient discomfort (COLLIER et al. 1987). In any case, the need for parenteral treatment is certainly a limiting factor.

As reviewed in Chaps. 11–13, this volume, glucagon is also used for diagnostic purposes such as assessment of residual insulinsecretion, assessment of the integrity of the glycogenolytic mechanisms, or stimulating catecholamine release (LEFÈBVRE and LUYCKX 1983), in metabolic investigations of insulinomas and in radiological investigations of the abdominal cavity because of its spasmolytic effects (DIAMANT and PICAZO 1983). With these indications, the parenteral route is obviously not a limiting factor.

The classic routes of administration of glucagon, with all the conditions mentioned above, are the subcutaneous, intramuscular and intravenous routes. As recently reviewed by LEFÈBVRE and SCHEEN (1995), novel routes have been recently investigated with nasal preparations, eye drops, or as suppositories so as to facilitate the management of hypoglycemia by the patients themselves in order to achieve the aim of avoiding hospital admittance. The aim of this contribution is to review the conventional routes of administration and to emphasize the potential of novel routes for the administration of this peptide.

B. Classic Routes of Administration

I. Intravenous Route

Intravenous glucagon administration is used for the management of severe hypoglycemia in intensive care units, but also to allow radiological gastrointestinal examinations (see Chap. 12, this volume), and for cardiac recovery in various types of poisoning (see Chap. 11, this volume). This route is also frequently used for the diagnosis of insulinomas and for the assessment of insulin secretion since glucagon is an insulin-releasing factor (see Chap. 13, this volume). The latter indication with measurements of insulin and/or C-peptide levels is a classic test for the evaluation of residual insulin secretion in both insulin-dependent and non-insulin-dependent diabetic patients (FABER and BINDER 1977).

When 1 mg glucagon is injected intravenously, its half-life in plasma is approximately 5 min, the peak value observed being greater than 4000 pg/ml. The metabolic clearance of glucagon is mainly processed by the liver and kidneys, which each contribute about 30% to the overall elimination of the peptide. The remaining fraction is degraded in the plasma or bound and inactivated by peripheral tissues. Results showing a slower disappearance from the blood with a half-life of about 12 min have been reported in insulin-treated diabetic patients. The clearance rate of glucagon over a wide range of plasma concentrations was found to be approximately 20 mg/kg per minute. The response of blood glucose levels to intravenous glucagon appeared very rapidly within a few minutes, peaked 15–20 min after injection and then decreased slowly to return to baseline after about 45 min (PONTIROLI et al. 1993).

II. Intramuscular Route

In addition to being used for the majority of injectable drugs, the intramuscular route is supposed to be more rapidly effective than the subcutaneous route and therefore preferable for the management of acute hypoglycemia. However, this has not been unanimously confirmed for glucagon (AMAN and WRANNE 1988). In a study of the pharmacokinetics and bioavailability of 1 mg glucagon injected intramuscularly in healthy volunteers in whom hypoglycemia was first induced by a 10-U injection of soluble insulin, MÜLHAUSER et al. (1985a) reported similar results: plasma glucagon peaked 10 min after injection, remaining at this level for about 1 h. With other conditions, comparing the pharmacokinetic profiles of different parenteral routes of administration (intravenous, intramuscular and nasal routes), PONTIROLI et al. (1993) recently reported an apparent elimination half-life of about 30 min after a 1-mg administration of glucagon intramuscular.

The increase in blood glucose concentration in MÜLHAUSER's protocol reached approximately 1 mmol/l 10 min after an intramuscular injection with maximum levels of about 2.4 mmol/l reached about 15 min later. Furthermore, in another study in 12 insulin-dependent diabetic patients with moderate hypoglycemia induced by intravenous insulin infusion, it was shown that the blood glucose increment was 2.5 ± 1 mmol/l after 15 min and 4.3 ± 1.8 mmol/l after 30 min, when 1 mg intramuscular glucagon was used; in this study, plasma glucagon levels rose to 3000 ± 1800 pg/ml (ROSENFALCK et al. 1992).

III. Subcutaneous Route

The subcutaneous route is the most-utilized route for the treatment of insulin-induced hypoglycemic episodes in ambulatory conditions because of its better accessibility for relatives and nurses and also for medical practitioners when the patient is agitated. The administration of 1 mg subcutaneous glucagon allows the patient to recover consciousness within about 15 min, when he or she is able again to ingest further carbohydrates.

The pharmacokinetics of subcutaneous glucagon is very close to that when the intramuscular route is used. Thus, when comparing intravenous and intramuscular routes of glucagon administration, in the same six subjects made hypoglycemic by 10 U of subcutaneous regular insulin, MÜLHAUSER et al. (1985a) reported the following results: plasma glucagon levels increased from a mean of 246 to 3233 pg/ml, which is a concentration slightly higher than that obtained after intramuscular injection (250–2638 pg/ml). Plasma glucagon levels remained high for about 60 min. Blood glucose profiles at all times were similar after subcutaneous or intramuscular glucagon injections, and the duration of the clinical effect seemed slightly longer with subcutaneous than with intramuscular administration. The conclusion of the authors was that the subcutaneous route was as effective as the intramuscular route in increasing blood glucose level.

In our experience, in 12 normoglycemic healthy volunteers, the immunoreactive plasma glucagon concentration reached 3500–5500 pg/ml in 15 min and remained at that level for at least 30 min after the injection of 1 mg glucagon subcutaneously. Plasma glucose levels consequently began to increase from basal values to 7 mmol/l between 5 and 30 min. In diabetic insulin-dependent patients submitted either to intravenous or subcutaneous insulin-induced hypoglycemia, a comparison of the response of plasma glucose increase and the increase in immunoreactive plasma glucagon levels showed very similar glycemic and glucagonemic curves with either subcutaneous or intravenous insulin. In these patients, plasma glucose rose progressively from 2.2 to 4.0 mmol/l about 15 min after the subcutaneous injection of 1 mg glucagon (SLAMA et al. 1990). In another study, the authors gradually induced hypoglycemia in diabetic children via a hyperinsulinemic hypoglycemic clamp. When blood glucose levels reached a value of 1.6–1.8 mmol/l, a dose of 0.5 mg glucagon was administered subcutaneously, increasing the plasma glucagon level to 2600 ± 300 pg/ml 15 min after injection, the values declining linearly and progressively thereafter. The blood glucose increment was 1.7 ± 0.2 mmol/l at 15 min and 2.2 ± 0.4 mmol/l at 45 min (STENNINGER and AMAN 1993).

C. New Routes of Glucagon Administration

In its main clinical indication, i.e., acute hypoglycemic states in insulin-treated diabetic patients, glucagon has to be considered as an emergency drug to be administered within minutes more frequently by patient's relatives. However, even if efforts to educate patients and families are made, relatives are in fact often reluctant to perform the injection, for various reasons: the storage conditions (+4°C) (not really convenient), the galenic form of the hormone, which often has to be solubilized immediately before injection, requiring a good level of skill, and most frequently, family member's fear of performing the injection.

I. Intranasal Route

Nasal absorption of drugs such as snuff or cocaine were in use long before WOODYATT (1923) investigated the administration of insulin by the nasal route. It was subsequently shown that insulin was poorly absorbed through the nasal route. PONTIROLI (1991) has clearly shown that the nasal absorption of peptides is inversely related to their molecular size: small peptides such as thyrotropin-releasing hormone (TRH), dDAVP, and luteinizing hormone releasing hormone (LH-RH) are absorbed whereas the bioavailability of pancreatic peptides, i.e., insulin and glucagon (51 and 29 aminoacids, respectively) is weak and unsatisfactory for clinical use. To enhance absorption, different approaches have been developed such as the use of acid solutions or the use of absorption promoters. The importance of the combination of promoters with insulin and then glucagon, in order to facilitate nasal absorption, has been widely investigated. More recently, different chemical classes of enhancing agent have been assessed for their effect on the nasal absorption of peptides. The feasibility of administering glucagon by the nasal route was reported for the first time by PONTIROLI et al. in 1983, and its efficacy has now been clearly demonstrated although glucagon nasal formulations are not commercially available at the time of writing.

Glucagon diluted in saline and supplemented with sodium glycocholate as the enhancer and administered intranasally to healthy volunteers by PONTIROLI et al. (1983) has been shown to increase blood glucose and glucagon immunoreactive plasma levels significantly: when 1 mg glucagon was instilled as a solution by the nasal route, glucagon plasma levels peaked at about 1300 pg/ml and returned to baseline within 60 min. Subsequently, the same authors showed that spray solutions containing the same enhancer were more effective than drops (PONTIROLI et al. 1985). FREYCHET et al. (1988) investigated and compared the intranasal efficacy of glucagon as a spray with that of the subcutaneous route in six insulin-dependent diabetics subjected to insulin-induced hypoglycemia and using desoxycholate as promoter. They concluded that the intranasal route was as effective as the subcutaneous route under their experimental conditions.

SLAMA et al. (1990) reported results from the use of an intranasal freeze-dried powder glucagon preparation containing sodium glycocholate as the absorption promoter. The powder was insufflated into one nostril by means of a rubber bulb flushing out the powder contained in a small glass tube. Intranasal glucagon 1 mg combined with 1 mg sodium glycocholate was tested vs. subcutaneous injection of the same dose in 12 healthy normoglycemic subjects (Fig. 1). Mean plasma glucose concentrations recorded were not significantly different concerning the peak values at 30 min and the areas under the curves.

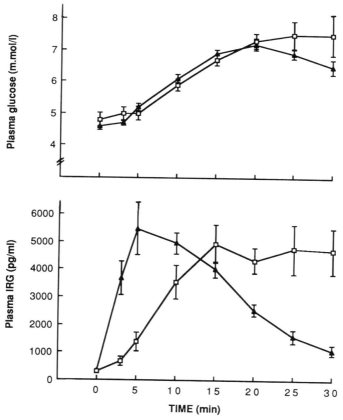

Fig. 1. Mean (SEM) plasma glucose and immunoreactive glucagon (*IRG*) levels in 12 normoglycemic healthy subjects in response to 1 mg glucagon given at t_0 either subcutaneously ($n = 6$, *squares*) or intranasally as a glycocholate-glucagon mixture ($n = 6$, *triangles*). (From SLAMA et al. 1990)

For immunoreactive plasma glucagon, peak values and areas under the curves were significantly different with a peak at 5 min for intranasal glucagon and at about 15 min for the subcutaneous form. However, at 30 min, plasma glucagon levels were almost at their maximal values with the subcutaneous glucagon, whereas those after administration of intranasal glucagon had already returned to their initial values. In the same report, the efficacy of this form of intranasal glucagon was also assessed in six type I diabetic patients subjected to intravenous or subcutaneous insulin-induced hypoglycemia. In moderate hypoglycemia (about 3 mmol/l) induced by an intravenous insulin injection and treated with 1 mg nasal glucagon at *t0*, plasma glucose levels increased steadily for 25 min, attaining at that point a mean of 4.4 mmol/l. The plasma immunoreactive glucagon levels attained a mean of 1822 pg/ml at about 8 min and then returned progressively to baseline within 30 min. In more severe hypoglycemia (about 2 mmol/l) induced by subcutaneous insulin added to the usual insulin morning dose, there was no significant difference in plasma glucose or in immunoreactive glucagon kinetics between 1 mg intranasal glucagon powder and 1 mg subcutaneous glucagon solution (Fig. 2). Nasal tolerance to sodium glycocholate was excellent in these acute conditions. PONTIROLI et al. (1989) showed that 1 mg glucagon combined with a 15-mg dose of sodium glycocholate as a spray was almost as effective as the same dose administered intramuscularly in increasing blood glucose level in healthy subjects, although an oral glucose challenge was significantly more efficient than glucagon. The bioavailability of intranasal 15 mg glycocholate/1 mg glucagon was significantly lower than that of an intramuscular injection. Also, by comparing 1 and 2 mg intranasal with 1 mg intramuscular glucagon, these authors showed that both plasma glucose and plasma immunoreactive glucagon levels were significantly higher after the injection although 2 mg intranasal glucagon was nonetheless more efficient than 1 mg. These results are in accordance with other reports as far as intranasal glucagon pharmacokinetics is concerned as well as regarding onset of activity and duration of effect in comparison to subcutaneous or intramuscular injections, which have a significantly more sustained hyperglycemic effect (SLAMA et al. 1990). As the intranasal glucagon form used by SLAMA et al. was shown to act more quickly than the subcutaneously administered solution, these authors experimented with the preparation in 20 diabetic children under pragmatic conditions at a summer camp, because simplicity of use is particularly important in diabetic children most exposed to severe hypoglycemia (SLAMA et al. 1992). Twenty severe hypoglycemic events occurred and were treated by either intranasal or subcutaneous administration of glucagon in random order. It was concluded that 1 mg intranasal glucagon with 1 mg sodium glycocholate as a powder form was as efficient in correcting severe hypoglycemic attacks as was the subcutaneous route. Both operating time (3 ± 1 vs. 7 ± 2 min for subcutaneous injection) and clinical recovery (18 ± 3 vs. 31 ± 5 min for subcutaneous glucagon) were in favor of the nasal form. Although blood glucose increments were globally similar 10 min after treatment, they lapsed for the nasal route at 30 min. These values were high enough

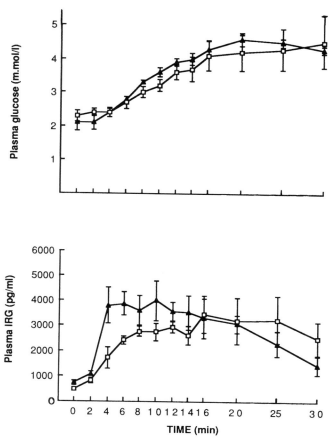

Fig. 2. Mean (SEM) plasma glucose and IRG levels in six insulin-dependent diabetic subjects subjected to subcutaneous-induced hypoglycemia in response to 1 mg glucagon either subcutaneously ($n = 3$, *squares*) or intranasally ($n = 3$, *triangles*) administered at t_0, (from SLAMA et al. 1990)

to allow clinical recovery and oral carbohydrate supplementation. This study was the first to demonstrate and quantify the efficacy of intranasal glucagon in the treatment of hypoglycemic attacks under emergency pragmatic conditions and all users found the nasal route much easier to operate.

In preliminary studies, JORGENSEN et al. (1991) reported efficacy of intranasal glucagon as a powder containing other absorption enhancers such as didecanoyl-L-α-phosphatidyl-choline or α-cyclodextrin in healthy volunteers. Similar results were reported in 12 C-peptide negative type I diabetic patients by ROSENFALCK et al. (1992), who used the same preparation (Hypogon Nasal) as an emergency kit. Recovery with 1 and 2 mg intranasal glucagon, or 1 mg intramuscular glucagon, were compared to spontaneous recovery from insulin-induced hypoglycemia (nadir 1.1–2.3 mmol/l). When evaluated as increments in blood glucose levels, all treatments were more effective than spontaneous recovery. There was no difference between nasal treatment with 2 mg glucagon and intramuscular treatment with 1 mg, both being more effective than 1 mg nasal treatment. This nasal formulation, however, showed side-effects in 80% of patients such as transient local irritation, rhinitis or sneezing, although none of the patients had such side-effects that this type of treatment had to be postponed. This preparation was further tested by STENNINGER and AMAN (1993) in 11 diabetic children undergoing intravenous insulin-induced controlled hypoglycemia. One milligram intranasal glucagon or 0.5 mg subcutaneous glucagon was compared and administered in random order. Although the efficiency of intranasal glucagon was confirmed, the activity shown by the intranasal route was greatly below that of the subcutaneous route due to significantly lower plasma immunoreactive glucagon levels. This could be explained by a lower enhancing effect of their absorption promoters and could also explain the lower occurrence of the side-effects noted such as nausea when compared with subcutaneous administration, i.e., 10 of the 11 children had severe nausea after subcutaneous glucagon versus only one with the intranasal form. Mild subjective nasal irritation was reported by four children and minimal signs of inflammation were observed when the nasal mucosae were inspected.

More recently, HVIDBERG et al. (1994) have compared the hyperglycemic effect of 2 mg intranasal glucagon with the intramuscular route in 12 healthy subjects subjected to insulin-induced hypoglycemia. If glucose recovery was slightly better with intramuscular injection than with the nasal administration, the differences between the two routes regarding blood glucose increments and time for incremental plasma glucose to exceed 3 mmol/l were not considered to be of major clinical importance.

II. Eye Drops

The conjunctival mucosae seem able to absorb pancreatic peptides such as glucagon and insulin; however, the need for an absorption promoter, at least for glucagon in order for an adequate activity to be reached, has not as yet been entirely established. Absorption promoters, however, seem to be necessary for ocular insulin administration since YAMAMOTO et al. (1989) have shown that the bioavailability calculated from the area under curves for insulin eye drops in albino rabbits was 4.9%–7.9% with sodium glycocholate compared with 0.7%–1.3% without a promoter. According to these authors, the nasal mucosae contributed four times more than conjunctival mucosae to the systemic absorption of applied insulin. This suggests that the effect of eye drops is due, at least partly, to nasal absorption as the solution reaches the nasal mucosae through the lacrymonasal canal.

Nevertheless, for glucagon, possibly because of its lower molecular mass, CHUANG et al. (1992) have reported significant absorption at 2.5%, 5.0% and 7.5% w/v (i.e., about 1.25, 2.5 and 3.75 mg glucagon/drop, respectively) in phosphate saline buffer without any promoter. One drop of buffer saline or one drop of the glucagon solutions was instilled in both eyes of insulin-dependent diabetic patients. Fifteen control subjects received the buffer saline, 10 received the 2.5% glucagon solution, 11 the 5.0% and 7 the 7.5% solution. In all subjects receiving glucagon eye drops, a significant increase in blood glucose concentrations was observed at 30 min. The rise in blood glucose concentration was 0.83, 1.76 and 1.91 mmol/l above baseline for the 2.5%, 5.0% and 7.5% solutions, respectively, at 30 min, and remained at 1.52 and 0.66 mmol/l for the 5.0% and 7.5% solutions at 60 min. No eye irritation was noted. In a previous study by the same team (Chiou and Chuang 1988), instillation of a 0.2% eye drop solution of glucagon in the rabbit raised the blood glucose level from 5.6 to 13.4 mmol/l. These findings indicate that it is feasible to induce a hyperglycemic effect with ocularly administered glucagon. However, more extensive studies are needed, principally in order to investigate the necessity for an absorption enhancer in the ocular solution, because of the possible irritant effect of the promoter on the conjunctival mucosae.

III. Rectal Route

The vascularization of the rectal mucosae is pharmacologically interesting because 30% is derived from the portal system. Hence, the rectal route, if possible, would be a more physiological approach for pancreatic peptide administration. However, data on intrarectal administration of glucagon are scarce. For insulin, Bruce et al. (1987) showed that rectal absorption was possible without an absorption enhancer in rats, but with a low bioavailability. With an added promoter, the absorption was improved, although remaining below that of nasal administration.

For glucagon, Parker et al. (1993) showed that it was absorbed by the rectal route when combined with indomethacin as an absorption promoter. This study was conducted in healthy volunteers in the postabsorptive state. Suppositories containing 0.05, 0.1, 1.0 or 5.0 mg glucagon combined with 100 mg indomethacin were inserted into the rectum. Three volunteers received each dose and another three received suppositories with 5 mg glucagon only. In the absence of promoter, glucagon showed no absorption. A linear dose-response relationship was observed when glucagon was administered in combination with indomethacin, with a blood glucose peak increment of at least 1.0 mmol/l after administration of the 0.05-mg suppositories and 2.8 mmol/l for the 5.0-mg ones. Plasma glucagon levels showed a similar dose-response relationship increment. Side-effects were noted and seemed more accentuated than with other routes of administration. These were nausea, abdominal pain and light-headedness felt by almost all the subjects. Side-effects thus limited treatment to the more acceptable dose of 0.1 mg glucagon. Further studies are needed to investigate whether side effects are strictly related to the rectal route of administration or to the nature of the absorption promoter used. In the case of similar efficacy, it seems probable that intranasal glucagon would be preferable for obvious reasons of comfort.

D. Conclusions

The conventional routes of glucagon administration remain the intravenous, intramuscular and subcutaneous routes. Of these, the subcutaneous mode of administration is the most used in the treatment of severe hypoglycemia in insulin-treated diabetic patients. Although the intravenous, but principally the intramuscular, routes can also be considered, they are only occasionally used because of the practical difficulties arising in emergency conditions when glucagon has to be administered by the patient's relatives. The intravenous route is sometimes used but is usually reserved for other conditions such as metabolic or radiological investigations. Although the subcutaneous mode of administration appears to be simpler to apply and needs minimal skill, the problem of preparation of the injectable solution has not been satisfactorily solved.

These considerations have orientated recent studies toward more suitable preparations for practical administration. Several studies have been performed with other nonconventional modes of administration such as the intranasal route, eye drops and suppositories. Preliminary data have shown that glucagon delivered in the form of eye drops and suppositories may give a significant increase in blood glucose levels. Nevertheless, besides the fact that these unusual preparations are not available and practical to use, additional investigations are needed to assess their potential.

The intranasal route has been much more thoroughly investigated during the past decade and appears to be the most suitable and practical route, with many definite advantages over the traditional subcutaneous mode. In fact, administration in this way is easy to perform with a ready-to-use device either as a powder, or as a spray or nasal drops. The nasal spray is preferable to nasal drops but the most efficient way of delivering glucagon into the nostril seems to be with the powdered forms, which can be administered by simple inhalation or insufflation almost instantly in this particular emergency condition. In addition to its feasible administration, the powder form has been shown to be more stable (for at least 6 months and probably more) than the solution form, which remains difficult to store. This also explains the need for the extemporaneous preparation used for the subcutaneous route. The bioavailability of intranasal glucagon is comparable to that of subcutaneous glucagon, with a more rapid onset of effect but with a sufficient, although reduced, duration of action due to the pharmacokinetics of nasal absorption. The enhancers used in these forms seem to be responsible for topical side-effects such as sneezing, burning and more or less pronounced reversible irritation. Although new enhancers could be better tolerated by nasal mucosae, those available, because of the acute nature of the treatment, could be considered safe and satisfactory. It seems therefore that, at present, intranasal glucagon represents a useful alternative to injectable glucagon forms or to intravenous glucose for rapid recovery in hypoglycemic states. In addition, it is also suitable for use in the treatment of minor hypoglycemic episodes with the supplementary advantage of avoiding secondary sustained hyperglycemia, which is frequent after oral glucose intake. Unfortunately, 10 years after its efficacy has been demonstrated, intranasal glucagon is still not marketed at the time of writing, and the reasons for this are incomprehensible.

References

Aman J, Wranne L (1988) Hypoglycaemia in childhood diabetes. Effect of subcutaneous or intramuscular injections of glucagon. Acta Paediatr Scand 77:548–553

Bartnicke BJ, Brown JJ (1993) Glucagon use in magnetic resonance imaging. Int Glucagon Monitor 3:1–3

Bruce JA, Nancy JR, Eli S (1987) Comparison of nasal, rectal, buccal, sublingual and intramuscular insulin efficacy and the effects of a bile salt absorption promoter. J Pharmacol Exp Ther 244 1:23–27

Chiou GCY, Chuang CY (1988) Treatment of hypoglycemia with glucagon eye drops. J Ocul Pharmacol 4:179–186

Chuang LM, Wu HP, Chiou GCY (1992) Increase of blood glucose concentrations in diabetic patients with glucagon eye drops. Chung Kuo Yao Li Hsuer Pao 13:193–197

Collier A, Steedman DJ, Patrick AW, Nimmo GR, Matthews DM, Macintyre CCA, Little K, Clarke BF (1987) Comparison of intravenous glucagon and dextrose in treatment of severe hypoglycemia in an accident and emergency department. Diabetes Care 10:712–715

DCCT Research Group (1991) The epidemiology of severe hypoglycemia in the Diabetes Control and Complications Trial. Am J Med 90:450–459

Diamant B, Picazo J (1983) Spasmolytic action and clinical use of glucagon. In: Lefèbvre PJ (ed) Glucagon II. Springer, Berlin Heidelberg New York Tokyo, pp 611–643 (Handbook of experimental pharmacology, vol 66/II)

Faber OK, Binder C (1977) C-peptide response to glucagon. A test for the residual β-cell function in diabetes mellitus. Diabetes 26:605–610

Freychet L, Desplanque N, Zirinis P, Riskalla SW, Basdevant A, Tchobroutsky G, Slama G (1988) Effect of intranasal glucagon on blood glucose levels in healthy subjects and hypoglycaemic patients with insulin-dependent diabetes. Lancet 1364–1366

Henry JA (1994) The use of glucagon in acute poisoning. Int Glucagon Monitor 4:7–14

Hvidberg A, Djurup R, Hilsted J (1994) Glucose recovery after intranasal glucagon during hypoglycaemia in man. Eur J Clin Pharmacol 46:15–17

Jorgensen S, Sorensen AR, Kimer LL, Mygind N (1991) A new non-invasive method for treating insulin reaction: intranasal lyophilized glucagon. Diabetes 40:A2190 (abstract)

Lefèbvre PJ, Luyckx AS (1983) Glucagon and catecholamines. In: Lefèbvre PJ (ed) Glucagon II. Springer, Berlin Heidelberg New York Tokyo, pp 537–543 (Handbook of experimental pharmacology, vol 66/II)

Lefèbvre PJ, Scheen AJ (1995) Routes of glucagon administration. Int Diabetes. Monitor 7:1–5

Marks V (1983) Glucagon in the diagnosis and treatment of hypoglycaemia. In: Lefèbvre PJ (ed) Glucagon II. Springer, Berlin Heidelberg New York Tokyo, pp 645–666 (Handbook of experimental pharmacology, vol 66/II)

Mülhauser I, Koch J, Berger M (1985a) Pharmacokinetics and bioavailability of injected glucagon: differences between intramuscular, subcutaneous, and intravenous administration. Diabetes Care 8:39–42

Mülhauser I, Berger M, Sonnenberg G, Koch J, Jörgens V, Scholz V (1985b) Incidence and management of severe hypoglycemia in 434 adults with insulin-dependent diabetes mellitus. Diabetes Care 8:268–273

Parker DR, Basdevant GD, Newrick PG, Corral RJM (1993) Rectal absorption of glucagon in man. Eur J Clin Invest 23:S1:A42

Pontiroli AE (1991) Present situation of work with nasal and oral administration of insulin and glucagon. In: Duchène D (ed) Buccal and nasal administration as an alternative to parenteral administration. Editions de Santé, Paris, pp 226–237

Pontiroli AE, Alberetto M, Pozza G (1983) Intranasal glucagon raises blood glucose concentrations in healthy volunteers. BMJ 287:462–463

Pontiroli AE, Alberetto M, Pozza G (1985) Metabolic effects of intranasally administered glucagon: comparison with intramuscular and intravenous injection. Acta Diabetol Lat 22:103–110

Pontiroli AE, Calderera A, Pajetta E, Alberetto M, Pozza G (1989) Intranasal glucagon as remedy for hypoglycemia. Studies in healthy subjects and type I diabetic patients. Diabetes Care 12:604–608

Pontiroli AE, Calderera A, Perfetti MG, Bareggi SR (1993) Pharmacokinetics of intranasal, intramuscular and intravenous glucagon in healthy subjects and diabetic patients. Eur J Clin Pharmacol 45:555–558

Rosenfalck AM, Bendtson I, Jorgensen S, Binder C (1992) Nasal glucagon in the treatment of hypoglycaemia in type I (insulin-dependent) diabetic patients. Diabetes Res Clin Pract 17:43–50

Slama G, Alamowitch C, Desplanque N, Letanoux M, Zirinis P (1990) A new non-invasive method for treating insulin reaction: intranasal lyophilized glucagon. Diabetologia 33:617–674

Slama G, Reach G, Cahane M, Quetin C, Villanove-Robin F (1992) Intranasal glucagon in the treatment of hypoglycaemic attacks in children: experience at a summer camp. Diabetologia 35:398

Stenninger E, Aman J (1993) Intranasal glucagon treatment relieves hypoglycaemia in children with type I (insulin-dependent) diabetes mellitus. Diabetologia 36:931–935

Vukmir RB, Paris PM, Yealy DM (1991) Glucagon: prehospital therapy for hypoglycemia. Ann Emerg Med 30:375–379

Woodyatt RT (1923) The clinical use of insulin. J Metab Res 2:793–801

Yamamoto A, Luo AM, Dodda-Kashi S, Lee VH (1989) The ocular route for systemic insulin delivery in the albino rabbit. J Pharmacol Exp Ther 1:249–255

CHAPTER 11
The Place of Glucagon in Emergency Medicine

M.M. JOHNSON and G.P. ZALOGA

Glucagon is a polypeptide hormone well known for its effects on glucose and lipid metabolism. Once thought to be primarily a hyperglycemic factor, glucagon is now known to alter physiologic function in many tissues. This chapter reviews the use of pharmacologic quantities of glucagon in the treatment of diseases which affect critically ill patients (Table 1).

A. Hypoglycemia

Glucagon raises blood glucose levels by stimulating hepatic glycogenolysis and gluconeogenesis. Simultaneously, it inhibits glycogen synthesis and glycolysis. These actions are mediated through cell surface glucagon receptors (see Chap. 5, this volume) which stimulate intracellular production of cyclic adenosine monophosphate (cAMP). Other metabolic actions of glucagon include the stimulation of lipolysis, proteolysis, and ketogenesis. These catabolic actions provide fuel for cellular energy.

Glucagon's metabolic actions relevant to glucose homeostasis are antagonized by insulin and potentiated by insulin deficiency. Glucagon also stimulates hepatic ketogenesis, an effect potentiated by insulin deficiency. Thus, the ketoacidosis of diabetes mellitus is driven by a low insulin/glucagon ratio. Diabetic ketoacidosis is usually treated with insulin administration. However, inhibition of glucagon secretion with somatostatin has also been shown to reverse diabetic ketoacidosis (LICKLEY et al. 1983; GERICH et al. 1975). Both treatments increase the insulin/glucagon ratio.

The integrity of glucagon and epinephrine secretion (i.e., counterregulatory hormones) is important for prevention of hypoglycemia in patients receiving insulin. Hypoglycemia is common in patients lacking these hormones (see Chap. 9, this volume).

Following a protein meal, insulin assures cellular uptake of nutrients (i.e., glucose) and utilization of amino acids. Glucagon secretion limits the degree of insulin-induced hypoglycemia. Glucagon is also important for maintaining plasma glucose levels and preventing hypoglycemia during starvation and stress (both common during critical illness).

Energy requirements during severe illness are met by glucose and fatty acids. Catecholamines, cortisol, and glucagon stimulate hepatic glucose and

Table 1. Clinical use of glucagon in emergency medicine

Indication	Dose
Hypoglycemia	Adult: 1 mg i.v., i.m., s.q.
	Children: 0.5 mg i.v., i.m., s.q.
Heart failure	1–10 mg/h intravenous infusion
Calcium-blocker overdose	1–5 mg i.v. bolus, 1–10 mg/h infusion
β-Blocker overdose	1–5 mg i.v. bolus, 1–10 mg/h infusion
Acute mesenteric ischemia	1–10 mg/h infusion
Acute renal colic	1–3 mg i.v. bolus, 1–5 mg/h infusion
Acute biliary colic	1–3 mg i.v. bolus, 1–5 mg/h infusion
Acute diverticulities	1–3 mg i.v. bolus, 1–5 mg/h infusion
Esophageal impaction	1 mg i.v., s.q., i.m.; may be repeated

fatty acid release and assure delivery of these nutrients during stress. In addition, stress hormones (i.e., epinephrine, growth hormone, cortisol, endorphins) stimulate glucagon secretion. Together, these hormones are responsible for the "stress hyperglycemia" of critical illness.

Maintenance of blood glucose (and cellular glucose supply) is essential for energy integrity of many tissues (especially brain, renal tissue, and white blood cells). When hypoglycemia occurs, carbohydrate administration is the treatment of choice. However, when glucose is not available or administration is not feasible, glucagon can effectively increase blood glucose levels (VUKMIR et al. 1991; MUHLHAUSER et al. 1985b; COLLIER et al. 1987; STEEL et al. 1992; MACCUISH et al. 1970). VUKMIR et al. (1991) evaluated the effect of glucagon on blood glucose levels and level of consciousness in 50 patients with prehospital hypoglycemia. Glucagon increased mean blood glucose in 49 of 50 patients from 33 to 133 mg/dl and improved level of consciousness. MUHLHAUSER et al. (1985b) evaluated home glucagon injections in adults with insulin-dependent diabetes mellitus. Forty-three severe hypoglycemic episodes were treated with intramuscular or subcutaneous injections of glucagon (1 mg). Therapy was successful in 98% of patients. COLLIER et al. (1987) compared intravenous glucagon (1 mg) with intravenous dextrose (25 g of D50) in hypoglycemic insulin-treated patients in the emergency room. Forty-nine patients were randomized. Both treatments effectively raised blood glucose and improved level of consciousness. However, the glucagon group was slower to achieve normoglycemia (median 6.5 vs. 4 min). PATRICK and colleagues (1990) compared intramuscular glucagon (1 mg) with 25 g intravenous D50 in 29 insulin-treated diabetics with hypoglycemia in an emergency department. In this randomized study, there was a more gradual return to normoglycemia and less severe hyperglycemia in the glucagon group. There was also a slower return to normal consciousness in the glucagon group (9 vs. 3 min). However, both treatments were effective in reversing hypoglycemia.

As reviewed in detail in Chap. 10, this volume, intranasal glucagon can also effectively reverse hypoglycemia in diabetic patients (ROSENFALCK et al. 1992; HVIDBERG et al. 1994; STENNINGER and AMAN 1993; SLAMA et al. 1992;

FREYCHET et al. 1988). STENNINGER and AMAN (1993) treated 11 hypoglycemic (insulin-induced) children with 1 mg intranasal glucagon or 0.5 mg subcutaneous glucagon. Blood glucose increased similarly with both treatments. In comparison with either the intramuscular or subcutaneous routes of administration, intranasal glucagon had a shorter time of onset (i.e., it was similar to the intravenous route) and was associated with less nausea (STENNINGER and AMAN 1993; SLAMA et al. 1992). Bioavailability of the intranasal form appears to be approximately 10% of that achieved with the parenteral route. In addition, CHIOU and CHUANG (1988) reported that 0.2% glucagon ophthalmic solution was effective in reversing hypoglycemia in rabbits and was as effective as intravenous glucagon. This route has not been yet evaluated in humans.

Glucagon has been used effectively to reverse hypoglycemia in patients following insulin overdose (VUKMIR et al. 1991; TAYLOR et al. 1978; MUHLHAUSER et al. 1985a,b; COLLIER et al. 1987; PATRICK et al. 1990; STEEL et al. 1992; MACCUISH et al. 1970; ELRICK et al. 1958). Because one of the physiologic effects of glucagon is stimulation of insulin release, glucagon is less effective in reversing the hypoglycemia due to sulfonylurea overdose and insulinoma. One author even suggests that glucagon my exacerbate the hypoglycemia resulting from sulfonylurea overdose (MARRI et al. 1968). Despite this possibility, glucagon may be useful in cases of sulfonylurea-induced hypoglycemia when glucose is not available or glucose alone is not effective in restoring normoglycemia (DAVIES et al. 1967; TAYLOR et al. 1978).

Glucagon is rarely used to treat hypoglycemia in the intensive care unit, where glucose is readily available. However, because it is available in a small volume and does not require intravenous administration, it is useful in situations where intravenous glucose is impractical. These include field emergencies, emergency rooms, ambulances, helicopters, airplanes, and homes. The drug can be safely administered by nonmedical personnel.

If glucagon fails to restore blood glucose levels, glucose must be given to avoid irreversible central nervous system injury. The etiology of hypoglycemia may persist beyond the duration of glucagon's antihypoglycemic actions. Thus, these patients should be monitored for recurrence of hypoglycemia. Recent evidence also suggests that hyperglycemia may exacerbate neurologic damage following cerebral ischemia. Infusion of hypertonic glucose (i.e., 50% dextrose) is invariably associated with hyperglycemia. Interestingly, although glucagon effectively reverses hypoglycemia it has not been associated with marked hyperglycemia.

The hyperglycemic action of glucagon is poorly maintained during continuous infusion due to insulin release. However, hypoglycemia does not occur due to increased hepatic gluconeogenesis. Glucagon's hyperglycemic actions are increased in patients with diminished insulin secretion (i.e., diabetes mellitus) unless exogenous insulin is administered.

Overall, the available data indicate that glucagon is a highly reliable agent for restoring normoglycemia in patients with hypoglycemic reactions. The usual dose for treating hypoglycemia in adults is 1 mg subcutaneously, intravenously, or intramuscularly. As reviewed in Chap. 10, this volume, the intravenous, subcutaneous, and intramuscular routes of administration of glucagon are all effective in raising blood glucose levels (TAYLOR et al. 1978; MUHLHAUSER et al. 1985a). The intramuscular and subcutaneous routes result in a slower increase in blood glucose but a more prolonged elevation compared to the intravenous route. Glucagon's onset of action is 1–3 min following intravenous injection and 8–10 min after intramuscular or subcutaneous administration. The duration of action is usually 15–30 min. The immediate hyperglycemic effect of glucagon results from its potent hepatic glycogenolytic properties. Glucagon's immediate hyperglycemic effect is diminished (or absent) in states of hepatic glycogen depletion (i.e., starvation, adrenal insufficiency, chronic hypoglycemia). It also works poorly in patients with hepatic insufficiency (i.e., cirrhosis, hepatitis, congestive heart failure, sepsis), glycogen storage disease, and alcohol abuse.

B. Cardiovascular Insufficiency

Endogenous glucagon does not appear to regulate cardiovascular activity. However, pharmacologic administration of glucagon (i.e., 1–5 mg/h) can alter cardiovascular performance.

Glucagon possesses both inotropic and chronotropic actions on the heart in animals and humans (FARAH 1983; FARAH and TUTTLE 1960; BODER and JOHNSON 1972; GLICK et al. 1968; LUCCHESI 1968; REGAN et al. 1964; MOURA and SIMPKINS 1975; MARSIGLIA et al. 1970; SIMAAN and FAWAZ 1976; KONES and PHILLIPS 1971; SMITHERMAN et al. 1978; ABEL 1983; MACLEOD et al. 1981; ZALOGA et al. 1988; CHERNOW et al. 1986; PURI and BING 1979). In the nonfailing heart, glucagon increases heart rate significantly more than cardiac output. On the other hand, in the failing heart, glucagon increases both heart rate and cardiac output to similar degrees (FARAH and TUTTLE 1960).

Glucagon's cardiotonic effects are not dependent on catecholamines or adrenergic receptors and they are not inhibited by α- or β-adrenergic receptor antagonists. Glucagon produces its cardiac effects by interacting with glucagon receptors, which increase intracellular cyclic AMP (via stimulation of adenylyl cyclase or inhibition of phosphodiesterase) (FARAH 1983; MOURA and SIMPKINS 1975; MACLEOD et al. 1981; CHERNOW et al. 1986; ROSS and GILMAN 1980; XENOPHONTOS et al. 1989; MERY et al. 1990) and improve calcium fluxes within the cell (FARAH 1983; MERY et al. 1990; PULLMAN et al. 1967; BARRITT and SPIEL 1981; NAYLER et al. 1970). Cyclic-AMP-dependent phosphorylation of L-type calcium channels is also felt to contribute to inotropic actions. MALCOLM et al. (1986) reported that glucagon could antagonize both the bradycardic and antinociceptive effects of morphine, indicating that endorphin or opiate receptors may play a role in the mechanism by which glucagon exerts its positive chronotropic effects.

As reviewed in detail in Chap. 5, this volume, glucagon is metabolized by endopeptidase into two fragments (PAVOINE et al. 1991). Glucagon-[19–29] is known as "miniglucagon." PAVOINE et al. (1991) reported that miniglucagon is essential for the positive inotropic actions of glucagon in cultured heart cells. Interestingly, miniglucagon alone is a negative inotrope but, when combined with glucagon, the net result is increased contractility. Glucagon's inotropic effects have also been enhanced by glutathione (TALESNIK and TSOPORIS 1984) and hypertonic glucose (DRUCKER et al. 1974). Maintenance of normal or elevated glucose levels is important in a number of glucagon's cardiovascular actions.

Glucagon's chronotropic actions are antagonized by ionized hypocalcemia and severe hypercalcemia (CHERNOW et al. 1987). Hypercalcemia has been shown to inhibit generation of cAMP by catecholamines (PRIELIPP et al. 1989; ABERNETHY et al. 1995). In addition, calcium influx through membrane calcium channels inhibits adenylate cyclase activity in myocardial cells (ABERNETHY et al. 1995). A similar inhibition by calcium may occur with glucagon, an agent whose cardiotonic actions also depend upon stimulation of adenylate cyclase.

Glucagon increases heart rate, rate of pressure change (dP/dt), cardiac output, stroke volume, and oxygen delivery in patients with cardiac disease. These effects have been demonstrated in patients with coronary artery disease, heart failure, and cardiogenic shock (BROGAN et al. 1969; LVOFF and WILCKEN 1972; PARMLEY et al. 1968, 1969; EDDY et al. 1969; WILCKEN and LVOFF 1970; SIEGEL et al. 1970; NORD et al. 1970; DIAMOND et al. 1971; VANDER-ARK and REYNOLDS 1979; TIMMUS et al. 1973). Glucagon acts synergistically with β-adrenergic receptor agonists in these disorders.

Maximal cardiac output effects from glucagon are obtained with 5–10mg intravenous glucagon per hour. Despite beneficial cardiotonic effects, glucagon is rarely used today to support cardiac function. β-Adrenergic agonists such as epinephrine, dobutamine, and dopamine have more potent cardiac-stimulating effects and are the agents of choice for treating cardiac insufficiency. However, due to glucagon's nonadrenergic inotropic actions, further research into cardiovascular applications of glucagon is warranted.

Glucagon increases conduction velocity, membrane responsiveness, and rate of rise of the zero phase of the action potential (PRASAD 1975). It has little effect on refractory period or duration/amplitude of the action potential. Glucagon improves conduction through the atrioventricular node and increases the rate of discharge of sinus and atrioventricular pacemaker cells. NIEMANN et al. (1987) reported that glucagon infusion was associated with an increased likelihood of rhythm restoration following induction of ventricular standstill in dogs. They attributed this effect to glucagon's ability to increase energy substrates within the myocardium. In a separate experiment, glucagon has been shown to increase myocardial carbohydrate uptake (KLINE et al. 1995).

Glucagon reverses the cardiotoxic (i.e., arrhythmogenic, hypotensive) effects of procainamide (PRASAD and WECKWORTH 1978), quinidine (PRASAD 1977; STEWART et al. 1969), ouabain (MADAN 1971; SINGH et al. 1980; COHN et al. 1970; PRASAD and DESOUSA 1972; GUPTA and PRASAD 1980), tricyclic antidepressants (SENER et al. 1995; RUDDY et al. 1972), and slow calcium channel blockers (JOLLY et al. 1987; SABATIER et al. 1991; ZARITSKY et al. 1988; ZALOGA et al. 1985). A portion of the antiarrhythmogenic effects of glucagon result from stimulation of heart rate (i.e., overdrive effect). Procainamide produces a dose-dependent decrease in heart rate, cardiac output, dP/dt, and left ventricular stroke volume in dogs (PRASAD and WECKWORTH 1978). It increases PR interval, QRS duration, and ventricular ectopic beats. These effects are antagonized by glucagon. Ouabain causes bradycardia and PR interval prolongation in dogs (GUPTA and PRASAD 1980). These effects are reversed by glucagon. In one study (PRASAD and DESOUSA 1972), glucagon abolished ouabain-induced ventricular toxicity in 71% of dogs. Quinidine depresses cardiac conduction (i.e., widens QRS interval, causes AV block) and myocardial contraction. Glucagon reverses the prolongation of the action potential and refractory period, improves conduction in quinidine-treated tissue (STEWART et al. 1969), and improves contractility in quinidine-treated hearts. Glucagon has little effect on the normal action potential. However, when it is depressed by drugs, glucagon can restore it toward normal (PRASAD and WECKWORTH 1978; PRASAD 1975, 1977; STEWART et al. 1969). There are no reports of glucagon use in human cases of quinidine, ouabain, and procainamide toxicity.

ZALOGA et al. (1985) evaluated the effect of glucagon on verapamil overdose in rats and found that glucagon reversed verapamil-induced hypotension and bradycardia. In the isolated heart, ZARITSKY et al. (1988) found that glucagon reversed the myocardial depressant actions of verapamil, diltiazem, and nifedipine. Glucagon also reversed the cardiovascular depression of verapamil in dogs (JOLLY et al. 1987; STONE et al. 1995; SABATIER et al. 1991) without causing sustained exacerbations of verapamil-induced hyperglycemia (THOMAS et al. 1995). Over the past few years glucagon has proved effective in reversing the hemodynamic effects of verapamil intoxication in humans (WALTER et al. 1993; WOLF et al. 1993; DOYON and ROBERTS 1993; QUEZADO et al. 1991; LINDEN and AGHABABIAN 1985).

KLINE et al. (1995) reported beneficial effects for insulin/glucose in reversing verapamil toxicity in the anesthetized dog. Insulin improved survival, hemodynamics, and cardiac function. In the Kline study, insulin was more effective than glucagon. It is interesting to note that glucagon was effective following initial administration but its effects dissipated as blood glucose levels decreased. This study suggests that verapamil toxicity renders the heart dependent upon carbohydrate metabolism. Inotropic actions of insulin, glucagon, and epinephrine were coincident with increased myocardial carbohydrate uptake. Thus, glucagon's cardiotonic actions may be dependent upon adequate carbohydrate (i.e., glucose) supply. It would be useful to evaluate the combined effects of glucagon/insulin/glucose during calcium blocker toxicity and in other states of myocardial contractile failure.

WOLF et al. (1993) suggested that glucagon (which stimulates cAMP production) may act synergistically with amrinone (a phosphodiesterase inhibitor) in the treatment of cardiovascular depression secondary to verapamil overdose. Calcium fails to antagonize the atrioventricular depressant effects of slow calcium channel antagonists (SABATIER et al. 1991; CRUMP et al. 1982), although it can reverse hypotension and myocardial depression. On the other hand, glucagon effectively reverses the conduction disturbances. Given glucagon's low toxicity, it appears that the agent should be considered a first-line agent for the treatment of slow calcium channel blocker overdose.

Glucagon's inotropic and chronotropic actions are independent of the β-adrenergic receptor. It effectively antagonizes heart block, bradycardia, hypotension, and diminished contractility resulting from overdose of β-adrenergic blockers in animals and humans (GLICK et al. 1968; LUCCHEST 1968; ABEL 1983; CHERNOW et al. 1986; WHITSITT and LUCCHEST 1968; PETERSON et al. 1978; AGURA et al. 1986; ZALOGA et al. 1986; PETERSON et al. 1984; TAI et al. 1990; WEINSTEIN 1984; KOSINSKI and MALINDZAK 1973; FRISHMAN et al. 1979; O'MAHONY et al. 1990; NEWMAN and SCHULTZ 1981). Glucagon is reported to reverse the toxic cardiovascular effects of β-adrenergic blockers even when sympathomimetic drugs and cardiac pacing have failed (AGURA et al. 1986; TAI et al. 1990). A portion of glucagon's beneficial effects during β-adrenergic blocker toxicity may result from its ability to increases hepatic blood flow and β-blocker (i.e., propranolol) clearance. WEINSTEIN (1984) reviewed 39 cases of β-blocker overdose reported in the literature. Patients were treated with atropine, β-adrenergic agonists, glucagon, and transvenous pacing. Glucagon was the most effective therapy and consistently produced a salutary effect on heart rate and blood pressure. Glucagon also improved outcome. Transvenous pacing consistently improved heart rate but not blood pressure. Epinephrine was the second most effective medical treatment for improving heart rate and blood pressure while atropine was the least effective.

LOVE et al. (1992) compared amrinone (a phosphodiesterase inhibitor) and glucagon in the treatment of propranolol cardiovascular toxicity in the dog. Amrinone and glucagon reversed propranolol-induced depression of cardiac contractility and cardiac output. Bradycardia did not respond well to amrinone while glucagon effectively increased heart rate. Overall. glucagon was more effective than amrinone in treating β-adrenergic blocker overdose. SATO et al. (1994) compared milrinone (a derivative of the phosphodiesterase inhibitor amrinone) with glucagon in a canine model of propranolol toxicity. Similar to LOVE et al. (1992), they found that glucagon had greater chronotropic effects than milrinone. However, improvement in cardiac output was similar in both groups with a trend toward greater improvements in stroke volume and pulmonary capillary wedge pressures in the milrinone group. In a subsequent study, LOVE et al. (1993) compared the combination of amrinone and glucagon to glucagon alone in a model of canine propranolol toxicity. Amrinone failed to significantly augment glucagon's cardiac-stimulating properties (heart rate, cardiac output, stroke volume, dP/dt_{max}). Mean arterial pressure was lower in the combined amrinone/glucagon group compared to glucagon alone. Overall, the combination of amrinone plus glucagon did not offer an advantage over glucagon alone in the treatment of β-blocker overdose. The available data support the use of glucagon as a first-line therapeutic agent for the treatment of β-adrenergic blocker overdose.

Glucagon's cardiac effects begin 1–5 min, peak in 5–15 min, and last 20–30 min after a single 5-mg intravenous dose. It may be administered in 1- to 5-mg boluses every 20–30 min or as a continuous infusion of 1–10 mg/h. In light of these recommendations, some Poison Centers recommend that hospitals maintain a supply of at least 50 mg glucagon. In a survey of 47 hospitals in the Washington, DC, area, Love and Tandy (1993) found that only 15% of hospitals met the suggested minimum. In none of the 47 hospitals did the emergency departments stock more than 20 mg glucagon. Vadhera (1992) also reported the inadequacy of hospital glucagon supplies.

C. Vascular Effects

Glucagon is a peripheral vasodilator (Chernow et al. 1986; Kazmers et al. 1981). However, its vascular effects are not uniform and vary between vascular beds. Glucagon increases aortic, superior mesenteric, and renal blood flow but has smaller effects on femoral and cerebral blood flow (Farah 1983; Kazmers et al. 1981; Tibblin et al. 1971; Kock et al. 1970; Madden et al. 1971; Okamura et a. 1986). Glucagon redistributes blood flow from the periphery to splanchnic area. Glucagon relaxes both arteries and veins. It antagonizes the vasoconstrictor effects of α-adrenergic agonists, vasopressin, and angiotensin.

Glucagon dilates the hepatic artery (D'Almeida and Lautt 1989) and antagonizes the vasoconstrictor effects of sympathetic stimulation, norepinephrine, angiotensin, and vasopressin on the hepatic vascular bed (Kock et al. 1971; Richardson and Withrington 1976). Glucagon dilates mesenteric arteries and inhibits norepinephrine-induced mesenteric artery vasoconstriction (D'Almeida and Lautt 1991; Wright et al. 1985). These effects are the basis for the use of glucagon in the treatment of nonocclusive acute mesenteric ischemia (Cronenwett et al. 1985; Kazmers et al. 1984; Boorstein et al. 1988; Cronenwett 1993a).

Glucagon increases mesenteric blood flow and survival in experimental models of acute mesenteric ischemia (Cronenwett et al. 1985; Kazmers et al. 1984; Boorstein et al. 1988; Oshima et al. 1990). Kazmers et al. (1984) reported that glucagon significantly improved the 48-h survival of rats with acute superior mesenteric artery occlusion from 54% to 85%. Dopamine, prostacyclin, superoxide dismutase, and heparin were without effect. Boorstein et al. (1988; Cronenwett 1993b) reported a 21% survival in control rats (receiving saline) following 85 min of superior mesenteric artery occlusion. Survival failed to improve with an angiotensin-converting enzyme inhibitor, allopurinol, or vasopressin antagonist. Intravenous glucagon, given following occlusion, improved survival to 86% while dopamine improved survival to 67%. When glucagon was given prior to mesenteric ischemia, it worsened reperfusion injury (Cronenwett 1993b; Clark and Gewertz 1990;

KAZMERS et al. 1984). CRONENWETT et al. (1985) varied the duration and timing of glucagon administration during and after 85 min of superior mesenteric artery (SMA) ischemia in rats and examined their subsequent survival. Survival was poor (33%) if glucagon was administered during SMA occlusion. In contrast, when administered after release of occlusion, survival improved to 83%. Injury may have resulted from glucagon stimulation of gut metabolic rate prior to the ischemic episode. Glucagon also redistributes blood flow away from the ischemic segment of the intestine when given before reperfusion (CRONENWETT 1993b). This "steal" from the ischemic region may contribute to injury. Thus, intravenous glucagon may be detrimental during occlusive ischemia by promoting vasodilation in adjacent nonischemic intestine, resulting in a reduction of collateral blood flow. However, after relief of occlusion, glucagon may be of benefit by improving mesenteric blood flow, especially in situations where vasoconstriction persists (CRONENWETT 1993b).

To determine whether glucagon improves outcome following mesenteric ischemia (i.e., intestinal shock) by vasodilating mesenteric vessels or via its cardiac inotropic effects, CRONENWETT (1993b) studied cardiac output in rats subjected to SMA occlusion. Following release of occlusion (85 min in duration), cardiac output decreased by 50% in control animals but only 11% in glucagon-treated animals. The relative distribution of radiolabeled microspheres to the intestine was not changed by glucagon (CRONENWETT 1993b; SARDELLA et al. 1990). Thus, a significant portion of glucagon's effects in this model result from maintenance of cardiac output and earlier restoration of mesenteric blood flow. Myocardial depressant factors are released into the circulation of animals subjected to mesenteric ischemia. Glucagon overcomes these depressant factors.

Glucagon improves blood flow to the intestine and has been shown to improve mucosal blood flow and gut viability in animal models of bowel strangulation (OSHIMA et al. 1990; BOND and LEVITT 1979; OSHIMA and KITAJIMA 1993). OSHIMA et al. (1990; OSHIMA and KITAJIMA 1993) evaluated the effect of glucagon on intestinal viability following 3.5 h of total ischemia. Glucagon was administered following reperfusion and improved intestinal viability from 0% (controls) to 67%. Glucagon was superior to heparin, urokinase, superoxide dismutase plus catalase, or a thromboxane inhibitor. Glucagon also improves gut blood flow during hypovolemia (BOND and LEVITT 1980; LINDBERG and DARLE 1977). Gut hypoperfusion and bacterial translocation are believed to play significant pathophysiologic roles in shock and ischemic states. Glucagon appears to have potential as an agent for improving perfusion and protecting the mucosa from injury during these conditions.

Glucagon has been shown to improve blood flow to normal human intestine (LILLY et al. 1989). Clinical experience with glucagon for the treatment of mesenteric ischemia has been limited to anecdotal reports. These reports suggest that glucagon may be beneficial for the treatment of nonocclusive mesenteric ischemia (CRONENWETT 1993a; BERK 1976; ROARK and SUZUKI

1987). However, additional clinical studies are required before glucagon can be recommended for routine treatment of nonocclusive mesenteric ischemia.

D. Renal/Urologic Effects

Glucagon causes renal vasodilation and a resulting diuresis and natriuresis (FARAH 1983; GAGNON et al. 1980; KATZ and LINDHEIMER 1977; DANFORD 1970). It increases both renal blood flow and glomerular filtration rate (PULLMAN et al. 1967; AKI et al. 1990). Glucagon preferentially dilates the afferent arteriole (AKI et al. 1990; UEDA et al. 1977) and inhibits renal tubular sodium reabsorption (PULLMAN et al. 1967; KIRSCHENBAUM and ZAWADA 1980).

Glucagon increases cardiac output and renal blood flow during hypovolemia in dogs (VANDERWALL et al. 1970) and subhuman primates (BOWMAN et al. 1972). The renal effects of glucagon are similar to but less potent than those of low-dose dopamine (i.e., 0.5–2.0 mg/kg per minute) (BOWMAN et al. 1972). Glucagon has the potential to improve renal blood flow during ischemic and hypovolemic states. Unfortunately, the renal actions of glucagon have not been studied in critically ill patients.

Glucagon relaxes ureteral smooth muscle (inhibits ureteral peristalsis) in unobstructed ureters (Ros 1993). Its effect on obstructed ureters requires further study. It also causes a diuresis. Thus, glucagon may be beneficial in the treatment of ureteral calculi. Some investigators have reported that glucagon relieves pain and helps the passage of renal calculi (LOWMAN et al. 1977; MORISHIMA and GHAED 1978; MURARO et al. 1982). Based upon these results, NEPPER-RASMUSSEN et al. (1984) performed a prospective randomized double-blind study of glucagon (single injection of 1 mg) in the treatment of acute ureteral colic. Compared with placebo, glucagon had no significant effect on the intensity of pain or movement of calculi. BAHN ZOBBE and coworkers (1986) evaluated glucagon (1 mg bolus plus 2 mg/h for 8 h) in a prospective randomized study of 37 patients with ureteral calculi. There was a trend toward faster passage of stones in the glucagon group. KAHNOSKI et al. (1987) examined the effects of glucagon on pain and passage of stone fragments following lithotripsy in a prospective, double-blind, placebo-controlled study. In a group of 18 patients, the investigators demonstrated that 10 mg glucagon given intravenously in two divided doses over 30 min was superior to placebo in reducing pain and increasing the passage of stone fragments. MINKOV et al. (1988) also found that glucagon enhanced passage of stone fragments after lithotripsy. Clearly, further large-scale prospective randomized studies are needed to evaluate the clinical efficacy of glucagon in the treatment of renal/ureteral stone disease.

E. Shock

Glucagon improves hemodynamics (i.e., cardiac output, hepatosplanchnic blood flow, renal blood flow) and hepatic glucose output in experimental endotoxin and hypovolemic shock (VANDERWALL et al. 1970; JAIN et al. 1978; BOWER et al. 1970; GUILLEN and PAPPAS 1972; BOWMAN et al. 1972). It increases glucose levels and improves survival (75% vs. 33%) in swine following endotoxin infusion (WEINGAND et al. 1986). Glucagon also improves liver and muscle glucose-glycogen stores and survival in rat hemorrhagic shock (JAIN et al. 1978).

Glucagon pretreatment demonstrated protective effects during the anaphylactic response in guinea-pig isolated hearts (ANDJELKOVIC and ZLOKOVIC et al. 1982). In these hearts, glucagon exerted antiarrhythmic activity, reduced histamine and CPK release, and improved coronary blood flow. Glucagon successfully reversed the hypotension and bronchospasm of anaphylactoid shock in patients receiving β-adrenergic blockers (ZALOGA et al. 1986; NEWMAN and SCHULTZ 1981).

DRUCKER et al. (1974) administered glucagon to 11 surgical patients in shock (i.e., from hemorrhage, sepsis, bowel infarction, spinal cord injury). Glucagon (5 mg) marginally increased heart rate, maintained blood pressure, and significantly increased cardiac index (from 3.77 to 5.13 l/min per square meter). The response to glucagon improved in patients receiving 50% glucose compared to patients receiving 5% glucose containing fluids. Glucagon also improved hemodynamics in patients with cardiogenic shock (LVOFF and WILCKEN 1972; WILCKEN and LVOFF 1970; SIEGEL et al. 1970).

F. Respiratory Effects

Glucagon relaxes smooth muscle by increasing muscle production of cAMP. This mechanism is similar to that of methylxanthines and β-adrenergic agonists, agents which promote bronchodilation. Glucagon relaxes bronchial smooth muscle (BLUMENTHAL and BRODY 1969; WARNER et al. 1971) and inhibits histamine (and other mediator) release from mast cells. Theoretically, glucagon may be useful in the treatment of bronchospasm (particularly in the setting of β-adrenergic blockade). Clinical studies demonstrate that glucagon produces mild to moderate improvement in spirometric tests (i.e., forced expiratory volume in 1 s or FEV-1, peak expiratory flow rate or PEFR) and subjective symptoms in patients with acute asthma (WILSON and NELSON 1990; SHERMAN et al. 1988; IMBRUCE et al. 1975; WILSON and FONTANAROSA 1993; LOCKEY et al. 1969; DIEZ-JARILLA et al. 1981). These effects are seen with both intravenous and aerosolized glucagon. In a prospective open label study, WILSON and NELSON (1990) evaluated the effect of glucagon (1 mg i.v.) in 14 young asthmatics with mild-moderate acute bronchospastic disease. Glucagon improved airflow and symptoms in 57% of the patients (mean increase in PEFR of 113 l/min). SHERMAN and colleagues (1988) studied intravenous glucagon (2 mg) in 11 patients using a double-blind, placebo, crossover protocol. Compared to placebo, glucagon significantly improved FEV-1 (glucagon 17.5% vs. placebo 2.4%). Clinically significant bronchodilation was apparent in 64% of patients. Glucagon was less effective than isoproterenol. LOCKEY et al. (1969) reported a modest improvement in FEV-1 (10%–15%) in nine of ten asthmatics given 2 mg intravenous glucagon. DIEZ-JARILLA et al. (1981) reported improvement in FEV-1 and forced vital capacity in eight of nine asthmatics given subcutaneous glucagon. Glucagon has also been reported to prevent bronchospasm during provocative inhalation testing (WILSON and FONTANAROSA 1993). Glucagon has not demonstrated benefit in the treatment of chronic obstructive pulmonary disease (WILSON and FONTANAROSA 1993; EL-NAGGAR and COLLINS 1974; IMBRUCE et al. 1975). Overall, preliminary clinical data suggest that glucagon may be useful in the treatment of acute bronchospasm. However, further studies are needed to verify these results and to determine whether glucagon provides benefits over current medications.

Glucagon, administered for short periods of time to patients with the adult respiratory distress syndrome and pulmonary hypertension, improved PO_2, with little change in pulmonary shunt (WEIGELT et al. 1982). Cardiac output, oxygen delivery, and pulmonary artery pressures increased slightly. In contrast, nitroprusside decreased PO_2, pulmonary artery pressures, cardiac output, and oxygen delivery. Others (DEMLING et al. 1978; MURTAUGH et al. 1970) reported that glucagon lowered pulmonary vascular resistance. Longer studies are needed to determine whether these effects of glucagon are maintained and whether they favorably influence the course of lung disease.

G. Gastrointestinal Effects

Glucagon relaxes smooth muscle in the lower esophageal sphincter (LES), stomach, small intestine, colon, gallbladder, and biliary ducts. Glucagon reduces LES pressure without interfering with esophageal peristalsis and a 1-mg dose can relieve obstruction from esophageal impaction (FERRUCCI and LONG 1977; HANDAL et al. 1980; MARKS and LOUSTEAU 1979; PILLARI et al. 1979; BERGGREEN et al. 1993; ROBBINS and SHORTSLEEVE 1994; RATCLIFF 1991). Glucagon in combination with effervescent agents and water has been particularly effective (ROBBINS and SHORTSLEEVE 1994; KASZAR-SEIBERT et al. 1990). Glucagon can also relieve elevated LES pressure and dysphagia in patients with achalasia.

Glucagon is frequently administered to relax smooth muscle during endoscopic study of the intestines and bile ducts (CARSEN and FINBY 1976; QVIGSTAD et al. 1979; FERRUCCI et al. 1976). It has been used to reduce bile duct pressure, pain, spasm, and help the passage of gallstones (DOMAN and GINSBERG 1981; STOWER et al. 1982; JONES et al. 1980, 1981; PAUL 1979; BRANDSTATTER and KRATOCHVIL 1979; GROSSI et al. 1986). PAUL (1979) administered glucagon (0.2–1.0 mg intravenously) to 31 patients with biliary pain. Pain was relieved in the majority of patients in less than 5 min. In another report (BRANDSTATTER and KRATOCHVIL 1979), glucagon (1 mg intravenous bolus) relieved biliary pain within 30 s. STOWER et al. (1982) studied 43 patients with acute cholecystitis randomized to glucagon or placebo. Glucagon treatment relieved pain faster than placebo. GROSSI et al. (1986) compared glucagon (1 mg intramuscularly) to a nonsteroidal anti-inflammatory agent and anticholinergic for the treatment of biliary colic. Glucagon and the nonsteroidal agent were equally effective in relieving pain. These effects may also be useful during operative cholangiography, surgery on the intestines and biliary tract, and to relieve narcotic-induced biliary spasm. In addition, glucagon can relieve spasm and pain resulting from diverticulitis (DANIEL et al. 1974; ALMY and HOWELL 1980) and intestinal surgery (HARFORD 1979). It appears to relieve pain more effectively than anticholinergics and may decrease the need for analgesics. Furthermore, glucagon does not block the pain response or alter the abdominal examination.

Based upon their metabolic, growth potentiating, and hemodynamic effects, glucagon and insulin were thought to be beneficial for the treatment of severe liver disease (MANABE and STEER 1979). These anabolic hormones were postulated to speed liver regeneration. Glucagon and insulin improve outcome in animal models of liver failure (FARIVAR et al. 1976; FUGIWARA et al. 1988). Unfortunately, glucagon and insulin therapy have not proved to be of clinical benefit in human liver disease. Some studies report short-term benefits in patients with hepatitis (FEHER et al. 1987; RADVAN et al. 1982; BAKER et al. 1981; OKA et al. 1989). However, the majority of prospective randomized trials fail to demonstrate survival benefits for glucagon and insulin administration in alcoholic liver disease and nonalcoholic liver failure (BIRD et al. 1991; TRINCHET et al. 1992; WOOLF and REDEKER 1991; BAKER et al. 1981; HARRISON et al. 1990). A major risk of this therapy is hypoglycemia and deaths from hypoglycemia have been reported. At present, glucagon is not recommended for the treatment of hepatic failure or alcoholic hepatitis (MORGAN 1993; ACHORD 1993; BAKER 1993).

Total parenteral nutrition (TPN) is widely used to provide metabolic support to patients with dysfunctional gastrointestinal tracts. Unfortunately, hepatic and biliary complications remain major adverse consequences of TPN. The etiologies of hepatic dysfunction during TPN are complex. Substrate imbalance is one proposed etiology. Excess administration of substrate is known to injure the liver. Elevated insulin levels and decreased glucagon levels during TPN lead to excess carbohydrate storage, decreased lipid removal, and hepatic steatosis. Hepatic steatosis is believed to be a precursor to more severe hepatic injury. Glucagon administration during TPN ameliorates the development of steatosis (LI et al. 1988).

Because of glucagon's suppressive actions on pancreatic secretions and gastrin (CARR-LOCKE 1993), it was hypothesized that the hormone would be beneficial for the treatment of acute pancreatitis. Early animal studies indicated that glucagon improved outcome in some animal models of acute pancreatitis (KNIGHT et al. 1972; PAPP et al. 1975; WATERWORTH et al. 1976; MANABE and STEER 1979). Unfortunately, in human controlled studies (STEINBERG and SCHLESSELMAN 1987; GILSANZ et al. 1978; CARR-LOCKE 1993; MEDICAL RESEARCH COUNCIL MULTICENTER TRIAL 1980; KALIMA and LEMPINEN

1980; DURR et al. 1978), glucagon has not significantly improved outcome and it is not recommended for the management of acute pancreatitis.

H. Radiographic Studies

As reviewed in Chap. 12, this volume, by Skucas, due to its smooth muscle relaxant properties, glucagon is used to aid visualization of the intestines and bile ducts during radiologic procedures. It helps to discriminate artifact from true disease. Glucagon has also been used to aid reduction of intussusception by barium enema.

I. Adverse Effects

Glucagon stimulates the release of epinephrine from the adrenal medulla (SCIAN et al. 1960). Although rare, hypertensive crisis may occur following administration of glucagon to patients with pheochromocytoma (LAWRENCE 1967; SCHORR and ROGERS 1987). Insulinoma patients can have an exaggerated release of insulin in response to glucagon. Initial hyperglycemia may be followed by rebound hypoglycemia. In addition, glucagon may precipitate flushing in patients with carcinoid syndrome (LEBOVICS and ROSENTHAL 1991).

Because glucagon improves atrioventricular conduction, it can increase the ventricular response in patients with atrial flutter/fibrillation. Glucagon also increases heart rate and myocardial oxygen consumption. Similar to other cardiotonic agents, it may induce myocardial ischemia in the setting of fixed coronary artery disease.

Rare adverse complications from glucagon administration include erythema multiforme, Stevens-Johnson syndrome, and anaphylaxis (EDELL 1980). However, the most common side effects from glucagon administration are tachycardia, hypokalemia, nausea/vomiting, weakness, dizziness, and hyperglycemia. Symptoms are lessened by slower administration of the agent and by use of continuous infusion. These side effects are rarely a significant clinical problem if patients are properly monitored and treated.

References

Abel FL (1983) Action of glucagon on canine left ventricular performance and coronary hemodynamics. Circ Shock 11:45
Abernethy WB, Butterworth JF, Prielipp RC, Leith JP, Zaloga GP (1995) Calcium entry attenuates adenylyl cyclase activity: a possible mechanism for calcium-induced catecholamine resistance. Chest 107:1420–1425
Achord JL (1993) Review of alcoholic hepatitis and its treatment. Am J Gastroenterol 88:1822–1831
Agura ED, Wexler LF, Witzburg RA (1986) Massive propranolol overdose: successful treatment with high-dose isoproterenol and glucagon. Am J Med 80:755
Aki Y, Shoji T, Hasui K, Fukui K, Tamaki T, Iwao H, Abe Y (1990) Intrarenal vascular sites of action of adenosine and glucagon. Jpn J Pharmacol 54:433

Almy TP, Howell DA (1980) Diverticular disease of the colon. N Engl J Med 302:324–331
Andjelkovic I, Zlokovic B (1982) Protective effects of glucagon during the anaphylactic response in guinea-pig isolated heart. Br J Pharmacol 76:483
Bahn Zobbe V, Rygaard H, Rasmussen D, Strandberg C, Krause S, Hartvig Hartsen S, Thomsen HS (1986) Glucagon in acute ureteral colic: a randomized trial. Eur Urol 12:28–31
Baker AL (1993) Glucagon for the management of hepatic emergencies. In: Picazo J (ed) Glucagon in acute medicine. Kluwer Academic, Dordrecht, p 131
Baker AL, Jaspan JB, Haines NW, Hatfield GE, Krager PS, Schneider JF (1981) A randomized clinical trial of insulin and glucagon infusion for treatment of alcoholic hepatitis: progress report in 50 patients. Gastroenterology 80:1410–1414
Barritt GJ, Spiel PF (1981) Effects of glucagon on ^{45}Ca outflow exchange in the isolated perfused rat heart. Biochem Pharmacol 30:1407
Berggreen PJ, Harrison E, Sanowski RA, Ingebo K, Noland B, Zierer S (1993) Techniques and complications of esophageal foreign body extraction in children and adults. Gastrointest Endosc 39:626–630
Berk JL (1976) Non-occlusive mesenteric vascular insufficiency. Arch Surg 111:829–830
Bird G, Lau JY, Koskinas J, Wicks C, Williams R (1991) Insulin and glucagon infusion in acute alcoholic hepatitis: a prospective randomized controlled trial. Hepatology 14:1097–1101
Blumenthal MN, Brody TM (1969) Studies on the mechanism of drug-induced bronchiolar relaxation in the guinea pig. J Allergy 44:63–69
Boder GB, Johnson IS (1972) Comparative effects of some cardioactive agents on the automaticity of cultured heart cells. J Mol Cell Cardiol 4:453
Bond JH, Levitt MD (1979) Use of microspheres to measure small intestinal villus blood flow in the dog. Am J Physiol 236:577
Bond JH, Levitt MD (1980) Effect of glucagon on gastrointestinal blood flow of dogs in hypovolemic shock. Am J Physiol 238:G434
Boorstein JM, Dacey LJ, Cronenwett JL (1988) Pharmacologic treatment of occlusive mesenteric ischemia in rats. J Surg Res 44:555–560
Bower MG, Okude S, Jolley WB, Smith LL (1970) Hemodynamic effects of glucagon following hemorrhagic and entotoxic shock in the dog. Arch Surg 101:411–415
Bowman HM, Cowan D, Kovach G Jr, Hook JB (1972) Renal effects of glucagon in rhesus monkeys during hypovolemia. Surg Gynecol Obstet 134:937–941
Brandstatter G, Kratochvil P (1979) Glucagon ber gallenkoliken. Therapiewoche 29:3362–3365
Brogan E, Kozonis MC, Overy DC (1969) Glucagon therapy in heart failure. Lancet I:482
Carr-Locke DL (1993) Biliary and pancreatic emergencies. In: Picazo J (ed) Glucagon in acute medicine. Kluwer Academic, Dordrecht, p 141
Carsen GM, Finby N (1976) Hypotonic duodenography with glucagon. A clinical comparison study. Radiology 118:529
Chernow B, Reed L, Geelhoed GW, Anderson M, Teich S, Meyerhoff J, Beardsley D, Lake CR, Holaday JW (1986) Glucagon: endocrine effects and calcium involvement in cardiovascular actions in dogs. Circ Shock 19:393
Chernow B, Zaloga GP, Malcolm D, Willey SC, Clapper M, Holaday JW (1987) Glucagon's chronotropic action is calcium dependent. J Pharmacol Exp Ther 241:833
Chiou GCY, Chuang CY (1988) Treatment of hypoglycemia with glucagon eye drops. J Ocular Pharm 4:179
Clark ET, Gewertz BL (1990) Glucagon potentiates intestinal reperfusion injury. J Vasc Surg 11:270–279
Cohn KE, Agmon J, Gamble OW (1970) The effect of glucagon on arrhythmias to digitalis toxicity. Am J Cardiol 25:683

Collier A, Steedman DJ, Patrick AW, Nimmo GR, Matthews DM, MacIntyre CC, Little K, Clarke BF (1987) Comparison of intravenous glucagon and dextrose in treatment of severe hypoglycemia in an accident and emergency department. Diabetes Care 10:712–715

Cronenwett JL (1993a) Acute mesenteric ischaemia. In: Picazo J (ed) Glucagon in acute medicine. Kluwer Academic, Dordrecht, p 83

Cronenwett JL (1993b) Mesenteric ischemia. In: Picazo J (ed) Glucagon in acute medicine. Kluwer Academic, Dordrecht, p 131

Cronenwett JL, Ayad M, Kazmers A (1985) Effect of intravenous glucagon on the survival of rats after acute occlusive mesenteric ischemia. J Surg Res 38:446–452

Crump BJ, Holt DW, Vales JA (1982) Lack of response to intravenous calcium in severe verapamil poisoning. Lancet I:939–940

D'Almeida MS, Lautt WW (1989) The effect of glucagon on vasoconstriction and vascular escape from nerve and norepinephrine-induced constriction of the hepatic artery of the cat. Can J Physiol Pharmacol 67:1418

D'Almeida MS, Lautt WW (1991) Glucagon pharmacodynamics and modulation of sympathetic nerve and norepinephrine-induced constrictor responses in the superior mesenteric artery of the cat. J Pharmacol Exp Ther 259:118

Danford RO (1970) The effect of glucagon on renal hemodynamics and renal arteriography. Am J Roentgenol Radium Ther Nucl Med 108:665

Daniel O, Basu PK, Al-Samarrae HM (1974) Use of glucagon in the treatment of acute diverticulitis. Br Med J 3:720–722

Davies DM, MacIntyre A, Millar EJ, Bell SM, Mehra SK (1967) Need for glucagon in severe hypoglycemia induced by sulphonylurea drugs. Lancet I:363–364

Demling RH, Manohar M, Will J (1978) The effect of glucagon on the pulmonary transvascular fluid filtration rate. Chest 74:196–199

Diamond G, Forrester J, Danzig R, Parmley WW, Swan HJC (1971) Acute myocardial infarction in man: comparative hemodynamic effects of norepinephrine and glucagon. Am J Cardiol 27:612–616

Diez-Jarilla JL, Gonzalez-Macias J, Laso-Guzman FJ, de Castro del Pozo S (1981) Beta-blockade in asthma (letter). Br Med J 283:309

Doman DB, Ginsberg AL (1981) Glucagon infusion therapy for biliary tree stones. Gastroenterology 80:1137

Doyon S, Roberts JR (1993) The use of glucagon in a case of calcium channel blocker overdose. Ann Emerg Med 22:1229–1233

Drucker MR, Pindyck F, Brown RS, Elwyn DH, Shoemaker WC (1974) The interaction of glucagon and glucose on cardiorespiratory variables in the critically ill patient. Surgery 75:487

Durr HK, Maroske D, Zelder O, Bode JC (1978) Glucagon therapy of acute pancreatitis - report of a double blind trial. Gut 19:175–179

Eddy JD, O'Brien ET, Singh SP (1969) Glucagon and haemodynamics of acute myocardial infarction. Br Med J 4:663

Edell SL (1980) Erythema multiforme secondary to intravenous glucagon. Am J Radiol 134:385

El-Naggar M, Collins VJ (1974) Spirometry following glucagon and isoproterenol in chronic obstructive pulmonary disease. Crit Care Med 2:82–85

Elrick H, Witten TA, Arai Y (1958) Glucagon treatment of insulin reactions. N Engl J Med 258:476

Farah A, Tuttle R (1960) Studies on the pharmacology of glucagon. J Pharmacol Exp Ther 129:49

Farah AE (1983) Glucagon and the circulation. Pharmacol Rev 35:181

Farivar M, Wands JR, Isselbacher KJ, Bucher NLR (1976) Effect of insulin and glucagon in fulminant murine hepatitis. N Engl J Med 295:1517–1519

Feher J, Cornides A, Romany A, Karteszi M, Szalay L, Gogl A, Picazo J (1987) A prospective multicenter trial of insulin and glucagon infusion therapy in acute alcoholic hepatitis. J Hepatol 5:224–231

Ferrucci JT, Long JA (1977) Radiologic treatment of esophageal food impaction using intravenous glucagon. Radiology 125:25
Ferrucci JT, Wittenberg J, Stone LB, Dreyfuss JR (1976) Hypotonic cholangiography wiht glucagon. Radiology 118:466
Freychet L, Rizkalla SW, Desplanque N, Basdevant A, Zirinis P, Tchobroutsky G, Slama G (1988) Effect of intranasal glucagon on blood glucose levels in healthy subjects and hypoglycaemic patients with insulin-dependent diabetes. Lancet I:1364–1366
Frishman W, Jacob H, Eisenberg E, Ribner H (1979) Clinical pharmacology of the new beta-adrenergic blocking drugs: part 8: self-poisoning with beta-adrenergic blocking agents: recognition and management. Am Heart J 98:798–811
Fugiwara K, Ogata I, Mishiro S, Ohta Y, Oka Y, Takatsuki K, Sato Y, Hayashi S (1988) Glucagon and insulin for the treatment of hepatic failure in dimethylnitrosamine intoxicated rats. Scand J Gastroenterol 23:567
Gagnon G, Regoli D, Rioux F (1980) Studies on the mechanism of action of glucagon in strips of rabbit renal artery. Br J Pharmacol 69:389
Gerich JE, Lorenzi M, Bier DM, Schneider V, Tsalikian E, Karam JH, Forshamm PH (1975) Prevention of human diabetic ketoacidosis by somatostatin. Evidence for an essential role of glucagon. N Engl J Med 292:985–989
Gilsanz V, Oteyza CP, Rebollar JL (1978) Glucagon vs anticholinergics in the treatment of acute pancreatitis: a double-blind controlled trial. Arch Intern Med 138:535–538
Glick G, Parmley WW, Wechsler AS, Sonnenblick EH (1968) Glucagon: its enhancement of cardiac performance in the cat and dog and persistence of its inotropic action despite beta-receptor blockade with propranolol. Circ Res 22:789
Grossi E, Broggini M, Quaranta M, Balestrino E (1986) Different pharmacologic approaches to the treatment of acute biliary colic. Curr Ther Res 40:876–882
Guillen J, Pappas G (1972) Improved cardiovascular effects of glucagon in dogs with endotoxin shock. Ann Surg 175:535
Gupta MM, Prasad K (1980) Studies of the effects of glucagon on ouabain-induced cardiac disorders using the PISA method. In: Tajuddin M, Das PK, Tariq M, Dhalla NS (eds) Advances in myocardiology, vol 1. University Press, Baltimore, p 313
Handal KA, Riordan W, Siese J (1980) The lower esophagus and glucagon. Ann Emerg Med 9:577
Harford FJ (1979) Use of glucagon in conjunction with end-to-end anastomosis (EEA) stapling device for low anterior anastomoses. Dis Colon Rectum 22:452
Harrison PM, Hughes RD, Forbes A, Portmann B, Alexander GJM, Williams R (1990) Failure of insulin and glucagon infusion to stimulate liver regeneration in fulminant hepatic failure. J Hepatol 10:1–5
Hvidberg A, Djurup R, Hilsted J (1994) Glucose recovery after intranasal glucagon during hypoglycemia in man. Eur J Clin Pharmacol 46:15
Imbruce R, Goldfedder A, Macguire W, Briscoe W, Nair S (1975) The effect of glucagon on airway resistance. J Clin Pharmacol 15:680–684
Jain KM, Rush BF Jr, Hastings OM, Ghosh A, Slotman G, Albousamra S (1978) Glucagon treatment of hemorrhagic shock: improved survival and metabolic parameters in a murine shock model. Adv Shock Res 1:149
Jolly SR, Kipnis JN, Lucchesi BR (1987) Cardiovascular depression by verapamil: reversal by glucagon and interactions with propranolol. Pharmacology 35:249–255
Jones RM, Fiddian-Green R, Knight PR (1980) Narcotic induced choledochoduodenal sphincter spasm reversed by glucagon. Anesth Analg 59:946–947
Jones RM, Detmer M, Hill AB, Bjoraker DG, Pandit U (1981) Incidence of choledochoduodenal sphincter spasm during fentanyl-supplemented anesthesia. Anesth Analg 60:638–640
Kahnoski RJ, Lingeman JE, Woods JR, Eckley R, Brooks-Brunn J, Coury TA (1987) Efficacy of glucagon in the relief of uretheral colic following treatment by

extracoporeal shock wave lithotripsy: a randomized double-blind trial. J Urology 137:1124

Kalima TV, Lempinen M (1980) The effect of zinc-protamine-glucagon in acute pancreatitis. Ann Chir 69:293–295

Kaszar-Seibert DJ, Korn WT, Bindman DJ, Shortsleeve MJ (1990) Treatment of acute esophageal food impaction with a combination of glucagon, effervescent agent, and water. Am J Radiol 154:533–534

Katz AI, Lindheimer MD (1977) Actions of hormones on the kidney. Annu Rev Physiol 39:97

Kazmers A, Whitehouse WM Jr, Lindenauer SM, Stanley JC (1981) Dissociation of glucagon's central and peripheral hemodynamic effects: mechanisms of reduction and redistribution of canine hindlimb blood flow. J Surg Res 30:384

Kazmers A, Zwolak R, Appelman HD, Whitehouse WM Jr, Wu SC, Zelenock GB, Cronenwett JL, Lindenauer SM, Stanley JC (1984) Pharmacologic interventions in acute mesenteric ischemia: improved survival with intravenous glucagon, methylprednisolone, and prostacyclin. J Vasc Surg 1:472–481

Kirschenbaum MA, Zawada ET (1980) The role of prostaglandins in glucagon-induced natriuresis. Clin Sci 58:393

Kline JA, Leonova E, Raymond RM (1995) Beneficial myocardial metabolic effects of insulin during verapamil toxicity in the anesthetized canine. Crit Care Med 23:1251–1263

Knight MJ, Condon JR, Day JL (1972) Possible role of glucagon in pathogenesis of acute pancreatitis. Lancet I:1097–1099

Kock NG, Tibblin S, Schenk WG Jr (1970) Hemodynamic responses to glucagon: an experimental study of central, visceral and peripheral effects. Ann Surg 171:373

Kock NG, Tibblin S, Schenk WG Jr (1971) Modification by glucagon of the splanchnic vascular responses to activation of the sympathicoadrenal system. J Surg Res 11:12

Kones RJ, Phillips JH (1971) Glucagon: present status in cardiovascular disease. Clin Pharmacol Ther 12:427

Kosinski EJ, Malindzak GS (1973) Glucagon and isoproterenol in reversing propranolol toxicity. Arch Intern Med 132:840–843

Lawrence AM (1967) Glucagon provocative test for pheochromocytoma. Ann Intern Med 66:1091

Lebovics E, Rosenthal WS (1991) Glucagon precipitation of carcinoid flush. Gastrointest Endosc 37:212–213

Li SJ, Nussbaum MS, McFadden DW, Gapen CL, Dayal R, Fischer JE (1988) Addition of glucagon to total parenteral nutrition prevents hepatic steatosis in rats. Surgery 104:350–357

Lickley HL, Kemmer FW, Doi K, Vranic M (1983) Glucagon suppression improves glucoregulation in moderate but not chronic severe diabetes. Am J Physiol 245:E424

Lilly MP, Harward TRS, Flinn WR, Blackburn DR, Astleford PM, Yao JST (1989) Duplex ultrasound measurement of changes in mesenteric flow velocity with pharmacologic and physiologic alteration of intestinal blood flow in man. J Vasc Surg 9:10–25

Lindberg B, Darle N (1977) The effect of glucagon and blood transfusion on hepatic circulation and oxygen consumption in hemorrhagic shock. J Surg Res 23:257–263

Linden CH, Aghababian RV (1985) Further uses of glucagon. Crit Care Med 13:248

Lockey SD, Reed CE, Ouellette JJ (1969) Bronchodilating effect of glucagon in asthma. J Allergy 43:177–178

Love JN, Tandy TK (1993) Beta-adrenergic anatgonist toxicity: a survey of glucagon availability. Ann Emerg Med 22:267

Love JN, Leasure JA, Mundt DJ, Janz TG (1992) A comparison of amrinone and glucagon therapy for cardiovascular depression associated with propranolol toxicity in a canine model. Clin Toxicol 30:399–412

Love JN, Leasure JA, Mundt DJ (1993) A comparison of combined amrinone and glucagon therapy to glucagon alone for cardiovascular depression associated with propranolol toxicity in a canine model. Am J Emerg Med 11:360

Lowman RM, Belleza NA, Goetsch JB, Finkelstein HI, Berneike RR, Rosenfield AT (1977) Glucagon (letter). J Urol 118:128

Lucchesi BR (1968) Cardiac actions of glucagon. Circ Res 22:777

Lvoff R, Wilcken DE (1972) Glucagon in heart failure and in cardiogenic shock. Experience in 50 patients. Circulation 45:534–542

MacCuish AC, Munro JF, Duncan LJ (1970) Treatment of hypoglycemic coma with glucagon, intravenous dextrose, and mannitol infusion in a hundred diabetics. Lancet II:946

MacLeod KM, Rodgers RL, McNeill JH (1981) Characterization of glucagon-induced changes in rate, contractility and cyclic AMP levels in isolated cardiac preparations of the rat and guinea pig. J Pharmacol Exp Ther 217:798

Madan BR (1971) Effect of glucagon on ventricular arrhythmias after coronary artery occlusion and on ventricular automaticity in the dog. Br J Pharmacol 43:279

Madden JJ Jr, Ludewig RM, Wagensteen SL (1971) Effects of glucagon on the splanchnic and the systemic circulation. Am J Surg 122:85

Malcolm D, Zaloga G, Chernow B, Holaday J (1986) Glucagon is an antagonist of morphine bradycardia and antinociception. Life Sci 39:399

Manabe T, Steer ML (1979) Experimental acute pancreatitis in mice. Protective effects of glucagon. Gastroenterology 76:529–534

Marks HW, Lousteau RJ (1979) Glucagon and esophageal meat impaction. Arch Otolaryngol 105:367

Marri G, Cozzolono G, Palumbo R (1968) Glucagon in sulphonylurea hypoglycemia? Lancet I:303

Marsiglia JC, Moreyra AE, Lardani H, Cingolani HE (1970) Glucagon: its effect upon myocardial oxygen consumption. Eur J Pharmacol 12:265

Medical Research Council Multicentre Trial (1980) Morbidity of acute pancreatitis – the effect of aprotinin and glucagon. Gut 21:334–339

Mery PF, Brechler V, Pavoine C, Pecker F, Fischmeister R (1990) Glucagon stimulates the cardiac Ca^{2+} current by activation of adenylyl cyclase and inhibition of phosphodiesterase. Nature 345:158

Minkov N, Shumleva V, Pironkov A, Gotsev G, Voinikova IN, Nicolov N (1988) A new method for the management of ureteral colic after extracorporeal shock wave lithotripsy. Int Urol Nephrol 20:251–255

Morgan TR (1993) Treatment of alcoholic hepatitis. Semin Liver Dis 13:384–394

Morishima MS, Ghaed N (1978) Glucagon and diuresis in the treatment of ureteral calculi. Radiology 129:807–809

Moura AM, Simpkins H (1975) Cyclic AMP levels in cultured myocardial cells under the influence of chronotropic and inotropic agents. J Mol Cell Cardiol 7:71

Muhlhauser I, Koch J, Berger M (1985a) Pharmacokinetics and bioavailability of injected glucagon: differences between intramuscular, subcutaneous, and intravenous administration. Diabetes Care 8:39–42

Muhlhauser I, Berger M, Sonnenberg G, Koch J, Jorgens V, Schernthaner G, Scholz V, Padagogin D (1985b) Incidence and management of severe hypoglycemia in 434 adults with insulin-dependent diabetes mellitus. Diabetes Care 8:268–273

Muraro GB, Ruin G, Giusti G, Mazzarino A, Torri T, Lattanzi P (1982) Glucagon and hyperdiuresis in the treatment of ureteral calculi. Min Urol 34:111–115

Murtaugh JG, Binnion PF, Lal S, Hutchinson KJ, Fletcher E (1970) Haemodynamic effects of glucagon. Br Heart J 32:307–310

Nayler WG, McInnes I, Chipperfield D, Carson V, Daile P (1970) The effect of glucagon on calcium exchangeability, coronary blood flow, myocardial function and high energy phosphate stores. J Pharmacol Exp Ther 171:265

Nepper-Rasmussen J, Storgaard Pedersen O, Anderson A, Dalssgaard J (1984) Glucagon and ureteral colic. Urol Res 12:23–24

Newman BR, Schultz LK (1981) Epinephrine resistant anaphylaxis in a patient taking propranolol hydrochloride. Ann Allergy 47:35–36

Niemann JT, Haynes KS, Garner D, Jagels G, Rennie CJ (1987) Postcountershock pulseless rhythms: hemodynamic effects of glucagon in a canine model. Crit Care Med 15:554–558

Nord HJ, Fontaines AL, Williams JF (1970) Treatment of congestive heart failure with glucagon. Ann Intern Med 72:649–653

Oka H, Fujiwara K, Okita K, Ishii H, Sakuma A (1989) A multi-centre double-blind controlled trial of glucagon and insulin therapy for severe acute hepatitis. Gastroenterol Jpn 24:332–336

Okamura T, Miyazaki M, Toda N (1986) Responses of isolated dog blood vessels to glucagon. Eur J Pharmacol 125:395

Oshima A, Kitajima M (1993) Intestinal strangulation. In: Picazo J (ed) Glucagon in acute medicine. Kluwer Academic, Dordrecht, p 107

Oshima A, Kitajima M, Sakai N, Ando N (1990) Does glucagon improve the viability of ischemic intestine? J Surg Res 49:524–533

O'Mahony D, O'Leary P, Molloy MG (1990) Severe oxprenolol poisoning: the importance of glucagon infusion. Hum Exp Toxicol 9:101–103

Papp M, Ribet A, Fodor I, Nemeth PE, Feher S, Horvath JE, Folly G (1975) Glucagon treatment of experimental acute pancreatitis. Acta Med Acad Sci Hung 32:105–116

Parmley WW, Glick G, Sonnenblick EH (1968) Cardiovascular effects of glucagon in man. N Engl J Med 279:12–17

parmley WW, Matloff JM, Sonnenblick EH (1969) Hemodynamic effects of glucagon in patients following prosthetic valve replacement. Circulation 39:I163

Patrick AW, Collier A, Hepburn DA, Steedman DJ, Clarke BF, Robertson C (1990) Comparison of intramuscular glucagon and intravenous dextrose in the treatment of hypoglycemic coma in an accident and emergency department. Arch Emerg Med 7:73–77

Paul F (1979) The role of glucagon in the treatment of biliary tract pathology. In: Picazo J (ed) Glucagon in gastroenterology. MTP Press, Lancaster, p 107

Pavoine C, Brechler V, Kervran A, Blache P, Le-Nguyen D, Laurent S, Bataille D, Pecker F (1991) Miniglucagon [glucagon-(19-29)] is a component of the positive inotropic effect of glucagon. Am J Physiol 260:C993

Peterson A, Lucchesi B, Kirsh MM (1978) The effect of glucagon in animals on chronic propranolol therapy. Ann Thorac Surg 25:340

Peterson CD, Leeder JS, Sterner S (1984) Glucagon therapy for β-blocker overdose. Drug Intell Clin Pharm 18:394

Pillari G, Bank S, Katzka I, Fulco JD (1979) Meat bolus impaction and the lower esophagus associated with paraesophageal hernia. Successful noninvasive treatment with intravenous glucagon. Am J Gastroenterol 71:287

Prasad K (1975) Electrophysiologic effects of glucagon on human cardiac muscle. Clin Pharmacol Ther 18:22

Prasad K (1977) Use of glucagon in the treatment of quinidine toxicity in the heart. Cardiovasc Res 11:53

Prasad K, DeSousa HH (1972) Glucagon in the treatment of ouabain-induced cardiac arrhythmias in dogs. Cardiovasc Res 6:333

Prasad K, Weckworth P (1978) Glucagon in procainamide-induced cardiac toxicity. Toxicol Appl Pharmacol 46:517

Prielipp RC, Hill T, Washburn D, Zaloga GP (1989) Circulating calcium modulates adrenaline-induced cyclic adenosine monophosphate production. Cardiovasc Res 23:838

Pullman TN, Lavender AR, Aho I (1967) Direct effects of glucagon on renal hemodynamics and excretion of inorganic ions. Metabolism 16:358

Puri PS, Bing RJ (1979) Effects of glucagon on myocardial contractility and hemodynamics in acute experimental myocardial infarction: basis for its possible use in cardiogenic shock. Am Heart J 78:660–668

Quezado Z, Lippmann M, Wertheimer J (1991) Severe cardiac, respiratory, and metabolic complications of massive verapamil overdose. Crit Care Med 19:436–438

Qvigstad T, Larsen S, Myren J (1979) Comparison of glucagon, atropine, and placebo as premedication for endoscopy of the upper gastrointestinal tract. Scand J Gastroenterol 14:231

Radvan G, Kanel G, Redeker A (1982) Insulin and glucagon infusion in acute alcoholic hepatitis. Gastroenterology 82:1154

Ratcliff KM (1991) Esophageal foreign bodies. Am Fam Physician 44:824–831

Regan TJ, Lehan PH, Henneman DH, Behar A, Hellems HK (1964) Myocardial, metabolic and contractile response to glucagon and epinephrine. J Lab Clin Med 63:638

Richardson PD, Withrington PG (1976) The inhibition by glucagon of the vasoconstrictor actions of noradrenaline, angiotensin and vasopressin on the hepatic arterial vascular bed of the dog. Br J Pharmacol 57:93

Roark KM, Suzuki NT (1987) Glucagon use in mesenteric ischemia. Drug Intell Clin Pharm 21:660–661

Robbins MI, Shortsleeve MJ (1994) Treatment of acute esophageal food impaction with glucagon, an effervescent agent, and water. Am J Radiol 162:325–328

Ros SP (1993) Genitourinary emergencies and glucagon. In: Picazo J (ed) Glucagon in acute medicine. Kluwer Academic, Dordrecht, p 37

Rosenfalck AM, Bendtson I, Jorgensen S, Binder C (1992) Nasal glucagon in the treatment of hypoglycemia in type 1 (insulin-dependent) diabetic patients. Diabetes Res Clin Prac 17:43–50

Ross EM, Gilman AG (1980) Biochemical properties of hormone-sensitive adenylate cyclase. Annu Rev Biochem 49:533

Ruddy JM, Seymour JL, Anderson NG (1972) Management of tricyclic antidepressant ingestion in children with special reference to the use of glucagon. Med J Aust 1:630–633

Sabatier J, Pouyet T, Shelvey G, Cavero I (1991) Antagonistic effects of epinephrine, glucagon and methylatropine but not calcium chloride against atrio-ventricular conduction disturbances produced by high doses of diltiazem, in conscious dogs. Fundam Clin Pharmacol 5:93–106

Sardella GL, Bech FR, Cronenwett JL (1990) Hemodynamic effects of glucagon after acute mesenteric ischemia in rats. J Surg Res 49:354–360

Sato S, Tsuji MH, Okuba N, Naito H (1994) Milrinone versus glucagon: comparative hemodynamic effects in canine propranolol poisoning. Clin Toxicol 32:277

Schorr RT, Rogers SN (1987) Intraoperative cardiovascular crisis caused by glucagon. Arch Surg 122:833

Scian LF, Westermann CD, Verdesca AS, Hilton JG (1960) Adrenocortical and medullary effects of glucagon. Am J Physiol 199:867

Sener EK, Gabe S, Henry JA (1995) Response to glucagon in imipramine overdose. J Toxicol Clin Toxicol 33:51–53

Sherman MS, Lazar EJ, Eichacker P (1988) A bronchodilator action of glucagon. J Allergy Clin Immunol 81:908–911

Siegel JH, Levine MJ, McConn R, DelGuercio LR (1970) The effect of glucagon infusion on cardiovascular function in the critically ill. Surg Gynecol Obstet 131:505

Simaan J, Fawaz G (1976) The cardiodynamic and metabolic effects of glucagon. Naunyn Schmiedebergs Arch Pharmacol 294:277

Singh J, Bala S, Kaur AH, Garg KN (1980) Effect of glucagon on arrhythmias induced by coronary artery occlusion and ouabain in dogs. Indian J Physiol 24:329

Slama G, Reach G, Cahane M, Quetin C, Villanove-Robin F (1992) Intranasal glucagon in the treatment of hypoglycemic attacks in children: experience at a summer camp (letter). Diabetologia 35:398

Smitherman TC, Osborn RC Jr, Atkins JM (1978) Cardiac dose response relationship for intravenously infused glucagon in normal intact dogs and men. Am Heart J 96:363

Steel JM, Allwinkle J, Moffat R, Carrington DJ (1992) Use of lucozade and glucagon by ambulance staff for treating hypoglycaemia. Br Med J 304:1283–1284

Steinberg WM, Schlesselman SE (1987) Treatment of acute pancreatitis: comparison of animal and human studies. Gastroenterology 93:1420–1427

Stenninger E, Aman J (1993) Intranasal glucagon relieves hypoglycemia in children with type 1 (insulin dependent) diabetes mellitus. Diabetologia 36:931–935

Stewart JW, Myerburg RJ, Hoffman BF (1969) The effect of glucagon on quinidine-induced changes in Purkinje fibers. Circulation 40:196

Stone CK, May WA, Carroll R (1995) Treatment of verapamil overdose with glucagon in dogs. Ann Emerg Med 25:369–374

Stower MJ, Foster GE, Hardcastle JD (1982) A trial of glucagon in the treatment of painful biliary tract disease. Br J Surg 69:591–592

Tai YT, Lo CW, Chow WH, Cheng CH (1990) Successful resuscitation and survival following massive overdose of metoprolol. Br J Clin Pract 44:746

Talesnik J, Tsoporis J (1984) Enhancement by glutathione of the inotropic actions of catecholamines and glucagon. J Mol Cell Cardiol 16:573

Taylor JR, Sherratt HS, Davies DM (1978) Intramuscular or intravenous glucagon for sulphonylurea hypoglycemia? Eur J Clin Pharmacol 14:125–127

Thomas S, Stone K, May WA (1995) Exacerbation of verapamil induced hyperglycemia with glucagon. Am J Emerg Med 13:27

Tibblin S, Kock NG, Schenk WG Jr (1971) Response of mesenteric blood flow to glucagon. Influence of pharmacological stimulation and blockade of adrenergic receptors. Arch Surg 102:65

Timmus GC, Lin R, Ramos RG, Gordon S (1973) Prolonged glucagon infusion in cardiac failure. J Am Med Assoc 223:293–296

Trinchet JC, Balkau B, Poupon RE, Heintzmann F, Callard P, Gotheil C, Grange JD, Vetter D, Pauwels A, Labadie H (1992) Treatment of severe alcoholic hepatitis by infusion of insulin and glucagon: a multicenter sequential trial. Hepatology 15:76–81

Ueda J, Nakanishi H, Miyazaki M, Abe Y (1977) Effects of glucagon on the renal hemodynamics of dogs. Eur J Pharmacol 41:209

Vadhera RB (1992) Propranolol overdose (letter). Anesthesia 47:279

Vander-Ark CR, Reynolds EW (1979) Clinical evaluation of glucagon by continuous infusion in the treatment of low cardiac output states. Am Heart J 79:481–487

VanderWall DA, Stowe NT, Spangenberg R, Hook JB (1970) Effect of glucagon in hemorrhagic shock. J Surg Oncol 2:177

Vukmir RB, Paris PM, Yealy DM (1991) Glucagon: prehospital therapy for hypoglycemia. Ann Emerg Med 20:375–379

Walter FG, Frye G, Mullen JT, Ekins BR, Khasigian PA (1993) Amelioration of nifedipine poisoning associated with glucagon therapy. Ann Emerg Med 22:1234–1237

Warner WA, Begley L, Penman RW (1971) Effect of glucagon on pulmonary airflow resistance in dogs. Fed Proc 30:55S

Waterworth MW, Barbezat GO, Hickman R, Terblanche J (1976) A controlled trial of glucagon in experimental pancreatitis. Br J Surg 63:617–620

Weigelt JA, Gewertz BL, Aurbakken CM, Snyder WH (1982) Pharmacologic alterations in pulmonary artery pressure in the adult respiratory distress syndrome. J Surg Res 32:243

Weingand KW, Fettman MJ, Phillips RW, Hand MS (1986) Metabolic effects of glucagon in endotoxemic minipigs. Circ Shock 18:289–300

Weinstein RS (1984) Recognition and management of poisoning with beta-adrenergic blocking agents. Ann Emerg Med 13:1123

Whitsitt LS, Lucchesi BR (1968) Effects of beta-receptor blockade and glucagon on the atrioventricular transmission system in the dog. Circ Res 23:585

Wilcken DE, Lvoff R (1970) Glucagon in resistant heart-failure and cardiogenic shock. Lancet I:1315

Wilson JE, Fontanarosa PB (1993) Glucagon for the treatment of respiratory emergencies. In: Picazo J (ed) Glucagon in acute medicine. Kluwer, Dordrecht, p 27

Wilson JE, Nelson RN (1990) Glucagon as a therapeutic agent in the treatment of asthma. J Emerg Med 8:127–130

Wolf LR, Spadafora MP, Otten EJ (1993) Use of amrinone and glucagon in a case of calcium channel blocker overdose. Ann Emerg Med 22:1225–1228

Woolf GM, Redeker AG (1991) Treatment of fulminant hepatic failure with insulin and glucagon. A randomized control trial. Dig Dis Sci 36:92–96

Wright CD, Kazmers A, Whitehouse WM Jr, Stanley JC (1985) Comparative hemodynamic effects of selective superior mesenteric arterial and peripheral intravenous glucagon infusions. J Surg Res 39:230

Xenophontos XP, Watson PA, Chua BH, Haneda T, Morgan HE (1989) Increased cyclic AMP content accelerates protein synthesis in rat heart. Circ Res 65:647

Zaloga GP, Malcolm D, Holaday J et al (1985) Glucagon reverses the hypotension and bradycardia of verapamil overdose in rats (abstract). Crit Care Med 13:273

Zaloga GP, Delacey W, Holmboe E, Chernow B (1986) Glucagon reversal of hypotension in a case of anaphylactoid shock. Ann Intern Med 105:65

Zaloga GP, Malcolm DS, Holaday JW, Chernow B (1988) Glucagon. In: Chernow B (ed) The pharmacologic approach to the critically ill patient, 2nd edn. Williams and Wilkins, Baltimore, p 659

Zaritsky AL, Horowitz M, Chernow B (1988) Glucagon antagonism of calcium channel blocker-induced myocardial dysfunction. Crit Care Med 16:246

CHAPTER 12
The Place of Glucagon in Medical Imaging

J. SKUCAS

A. Introduction

Bowel hypotonia is useful in a number of imaging modalities. Bowel tonicity, however, is not the same as peristalsis, although in general those pharmacological agents that decrease bowel tonicity also result in decreased peristalsis. Currently several spasmolytic pharmacological agents are available; the two most often employed are glucagon, a hormone, or one of the anticholinergic agents.

In radiology, the primary use of exogenously administered pharmacological doses of glucagon is for its spasmolytic action on smooth muscle. The various smooth muscle target organs in the gastrointestinal tract have different sensitivities to glucagon and such differences can be exploited to achieve an optimal hypotonic effect. Thus, the individual target organs should be considered when determining the glucagon dose to be employed.

Considerable controversy surrounds the optimal dose of glucagon to be used and there is a wide variation in the dose employed (SKUCAS 1992). In the 1970s many radiologists used 2 mg glucagon i.m.; as the late Dr. Roscoe MILLER et al. (1974a,b) have shown, this dose is effective throughout the gastrointestinal tract and is associated with few side effects. The escalating cost of glucagon, however, has converted most radiologists to the i.v. route, where a smaller dose can be employed.

The onset of hypotonicity with an i.m. dose is between 5 and 10 min, while the onset with an i.v. injection is within 1 min, regardless of dose (MILLER et al. 1979, 1982). A certain minimal dose is required to achieve hypotonicity or atonicity and any further increase in dose does not result in an increase in the degree of hypotonicity, but simply prolongs the duration of hypotonia.

B. Upper Gastrointestinal Tract

I. Esophagus

The action of glucagon on the esophagus can be studied both by its effect upon smooth muscle peristalsis and lower esophageal sphincter (LES) tonicity. Regardless of dose, glucagon does not abolish esophageal peristalsis; although early studies concluded that glucagon had no effect on esophageal peristalsis, more sensitive studies by ANVARI et al. (1989) showed that pharmacological doses (0.5 mg or larger, i.v.) do lower the peristaltic amplitude in the distal esophagus, namely in that segment of esophagus containing smooth muscle.

Glucagon relaxes the LES and can thus induce gastroesophageal reflux, a finding that should be interpreted with caution if glucagon was administered during the examination (HAGGAR et al. 1982; FECZKO et al. 1983). With glucagon doses from 0.025 to 0.125 mg i.v., whether or not reflux occurs is not dose dependent (HAGGAR et al. 1982).

Because glucagon decreases the LES pressure, it can aid clearing of lower esophageal food impaction. Indeed, ROBBINS and SHORTSLEEVE (1994), RATCLIFF (1991), KASZAR-SEIBERT et al. (1990) have shown such an effect clinically. A dose greater than 1.0 mg i.v. is not needed. Water, a barium sulfate suspension, an effervescent agent, or a combination can be used to propel the foreign material bolus distally.

II. Stomach and Duodenum

A basic question is whether hypotonia of the stomach and duodenum improves one's ability to detect a lesion. The answer is controversial and a number of radiologists do not induce hypotonia. The hypotonic agents that have been evaluated include glucagon, morphine, pro-pantheline bromide, atropine, and other related compounds. With some, after initial enthusiasm the recognition of toxicity and undesirable side effects led to their abandonment. In the United States, for gastroduodenal hypotonia glucagon is the pharmacological agent of choice. In some countries, one of the anticholinergic agents is used. The main advantage of glucagon over the anticholinergic agents is a lack of side effects. Although MAGLINTE et al. (1982) found that glucagon improved mucosal coating, others found no such improvement (MILLER et al. 1974a; HAGGAR et al. 1982). ROTHE et al. (1987) found that the diagnostic quality did not differ significantly whether glucagon was used or not.

Both glucagon (LOUD 1989) and the anticholinergics decrease gastric secretions. TAKEMOTO et al. (1991) found that glucagon decreases the intragastric and intraduodenal mean pressures. There is also a delay in gastric emptying (JONDERKO et al. 1988; KAWAMOTO et al. 1985). The effect of glucagon upon the pylorus is still controversial; although some believe that glucagon induces pylorospasm, the delayed gastric emptying following administration of glucagon may also have a role.

The duration of atonicity and hypotonicity in the stomach and duodenum are dose related (Table 1). Even in normal volunteers there is some variability in the duration of hypotonicity. MILLER et al. (1982) found that a dose of 0.05 mg glucagon i.v. will result in hypotonicity in most patients. In the average individual this dose results in gastroduodenal hypotonia for approximately 3–5 min. If glucagon is administered immediately prior to the examination, the appropriate hypotonic stomach radiographs can be obtained within this time

Table 1. Effect of glucagon on the upper gastrointestinal tract

Location and response (min)	Glucagon (mg) i.v.							
	0.025	0.05	0.1	0.2	0.25	0.5	1.0	2.0
Stomach								
Atonicity	0.1	1.2	2.1	3.5	4.9	8.7	10.1	15.1
Moderate hypotonicity	0.8	3.3	4.8	7.4	8.5	13.1	14.7	21.7
Duodenal bulb								
Atonicity					7.5	10.1	12.5	16.7
Moderate hypotonicity					12.5	16.2	18.3	24.0
Duodenal loop								
Atonicity	0.5	1.5	3.1	5.2	7.8	10.1	12.5	16.1
Moderate hypotonicity	2.5	4.9	6.0	8.8	12.2	16.5	18.7	23.7

Note: Response to 0.025 to 0.2-mg doses adopted from Miller et al. (1982). Response to 0.25 to 2.0-mg adopted from Miller et al. (1978a)

period in most patients. The subsequent gradual onset of peristalsis then propels the barium bolus into the duodenum and the study can then be completed in the usual manner. If a longer period of gastric atonicity is desired, the glucagon dose can be increased accordingly. For instance, a dose of 0.25 mg induces gastric atonicity for almost 5 min and moderate hypotonicity for 8.5 min, times that should be sufficient for the radiographs even in patients with limited mobility. The side effects with such a small dose are minimal.

The duodenal bulb and duodenal sweep are more sensitive to glucagon than the stomach. The duodenum is usually the first region to show a drug effect and the last to recover. In most patients, a glucagon dose of 0.25 mg results in duodenal atonicity lasting 7–8 min. Obviously even smaller doses can be used.

Miller et al. (1978b) found that a dose of 0.25 mg glucagon i.v. is approximately equivalent in magnitude of response and duration of effect to 1.0 mg given i.m. The onset of drug effect with the i.m. dose, however, is considerably delayed.

Following gastric surgery, especially if part of the stomach has been resected, adequate gastric distension can be difficult to achieve because of rapid emptying. Induction of bowel hypotonia appears helpful in these patients and a dose of 1 mg glucagon i.v. has been recommended by Kovacs and Forgon (1983). A smaller dose undoubtedly can be used. In particular, following an antrectomy and gastrojejunostomy (Billroth II operation), visualization of the afferent loop is improved if bowel hypotonia is induced.

There has been limited study on the glucagon dose in infants and children. For the upper gastrointestinal tract, Ratcliffe (1980) recommends 0.5–1.0 mg/kg body weight, i.v. Such a dose results in atonia for approximately 3–5 min. The author injects the glucagon after barium opacifies the duodenal bulb.

C. Small Bowel

Glucagon is not used in a conventional small bowel follow through study because the resultant hypotonia simply prolongs the examination. The small doses of glucagon recommended by FECZKO et al. (1983) for the upper gastrointestinal tract (up to 0.125 mg i.v.) do not significantly prolong transit time, because the induced hypotonia is short enough that it does not interfere with a subsequent small bowel examination.

I. Enteroclysis

During enteroclysis barium is instilled directly into the jejunum, thus bypassing the pylorus. Either a single contrast or double contrast study can be performed and glucagon is generally not employed. In an occasional patient, where a suspicious lesion is detected and additional views are desired, injection of glucagon delays the forward propulsion of barium and allows a more leisurely study of the area in question. A radiographically suspicious area probably can be adequately evaluated with a dose as low as 0.1 mg i.v. In most patients a dose of 0.25 mg i.v. should result in atonicity for more than 8 min (Table 2); larger doses simply prolong the atonicity and hypotonicity.

II. Retrograde Ileography

Retrograde ileography is an underused technique for study of the distal small bowel. Either a single contrast or a double contrast study can be performed. The key to this study, however, is relaxation of the ileocecal valve, thus allowing reflux of contrast into the terminal ileum. A number of investigators have documented that pharmacological doses of glucagon result in considerably more ileal reflux as compared to placebo (MILLER et al. 1974b; VIOLON et al. 1981; STONE and CONTE 1988; MONSEIN et al. 1986). No specific minimal effective dose glucagon has been established, with many investigators using up to 1.0 mg i.v. Thus if Crohn's disease or some other abnormality of the distal ileum is suspected, glucagon should be employed prior to a barium enema.

Table 2. Effect of glucagon on the small bowel

Location and response (min)	Glucagon (mg) i.v.			
	0.25	0.5	1.0	2.0
Jejunum				
Atonicity	8.3	9.4	13.7	19.7
Moderate hypotonicity	11.5	14.9	19.7	25.0
Ileum				
Atonicity	8.6	9.4	14.0	19.7
Moderate hypotonicity	11.9	14.7	20.3	24.3

Note: Adopted from MILLER et al. (1978a)

Glucagon may have a role in distal small bowel obstructions secondary to a foreign body. The most common such obstruction is due to a gallstone (gallstone ileus). Although most gallstones impact proximal to the ileocecal valve, glucagon also induces ileal hypotonia and, theoretically at least, may be useful.

SLAPNIK and DOUNIES (1992) reported a swallowed coin producing distal small bowel obstruction that resolved following glucagon administration. Likewise in the therapy of meconium ileus and the ileal plug syndrome glucagon may have a role (MANDELL and TEPLICK 1982).

III. Peroral Pneumocolon

In selected patients, a peroral pneumocolon may be the only method possible in visualizing the terminal ileum and cecum. Because of its ability to allow reflux through the ileocecal valve, glucagon should aid in this examination. FITZGERALD et al. (1985) suggest doses of 0.25–0.5 mg i.v., while KELVIN et al. (1982) use 0.5–1.0 mg i.v.

D. Large Bowel

I. Barium Enema

There is variability in whether glucagon is employed during a barium enema. In general, the elderly and hospitalized patients tend to be premedicated more often with glucagon than younger outpatients. In some practices glucagon is used routinely, while in others it is employed only if significant spasm is encountered or the patient is not able to retain the enema. Some radiologists use it routinely with double contrast barium enemas but selectively with single contrast studies. The primary purpose of glucagon is simply to decrease bowel peristalsis and associated spasm and increase patient comfort (PIETILÄ 1992; SKUCAS 1994).

In their initial studies, Dr. MILLER et al. (1974b) used 2 mg glucagon i.m. There has been no well-controlled double-blind study evaluating the minimal effective i.v. dose of glucagon on the large bowel and thus the various doses have been empirically derived. A dose of 0.25–0.5 mg i.v. produces hypotonia almost immediately and lasts for 10–15 min. Because the main part of a barium enema can be completed within 10–15 min, it is believed that such a dose is adequate in most patients. In the occasional patient where subsequent spasm is encountered, a second such dose can be given. With such doses, nausea and vomiting is rarely encountered. Others use a dose up to 1 mg i.v. Such a relatively large dose is more than adequate in inducing hypotonia and abolishing most spasm, yet it is probably neither warranted nor necessary.

In infants and children RATCLIFFE (1980) uses a dose of 0.8–1.25 mg/kg i.v. The resultant hypotonia lasts approximately 5–10 min.

Whether glucagon aids filling the appendix is debatable. STONE and CONTE (1988) showed no statistical difference in visualizing the appendix whether glucagon was given or not, although there was a trend toward increased visualization.

Does use of glucagon result in a more "accurate" diagnosis? One would expect that the increased patient comfort and decreased spasm should improve the study. THOENI et al. (1984) found that the results were not statistically different whether glucagon was used or not. The authors recommended glucagon only if the patient has considerable discomfort during the examination, if spasm is encountered, if the patient has difficulty retaining the enema, or if there is clinical suspicion of colitis or diverticulitis. Extension of this study to 120 patients in each category added validity to the original conclusions (THOENI 1984). Similar findings were reached by ROTHE et al. (1987).

Because of increased reflux of air into the small bowel, STONE and CONTE (1988) found that glucagon resulted in less colon distension during a double-contrast barium enema. They added no further air to compensate for the air lost into the small bowel. The effect of glucagon on spasm, overall colon tonicity, and patient comfort were not addressed.

Some diabetic patients have relatively high glucagon plasma levels. Likewise, some develop autonomic neuropathy and, even with large doses of glucagon, some continue to have persistent spasm.

II. Intussusception Reduction

Because of the relaxation effect upon the ileocecal valve, it has been postulated that glucagon should have a role in the reduction of pediatric intussusceptions. Indeed a number of earlier anecdotal reports did suggest such a role, but both FRANKEN et al. (1983) and HSIAO et al. (1988) in subsequent well-controlled studies did not find any statistical significance in the success rate between glucagon and placebo. A dose of 0.05 mg/kg body weight, up to a maximum of 1 mg, was used, which is more than recommended by RATCLIFFE (1980) for barium enema examinations.

The role of glucagon in intussusception reduction is not settled. Some radiologists attempt hydrostatic reduction without glucagon first; if this attempt fails, glucagon is administered. Such empiric use of glucagon does not imply that eventual successful reduction can be attributed to glucagon, because even a second or third attempt at simple hydrostatic reduction improves the overall success rate. MORTENSSON et al. (1984) administered 0.05 mg/kg glucagon i.m. in a random fashion before attempted intussusception reduction. There was no statistical difference in the success rate nor in the time required for successful reduction whether glucagon was used or not. Of interest is that patients in the control group, who had three unsuccessful attempts at hydrostatic reduction, then received glucagon; the fourth attempt at reduction resulted in a success rate of 59%. Their recommendation was that after two unsuccessful attempts at hydrostatic reduction using conventional technique, glucagon should be employed prior to the third attempt.

In some patients even intraoperative reduction of an intussusception is not possible without bowel resection. MEULI et al. (1988) suggest that use of glucagon may avoid some bowel resections; they used $0.5\,mg/m^2$ i.v.

E. Biliary Tract

Pharmacological doses of glucagon relax the gallbladder, increase bile flow, and relax the papilla of Vater (BRANUM et al. 1991). Glucagon has been used to help relieve the pain of biliary colic (JACOBSON et al. 1984; GROSSI et al. 1986).

Glucagon has a limited role in percutaneous transhepatic cholangiography. In the patient with a narrowed segment in the distal common bile duct, glucagon can aid in differentiating between tumor, impacted stone, or spasm (CANNON and LEGGE 1979). A similar application had been claimed in operative cholangiography, although in controlled studies by COFER et al. (1988) and FOSTER and FOSTER (1984) glucagon did not improve the quality of intraoperative cholangiography.

Spasm can be encountered during manipulation of retained bile duct stones through a previously surgically placed T-tube tract. An impacted stone in the distal common bile duct can be difficult to dislodge; it can also be difficult to distinguish it from a papilloma. In an occasional patient glucagon may aid sphincter relaxation (MANDELL and TEPLICK 1982; LATSHAW et al. 1981). Most such reports have been anecdotal. Neither the dose, nor whether indeed glucagon is advantageous, has been adequately studied.

F. Other Applications

I. Computed Tomography

Some of the older scanners produce significant motion artifacts because of peristalsis. MARKS et al. (1980) used glucagon with scanning times longer than 6 seconds and injected 0.25 mg i.v. over 15 s. With most fast scanners few motion artifacts are encountered and there is a limited role for glucagon in the current practice of computed tomography.

The dose of glucagon employed should be tailored to the length of the examination. One option is to place an i.v. catheter, inject 0.25–0.5 mg glucagon, and reinject 0.25 mg every 10 min or so as needed. Three such injections should result in moderate hypotonicity for approximately 30 min. JEHENSON (1991) proposed that such spaced injections would result in fewer side effects.

II. Ultrasonography

At times ultrasonography of the upper abdomen can be enhanced by bowel atonia. A water-filled stomach and duodenum, rendered hypotonic by glucagon, can be a sonic window to the pancreas and bile ducts (OP DEN ORTH 1985, 1987).

Doppler sonography can be used to evaluate portal blood flow. Studies before and after a test meal yield quantitative data on portal blood flow velocity. RENDA et al. (1994) proposed substituting glucagon for the test meal, noting that after glucagon the portal blood velocity increased by over 30% in normal volunteers.

III. Angiography

In abdominal digital and subtraction angiography, bowel peristalsis may result in a poor or nondiagnostic study. RABE et al. (1982) performed digital abdominal arteriography before and after 1.0 mg glucagon i.v. In all patients peristalsis artifacts were present in the pre-glucagon studies, while in the post-glucagon studies there was a reduction or absence of artifacts.

Glucagon is a potent mesenteric vasodilator, although its efficacy during mesenteric angiography has not been adequately explored. ROARK and SUZUKI (1987) have proposed glucagon in the therapy of mesenteric ischemia; encouraging results were seen in animals (SARDELLA et al. 1990; OSHIMA et al. 1990). In some animal models the enhanced survival seen with postischemic glucagon may be secondary to the systemic effects of glucagon, rather than to the direct mesenteric effects (SARDELLA et al. 1990; SCHNEIDER et al. 1994).

IV. Magnetic Resonance Imaging

Most magnetic resonance (MR) examinations of the abdomen are prolonged and both respiration and peristalsis tend to degrade the resultant images. WINKLER and HRICAK (1986) have proposed use of glucagon to decrease peristaltic artifacts. Currently some centers use glucagon almost routinely in abdominal imaging, while others do not use it at all. Glucagon doses of 1–2 mg i.m. result in prolonged bowel hypotonia. Another option is placement of a small i.v. catheter and repeated small sequential injections every 10 min or so or the continuous infusion of glucagon for most of the duration of the study.

The contrast agent perflubron, when given orally, results in a signal void. It is immiscible with water and readily outlines bowel lumen. BROWN et al. (1991) found that adding glucagon significantly increased bowel wall visualization. The use of glucagon in MR imaging has been summarized by BARTNICKE and BROWN (1993).

V. Hysterosalpingography

Most radiologists performing hysterosalpingography have encountered what appears to be tubal obstruction, but with subsequent gentle pressure or during another examination the fallopian tubes appear to be normal. Spasm of the fallopian tubes has been proposed as the mechanism for this finding, although HUGH (1993) believes that kinking is a more likely explanation.

WINFIELD et al. (1982) and MALEEV et al. (1988) showed that glucagon relieves tubal spasm. YUNE (1989) routinely used a "liberal" dose of glucagon (0.5 mg i.v.) unless there was a contraindication. Pelvic cramps and tubal spasms are claimed to be rare. One multi-institutional study found that glucagon played no significant role in filling normal fallopian tubes (WORLD HEALTH ORGANIZATION 1983); however, this study did not evaluate spasm per se and because glucagon was administered "within 20 min" of the study its effect may have already worn off.

The role of glucagon in relieving spasm in hysterosalpingography and in improving the accuracy of diagnosis is yet to be established.

VI. Scintigraphy

During cholescintigraphy, glucagon results in faster visualization of the gallbladder (MATSUMOTO 1986). The maximum gallbladder relaxation was shorter in those patients receiving glucagon simultaneously with the radioisotope, than in those in a control group.

Technetium-99m-labeled red blood cell scintigraphy is used to detect gastrointestinal hemorrhage. Because glucagon increases the mesenteric blood circulation and induces hypomotility and hypotonia, theoretically at least, it should aid detection of such bleeding. FROELICH and JUNI (1984) found that in bleeding patients undergoing scintigraphy, who had a nonspecific increase in abdominal activity at 30 min, half of the patients had a focal accumulation that was not evident prior to glucagon. They concluded that glucagon may aid recognition of small bowel bleeding sites. FAWCETT et al. (1986) studied bleeding patients who had no evidence of bleeding within 30 min after technetium injection. Glucagon was then injected and imaging extended for 1 h. None of these patients had additional evidence of bleeding and the conclusion was that if initial scintigraphy does not detect the bleeding, addition of glucagon does not improve the detection rate.

Ectopic gastric mucosa in a Meckel's diverticulum can concentrate technetium-99m-labeled pertechnetate. False negative scans are not uncommon and, in order to improve accuracy, a number of pharmacological agents have been used. For instance, the hormone pentagastrin increases pertechnetate uptake both in the stomach and a Meckel's diverticulum. Unfortunately, pentagastrin also increases pertechnetate secretion and stimulates bowel motility so that the radioactivity is carried away from the diverticulum and diluted. Glucagon decreases pertechnetate uptake in the stomach and reduces bowel motility and thus the use of both pentagastrin and glucagon appears advantageous. Animal studies support such an approach in scanning Meckel's diverticula but there are few clinical data available (DATZ et al. 1991).

VII. Urography

Although glucagon relaxes the ureters, there is also diminution of contrast in the collecting systems (KINNUNEN et al. 1990; HILLMAN 1983; THOMSEN et al. 1983). Such decreased contrast density presumably is secondary to increased renal blood flow and glomerular filtration and resultant contrast dilution. Glucagon thus does not improve pyeloureteral visualization during urography.

G. Side Effects and Contraindications

With the small doses used in medical imaging, glucagon is a very safe drug. It produces few, mild, and transient side effects not much greater than those of placebo (CHERNISH et al. 1975). The side effects with glucagon are less than those encountered with atropine or propantheline bromide (MILLER et al. 1974a,b). In one center in Japan, anticholinergic agents were not used in 7.1% of patients undergoing endoscopy because of contraindications. (TAKEMOTO et al. 1991). In the elderly, this number increased to 19.5%.

The side effects of glucagon are dose dependent. With a dose of 0.25–0.5 mg i.v., injected slowly, nausea and vomiting are uncommon, while i.v. doses greater than 1 mg lead to increased nausea and vomiting. Whether the nausea and vomiting are direct effects of glucagon on the central nervous system, as proposed by CHERNISH and MAGLINTE (1991), or whether they are associated with gastric dysrhythmia and thus a direct action on the stomach as suggested by ABELL and MALAGELADA (1985) is speculative.

Thus, besides the cost, the incidence of these side effects should encourage radiologists to use the smallest dose of glucagon that consistently produces the desired hypotonia for the study in question.

Following injection of a pharmacological dose of glucagon, the serum glucose level rises to a peak in approximately 15–30 min and then gradually returns to a basal value or slightly below basal (CHERNISH et al. 1975, 1988). It has been postulated that an occasional susceptible patient may develop delayed mild hypoglycemia. Orange juice or any drink containing sugar would be useful in such a patient following completion of the imaging study (CHERNISH and MAGLINTE 1990), a practice not generally followed by most radiologists.

Unfortunately, some radiologists still believe that glucagon should be avoided in diabetic patients. Little substantiation is provided for such an approach. In fact, glucagon was specifically developed commercially for use in diabetic individuals. Numerous brittle diabetic patients have received glucagon. In a setting of hypoglycemia, glucagon should make the patient feel better. If hyperglycemia is present, the resultant temporary additional serum glucose elevation should be of little consequence. Diabetologists routinely give glucagon to evaluate residual insulin reaction in diabetic patients and these patients can have elevated blood glucose levels (see Chap. 14, this volume). Symptoms generally occur because of keto-acidosis and not the glucose elevation. If hyperglycemia and ketoacidosis are present, therapy of the underlying ketoacidosis should be undertaken. A patient in diabetic coma should not be undergoing any prolonged studies in radiology; the basic underlying condition should be stabilized first and only then should diagnostic studies be performed. On the other hand, the diabetic patient who presents for a radiographic study can safely receive glucagon whenever clinically indicated.

Although uncommon, hypersensitivity reactions to glucagon can occur. In very few patients glucagon resulted in a skin rash, erythema multiforme, periorbital edema, loss of consciousness, respiratory distress, hypotension, and anaphylaxis (EDELL 1980; ZAVRAS et al. 1990).

Anaphylactic reactions have also occurred after the administration of a barium sulfate suspension prior to the injection of glucagon (GELFAND et al. 1985). A paraben derivative previously used as a preservative in barium has been associated with hypersensitivity reactions (SCHWARTZ et al. 1984; KAINBERGER et al. 1986; JAVORS et al. 1984). In some reports it is not clear whether a patient received glucagon, barium, or both. Latex has also been incriminated in a number of hypersensitivity reactions. The cause of death may not be apparent even on postmortem examination (HARRINGTON and KAUL 1987). In general, the incriminating agent has not been sought or detected in most instances when a reaction has occurred. The incidence of such reactions is low.

The contraindications to glucagon are a suspected pheochromocytoma, insulinoma, or a known prior glucagon sensitivity. Severe hypertension can develop in a setting of a pheochromocytoma (McLOUGHLIN et al. 1981). Some of these patients have a normal baseline blood pressure (YOSHIDA et al. 1990; ELLIOTT et al. 1989). If hypertension develops following glucagon injection, 5mg phentolamine, a short-acting α-adrenergic blocking agent, as indicated, may be helpful (see LEFÈBVRE and LUYCKX 1983).

In a setting of an insulinoma, the additional release of insulin after glucagon injection may lead to hypoglycemia with its resultant symptoms. Glucose should be administered in such a clinical setting.

Glucagon may precipitate facial flushing in a patient with a carcinoid tumor (LEBOVICS and ROSENTHAL 1991). Such a facial flush lasting for several minutes should alert the radiologist to a possible underlying carcinoid tumor.

References

Abell TL, Malagelada JR (1985) Glucagon-evoked gastric dysrhythmias in humans shown by an improved electrogastrographic technique. Gastroenterology 88:1932–1940

Anvari M, Richards D, Dent J, Waterfall WE, Stevenson GW (1989) The effect of glucagon on esophageal peristalsis and clearance. Gastrointest Radiol 14:100–102

Bartnicke BJ, Brown JJ (1993) Glucagon use in magnetic resonance imaging. Int Glucagon Monitor 3:1–4

Branum GD, Bowers BA, Watters CR, Haebig J, Cucchiaro G, Farouk M, Meyers WC (1991) Biliary response to glucagon in humans. Ann Surg 213:335–340

Brown JJ, Duncan JR, Heiken JP, Balfe DM, Corr AP, Mirowitz SA, Eilenberg SS, Lee JKT (1991) Perfluoroctylbromide as a gastrointestinal contrast agent for MR imaging: use with and without glucagon. Radiology 181:455–460

Cannon P, Legge D (1979) Glucagon as a hypotonic agent in cholangiography. Clin Radiol 30:49–52

Chernish SM, Maglinte DDT (1990) Glucagon: common untoward reactions – review and recommendations. Radiology 177:145–146

Chernish SM, Maglinte DDT (1991) Reducing doses of glucagon used in radiologic examinations. Radiology 179:286–287

Chernish SM, Davidson JA, Brunelle RL, Miller RE, Rosenak BD (1975) Response of normal subjects to a single 2-milligram dose of glucagon administered intramuscularly. Arch Int Pharmacodyn Ther 218:312–327

Chernish SM, Maglinte DDT, Brunelle RL (1988) The laboratory response to glucagon dosages used in gastrointestinal examinations. Invest Radiol 23:847–852

Cofer JB, Barnett RM, Major GR, Lafon ED, Redish E (1988) Effect of intravenous glucagon on intraoperative cholangiography. South Med J 81:455–456

Datz FL, Christian PE, Hutson WR, Moore JG, Morton KA (1991) Physiological and pharmacological interventions in radionuclide imaging of the tubular gastrointestinal tract. Semin Nucl Med 21:140–152

Edell SL (1980) Erythema multiforme secondary to intravenous glucagon. AJR 134:385–386

Elliott WJ, Murphy MB, Straus FH II, Jarabak J (1989) Improved safety of glucagon testing for pheochromocytoma by prior α-receptor blockade. Arch Intern Med 149:214–216

Fawcett HD, Morettin LB, Nusynowitz ML (1986) Failure of glucagon to improve detection of acute gastrointestinal bleeding using technetium-99m red blood cells. J Nucl Med 27:1941–1942

Feczko PJ, Simms SM, Iorio J, Halpert R (1983) Gastroduodenal response to low-dose glucagon. AJR 140:935–940

Fitzgerald EJ, Thompson GT, Somers SS, Franic SF (1985) Pneumocolon as an aid to small-bowel studies. Clin Radiol 36:633–637

Foster ME, Foster DR (1984) Glucagon in operative cholangiography. J R Coll Surg (Edinb) 29:301–302

Franken EA Jr, Smith WL, Chernish SM, Campbell JB, Fletcher BD, Goldman HS (1983) The use of glucagon in hydrostatic reduction of intussusception: a double-blind study of 30 patients. Radiology 146:687–689

Froelich JW, Juni J (1984) Glucagon in the scintigraphic diagnosis of small-bowel hemorrhage by Tc-99m-labelled red blood cells. Radiology 151:239–242

Gelfand DW, Sowers JC, DePonte KA, Sumner TE, Ott DJ (1985) Anaphylactic and allergic reactions during double-contrast studies: is glucagon or barium suspension the allergen? AJR 144:405–406

Grossi E, Broggini M, Quaranta M, Balestrino E (1986) Different pharmacological approaches to the treatment of acute biliary colic. Curr Ther Res 40:876–882

Haggar AM, Feczko PJ, Halpert RD, Simms SM (1982) Spontaneous gastroesophageal reflux during double-contrast upper gastrointestinal radiography with glucagon. Gastrointest Radiol 7:319–321

Harrington RA, Kaul AF (1987) Cardiopulmonary arrest following barium enema examination with glucagon. Drug Intell Clin Pharm 21:721–722

Hillman BJ (1983) Excretory urography with glucagon. Radiology 149:326–327

Hsiao J-Y, Kao H-A, Shih S-L (1988) Intravenous glucagon in hydrostatic reduction of intussusception: a controlled study of 63 patients. Acta Paediatr Sin 29:242–247

Hugh AE (1993) The role of hysterosalpingography in modern gynaecological practice. Br J Radiol 66:278

Jacobson G, Nilsonn B, Nordgren CE, Selking O (1984) Glucagon-(1-21)-peptide to prevent biliary colic pain. Lancet II:1149

Javors BR, Applbaum Y, Gerard P (1984) Severe allergic reaction: an unusual complication of barium enema. Gastrointest Radiol 9:357–358

Jehenson PM (1991) Reducing doses of glucagon used in radiologic examinations (letter). Badiology 179:286–287

Jonderko G, Golab T, Jonderko K (1988) A pharmacological dose of glucagon suppresses gastric emptying of a radiolabelled solid meal in humans. Scand J Clin Lab Invest 48:743–746

Kainberger F, Lindemayr H, Frühwald F, Schwaighofer B (1986) Intoleranzreaction nach Glukagonapplikation in der Doppelkontrastuntersuchung. Radiologe 26:531–533

Kaszar-Seibert DJ, Korn WT, Bindman DJ, Shortsleeve MJ (1990) Treatment of acute esophageal food impaction with a combination of glucagon, effervescent agent, and water. AJR 154:533–534

Kawamoto H, Yamamura H, Tatsuta M, Okuda S (1985) Effect of glucagon on gastric motility examined by the acetaminophen absorption method and the endoscopic procedure. Arzneimittelforschung 35:1475–1477

Kelvin FM, Gedgaudas RK, Thompson WM, Rice RP (1982) The peroral pneumocolon. Its role in evaluating the terminal ileum. AJR 139:115–121

Kinnunen J, Saarinen O, Ahovuo J, Edgren J, Laasonen L, Pietilä J (1990) Urography with spasmolytics. Rontgenblatter 43:20–23

Kovacs K, Forgon J (1983) Doppelkontrastuntersuchung des operierten Magens in Glucagonhypotonie. Rontgenblatter 36:374–378

Latshaw RF, Kadir S, Witt WS, Kaufman SL, White RI Jr (1981) Glucagon-induced choledochal sphincter relaxation. Aid for expulsion of impacted calculi into the duodenum. AJR 137:614–616

Lebovics E, Rosenthal WS (1991) Glucagon precipitation of carcinoid flush. Gastrointest Endosc 37:212–213

Lefèbvre PJ, Luyckx AS (1983) Glucagon catecholamines. In: Lefèbvre PJ (ed) Glucagon II. Springer, Berlin Heidelberg New York, p 537 (Handbook of experimental pharmacology, vol 66/II)

Loud FB (1989) Glucagon and gastric secretion in man. Dan Med Bull 36:532–537

Maglinte DDT, Caudill LD, Krol KL, Chernish SM, Brown DL (1982) The minimum effective dose of glucagon in upper gastrointestinal radiography. Gastrointest Radiol 7:119–122

Maleev N, Červenyashki S, Vasilev D, Rachev E (1988) Overcoming of the spasm of uterine tubes during hysterosalpingography with application of glucagon (in Russian). Akus i Ginekol 27:55–57

Mandell GA, Teplick SK (1982) Glucagon – its application to childhood gastrointestinal radiology. Gastrointest Radiol 7:7–13

Marks WM, Goldberg HI, Moss AA, Koehler FR, Federle MP (1980) Intestinal pseudotumors: a problem in abdominal computed tomography solved by directed techniques. Gastrointest Radiol 5:155–160

Matsumoto T (1986) Glucagon-dosed cholescintigraphy. Radiat Med 4:31–38

McLoughlin MJ, Langer B, Wilson DR (1981) Life-threatening reaction to glucagon in a patient with pheochromocytoma. Radiology 140:841–842

Meuli M, Hirsig J, Briner J (1988) Intraoperative Spasmolyse mit Glucagon bei schwer reduzierbarer Invagination. Z Kinderchir 43:218–219

Miller RE, Chernish SM, Skucas J, Rosenak BD, Rodda BE (1974a) Hypotonic roentgenography with glucagon. AJR 121:264–274

Miller RE, Chernish SM, Skucas J, Rosenak BD, Rodda BE (1974b) Hypotonic colon examination with glucagon. Radiology 113:555–562

Miller RE, Chernish SM, Brunelle RL, Rosenak BD (1978a) Double-blind radiographic study of dose response to intravenous glucagon for hypotonic duodenography. Radiology 127:55–59

Miller RE, Chernish SM, Brunelle RL, Rosenak BD (1978b) Dose response to intramuscular glucagon during hypotonic radiography. Radiology 127:49–53

Miller RE, Chernish SM, Brunelle RL (1979) Gastrointestinal radiography with glucagon. Gastrointest Radiol 4:1–10
Miller RE, Chernish SM, Greenman GF, Maglinte DDT, Rosenak BD, Brunelle RL (1982) Gastrointestinal response to minute doses of glucagon. Radiology 143:317–320
Monsein LH, Halpert RD, Harris ED, Feczko PJ (1986) Retrograde ileography: value of glucagon. Radiology 161:558–559
Mortensson W, Eklöf O, Laurin S (1984) Hydrostatic reduction of childhood intussusception. Acta Radiol Diagn 25:261–264
Op den Orth JO (1985) Tubeless hypotonic duodenography with water: a simple aid in sonography of the pancreatic head. Radiology 154:826
Op den Orth JO (1987) Sonography of the pancreatic head aided by water and glucagon. Radiographics 7:85–100
Oshima A, Kitajima M, Sakai N, Ando N (1990) Does glucagon improve the viability of ischemic intestine? J Surg Res 49:524–533
Pietilä JA (1992) Glukagon-unterstützte Reinigungseinläufe vor einem Doppelkontrast-Bariumeinlauf. Aktuelle Radiol 2:36–38
Rabe FE, Yune HY, Klatte EC, Miller RE (1982) Efficacy of glucagon for abdominal digital angiography. AJR 139:618–619
Ratcliff KM (1991) Esophageal foreign bodies. Am Fam Physician 44:824–831
Ratcliffe JF (1980) Glucagon in barium examinations in infants and children: special reference to dosage. Br J Radiol 53:860–862
Renda F, Olivieri A, Migliorato L, Sanita diToppi G, Colagrande C (1994) Valutazione con eco-Doppler degli effetti del glucagone sul flusso portale nei soggetti normali. Radiol Med (Torino) 87:447–451
Roark KM, Suzuki NT (1987) Glucagon use in mesenteric ischemia. Drug Intell Clin Pharm 21:660–661
Robbins MI, Shortsleeve MJ (1994) Treatment of acute esophageal food impaction with glucagon, an effervescent agent, and water. AJR 162:325–328
Rothe AJ, Young JWR, Keramati B (1987) The value of glucagon in routine barium investigations of the gastrointestinal tract. Invest Radiol 22:786–791
Sardella GL, Bech FR, Cronenwett JL (1990) Hemodynamic effects of glucagon after acute mesenteric ischemia in rats. J Surg Res 49:354–360
Schneider JR, Foker JE, Macnab JR, Marquardt CA, Cronenwett JL (1994) Glucagon effect on postischemic recovery of intestinal energy metabolism. J Surg Res 56:123–129
Schwartz EE, Glick SN, Foggs MB, Silverstein GS (1984) Hypersensitivity reactions after barium enema examination. AJR 143:103–104
Skucas J (1992) Optimal dose of glucagon. Radiology 183:325
Skucas J (1994) The use of antispasmodic drugs during barium enemas. AJR 162:1323–1325
Slapnik K, Dounies R (1992) Intravenous glucagon to dislodge an intestinal foreign body. Am Fam Physicians 45:55–58
Stone EE, Conte FA (1988) Glucagon-induced small bowel air reflux: degrading effects on double-contrast colon examination. Gastrointest Radiol 13:212–214
Takemoto T, Okita K, Tada M, Kawano H, Yoshida T, Akiyama T (1991) Glucagon in digestive endoscopy – its usefulness for premedication. Int Glucagon Monitor 1:15–19
Thoeni RF (1984) Importance of sample size for statistical significance. AJR 143:924
Thoeni RF, Vandeman F, Wall DS (1984) Effect of glucagon on the diagnostic accuracy of double-contrast barium enema examinations. AJR 142:111–114
Thomsen HS, Hartelius H, Jensen R, Flindt-Nielsen N, Ohlhues L (1983) Excretory urography with glucagon: a randomized trial. Radiology 147:335–338
Violon D, Steppe R, Potvliege R (1981) Improved retrograde ileography with glucagon. AJR 136:833–834

Winfield AC, Pittaway D, Maxson W, Daniell J, Wentz AC (1982) Apparent cornual occlusion in hysterosalpingography: reversal by glucagon. AJR 139:525–527

Winkler ML, Hricak H (1986) Pelvis imaging with MR: technique for improvement. Radiology 158:848–849

World Health Organization (1983) A new hysterographic approach to the evaluation of tubal spasm and spasmolytic agents. Fertil Steril 39:105–107

Yoshida K, Sasaguri M, Kinoshita A, Ideishi M, Ikeda M, Arakawa K (1990) A case of a clinically "silent" pheochromocytoma. Jpn J Med 29:27–31

Yune HY (1989) Hysterosalpingography. In: Skucas J (ed) Radiographic contrast agents, 2nd edn. Aspen, Rockville, p 282

Zavras GM, Papadaki PJ, Kounis NG, Dimopoulos JA (1990) Glucagon-induced severe anaphylactic reaction. Fortschr Rontgenstr 152:110

CHAPTER 13
The Glucagon Test for Evaluation of Insulin Secretion

A.J. Scheen

A. Introduction

Insulin secretion plays a key role in the homeostasis of blood glucose levels and may be altered in numerous pathological states. It is thus important to have appropriate methods for evaluating insulin secretion in research and clinical practice (Hovorka and Jones 1994; Scheen et al. 1995). In most cases, stimulated insulin secretion provides more information than basal insulin secretion, which is why several stimulatory tests have been proposed during the last 30 years to better assess the secretory capacity of the B cells of the pancreatic islets of Langerhans (Scheen et al. 1995).

Since the original observation that glucagon stimulates insulin secretion (Samols et al. 1965; review in Samols 1983), the glucagon test has become popular among diabetologists for routine evaluation of residual insulin secretion (Faber and Binder 1977). In the present review, we will first describe the classical test and consider several methodological aspects of the experimental procedure. The glucagon test will then be compared to other classical stimulatory tests alternatively used to assess B-cell function. Finally, several clinical applications of the glucagon test will be briefly described, essentially to evaluate insulin secretion in diabetic patients.

B. Methodological Aspects

I. Classical Test

The glucagon test for evaluating residual B-cell function in diabetes mellitus was first proposed in 1977 (Faber and Binder 1977). The C-peptide responses to glucagon were found to be well correlated with those of a standardized mixed meal, indicating a high predictive value of the glucagon test as to how the B cell will respond during normal daily life (Faber and Binder 1977). Subsequently, post-glucagon C-peptide measurements were shown to be more reproducible than insulin measurements in the determination of B-cell function (Gjessing et al. 1987a; Gottsäter et al. 1992).

The original test consisted of an intravenous bolus injection of 1 mg porcine glucagon in overnight fasted individuals, with serial measurements of plasma C-peptide levels during the following 20 min (FABER and BINDER 1977). However, the majority of the subjects showed maximum concentrations of plasma C-peptide 6 min after glucagon and prolongation of the test until 20 min provided no additional information. A good correlation between the routine 6-min C-peptide values and areas under the curve (AUCs) of plasma C-peptide concentrations after intravenous glucagon injection has been reported in non-insulin-dependent diabetes mellitus (NIDDM) patients (KOLENDORF et al. 1982). We were able to verify, in normal, obese, insulin-dependent diabetes mellitus (IDDM) and NIDDM diabetic subjects, the existence of an excellent correlation between a single post-glucagon C-peptide value and an integrated measurement performed on a blood sample collected continuously during the 30 min following the glucagon injection (CASTILLO et al. 1995).

II. Dose-Response Curve

Several studies have investigated the dose-response relationship between the amount of glucagon injected intravenously and the amplitude of the insulin secretory response. In non-diabetic subjects, the dose of 0.25 mg has been shown to exert a lower effect than the dose of 0.5 mg, while the latter induced a similar response to that provoked by a dose of 1 mg (AHREN et al. 1987). Several weight-adjusted doses (1, 2, 5, 10 µg/kg) have also been compared to the classical dose of 1 mg in healthy subjects and NIDDM patients (SNORGAARD et al. 1988). The maximal insulin response was observed with the dose of 10 µg/kg and there is no doubt that a dose of 1 mg glucagon induces a maximal response of the insulin secretion in both normal and diabetic subjects.

III. Reproducibility

The repeatability of C-peptide response to glucagon was shown to be very good both in 10 elderly non-diabetic subjects and in 20 elderly NIDDM patients in whom the glucagon test was repeated four times once a week during consecutive weeks (SARLUND et al. 1987a). In NIDDM subjects submitted to two glucagon tests several days apart, the intra-test coefficients of variation (CV) averaged 24.1% for post-glucagon C-peptide increments and 14.8% for absolute 6-min C-peptide values (34.8% and 24.8% for corresponding plasma insulin levels, respectively (GJESSING et al. 1987a). The insulin/C-peptide responses to three consecutive injections of 1 mg glucagon, at 2-h intervals on the same morning, were studied by our group in six NIDDM patients (SCHEEN et al. 1989; CASTILLO et al. 1996). Mean intra-test CVs averaged 11.2% for 6-min C-peptide levels, 17.9% for 6-min insulin concentrations, and 12.8% for 0- to 30-min integrated insulin secretion rates.

The reproducibility may be influenced by the population tested, probably due to the different amplitudes of the post-glucagon insulin secretory responses. Indeed, plasma C-peptide values measured 6 min after glucagon administration have been shown to have a mean intra-test CV of 9% in normal subjects, 15% in NIDDM patients and 31% in IDDM patients (ARNOLD-LARSEN et al. 1987). In the latter study, mean CVs were not significantly improved when considering the AUCs of the plasma C-peptide levels instead of the 6-min values: 9%, 17% and 26%, respectively.

Comparison of the reproducibility of C-peptide acute responses to 0.5 g/kg intravenous glucose (1 + 3 min post-glucose measurements) or 1 mg intravenous glucagon (6-min post-glucagon measurement) in healthy subjects showed similar intrasubject CVs, 15% vs. 14%, respectively (GOTTSÄTER et al. 1992).

IV. Influence of Prevailing Glucose Level

Soon after the demonstration of the insulinogenic effect of glucagon (SAMOLS et al. 1965), several authors pointed out the fact that the glucagon-induced insulin secretion may be markedly affected by the basal blood glucose concentration: hypoglycemia inhibits the insulin response to glucagon while hyperglycemia potentiates it (SAMOLS et al. 1966; OAKLEY et al. 1972; review in SAMOLS 1983). Similarly, fasting inhibits post-glucagon insulin secretion while feeding potentiates it (review in SAMOLS 1983). The confounding effect of the prevailing blood glucose level on the insulin response to glucagon has been further studied during the last 10 years in both IDDM and NIDDM subjects.

In insulin-treated subjects, comparison of the outcome of 97 paired glucagon and meal tests showed similar C-peptide responses at blood glucose concentrations >7 mmol/l, but blunted post-glucagon responses at levels <7 mmol/l (MADSBAD et al. 1987). In IDDM subjects, low but positive glucagon-induced insulin responses were found at a basal blood glucose concentration of 7.7 mmol/l, but not at 3.2 mmol/l (ARNOLD-LARSEN et al. 1987). Comparison of glucagon tests performed in IDDM individuals at basal blood glucose levels of 5, 12 and 20 mmol/l showed a significant potentiation of the C-peptide response from 5 to 12 mmol/l, but no further potentiation from 12 to 20 mmol/l (GJESSING et al. 1991).

In NIDDM patients, a progressive potentiation of insulin secretory response was reported by increasing basal blood glucose levels from 6–7 to 12–13 mmol/l and from 12–13 to 18–20 mmol/l (GJESSING et al. 1989b; SCHEEN et al. 1989; CASTILLO et al. 1996). Post-glucagon C-peptide response was found to be dramatically blunted in insulin-treated NIDDM subjects studied with a basal blood glucose concentration below 3.5 mmol/l, and this response was enhanced when the test was repeated 90 min after a meal, increasing the blood glucose concentrations (RÖNNEMAA 1986). Conversely, in hyperglycemic NIDDM subjects, acute glycemic normalization significantly reduced post-glucagon C-peptide release (NOSARI et al. 1992; GJESSING et al. 1992).

All these observations demonstrate that prevailing blood glucose levels profoundly modulate the B-cell response to glucagon. The optimal basal glucose level before glucagon injection seems to be around 7 mmol/l. Indeed, in

IDDM subjects, the predictive value of the glucagon test as to how the B cells will respond to a meal during everyday life may be low when fasting blood glucose is <7 mmol/l (MADSBAD et al. 1987). In contrast, in hyperglycemic NIDDM patients, measurements of C-peptide response to glucagon might overestimate B-cell function, a relevant finding with respect to the therapeutic implications of the glucagon test, especially when a secondary failure to antidiabetic oral drugs is suspected (NOSARI et al. 1992).

V. Combined Stimulation

1. Glucagon-Glucose Test

A combined injection of glucagon and glucose has been proposed in order to potentiate the insulin secretion observed after glucagon alone and to better differentiate several groups of diabetic patients (MIKI et al. 1992). When comparing the combined test with the glucagon test in individual cases, a greater C-peptide response was seen with the combined test in all cases except for IDDM patients. Similarly, a test consisting of sequential intravenous challenges with glucose and glucagon (10 min after the end of glucose injection) has been used to improve the detection of subjects at risk of developing type I diabetes (VIALETTES et al. 1993).

2. Glucagon-Meal Test

When a glucagon test was performed 1.5 h after a standardized breakfast in 15 insulin-treated diabetic patients with onset of diabetes after the age of 30 years, the mean concentration of plasma C-peptide was 62% higher than that measured in the test performed in the fasting state, thus demonstrating that the stimulatory actions of glucagon and breakfast on the secretion of insulin are additive (RÖNNEMAA 1986). Such a combination (standardized breakfast followed 60 min later by a glucagon injection) was used in a large population of NIDDM patients and control subjects (SARLUND et al. 1987). These authors preferred the glucagon stimulation after breakfast in order to avoid hypoglycemia, which may inhibit the effects of glucagon on insulin secretion (review in SAMOLS 1983).

C. Comparison with Other Stimuli

I. Oral Glucose Tolerance Test

The oral glucose tolerance test (OGTT) is the reference test to diagnose impaired glucose tolerance and diabetes, but it can also be used to evaluate the B-cell response to a glucose challenge. In healthy men, a 75-g oral glucose load induced higher insulin and C-peptide responses than a 1-mg intravenous glucagon bolus and was proposed as the test of choice for screening populations of normal subjects for adequacy of B-cell function (SMALL et al. 1985). In diabetic subjects, however, there is an early and specific loss of B-cell responsiveness to glucose, so that the glucagon test may provide valuable information in this population (HOEKSTRA et al. 1982; GANDA et al. 1984; HSIEH et al. 1987). In cystic fibrosis patients with normal, impaired and diabetic glucose tolerance, the 6-min post-glucagon C-peptide concentration was positively correlated with the initial insulin response to oral glucose (LANNG et al. 1993). When compared to the glucagon test, the OGTT has the disadvantage of a longer duration and lower reproducibility, essentially due to the variability of gastric emptying (GANDA et al. 1978).

II. Intravenous Glucose Tolerance Test

When compared to the OGTT, the intravenous glucose tolerance test (IVGTT) has the advantage of being a more reproducible stimulus and of inducing an earlier increase in insulin secretion (GANDA et al. 1978). The reproducibility of C-peptide acute responses to intravenous glucose and glucagon was found to be similar in healthy subjects (GOTTSÄTER et al. 1992). Only small differences in insulin/C-peptide responses to intravenous glucose and glucagon have been reported in obese and non-obese control and NIDDM subjects (JAYYAB et al. 1982) although some NIDDM subjects may have totally abolished early insulin/C-peptide secretory responses to intravenous glucose (AHREN et al. 1981). B-cell responses to intravenous glucose and glucagon were also found to be similar in patients with chronic pancreatitis and in nondiabetic twins of IDDM patients (HEATON et al. 1989). However, the situation may be different in "early" IDDM patients as the response to glucagon has been shown to be preserved during a longer time period than that to intravenous glucose (MIREL et al. 1980, GANDA et al. 1984; MCCULLOCH and PALMER 1991).

III. Meal

Almost similar results of insulin/C-peptide plasma levels have been reported in various groups of diabetic and nondiabetic subjects after stimulation with glucagon or with a standardized breakfast (FABER and BINDER 1977; ESCOBAR-JIMENEZ et al. 1990; RÖNNEMAA 1986; GJESSING et al. 1988a). However, in NIDDM subjects, the incremental C-peptide AUC after the meal correlated only slightly with the incremental C-peptide AUC after glucagon administration (GJESSING et al. 1988a). Nevertheless, equal discrimination between insulin-requiring and non-insulin-requiring NIDDM patients was found by glucagon-stimulated and 2-h post-breakfast C-peptide concentrations (KOSKINEN et al. 1988). Therefore, the standard breakfast test may be considered as an alternative, useful and practical approach to the study of residual B-cell function (ESCOBAR-JIMENEZ et al. 1990).

IV. Other Tests

In a few studies, the insulin/C-peptide response to glucagon has been compared to that to other stimuli of the B cell, such as arginine (SEINO et al. 1975; NUMATA et al. 1993), terbutaline (AHREN et al. 1981) or tolbutamide (MIREL et al. 1980), in nondiabetic and diabetic subjects. A positive, although slight, correlation of the glucagon-induced stimulation of insulin secretion with 24-h C-peptide urinary excretion has been reported in NIDDM subjects (GJESSING et al. 1987b, 1988b; 1989a; MATSUDA et al. 1985).

D. Clinical Applications

Before the development of genetic, immunological and molecular biological markers, C-peptide response to glucagon was an easy, although not completely specific, tool for separating IDDM from NIDDM subjects (POULSEN et al. 1985; GJESSING et al. 1989a; VAHLKAMP et al. 1990). However, other applications of the glucagon test have been developed during the last few years both in diabetic populations (SCHEEN et al. 1996a) and in subjects with other pathological states potentially affecting insulin secretion.

I. Insulin-Dependent Diabetes Mellitus

Most IDDM patients no longer have residual insulin secretion. They are the so-called "C-peptide negative diabetic patients", i.e. individuals with no detectable C-peptide levels even after various stimuli of the B cell, one of which is glucagon (HENDRIKSEN et al. 1977). Nevertheless, in the IDDM population, assessment of residual insulin secretion has become more and more important in recent years, not only to simply discriminate between C-peptide-positive and -negative individuals.

The clinical diagnosis of IDDM is preceded by a long silent phase characterized by an ongoing immunological destruction of the islet B cells (MCCULLOCH and PALMER 1991). At this early stage of the disease, blunted acute insulin response to intravenous glucose seems to be a more discriminant marker than that to glucagon, because of the early and specific desensitization of the B cell to glucose (MCCULLOCH and PALMER 1991).

On clinical diagnosis, a significant residual insulin secretion is still present and may persist during the first months or even the first few years following initiation of exogenous insulin therapy. Because the acute insulin response to glucose has already been abolished, the glucagon test may offer some advantage at this stage (GANDA et al. 1984) and has become the most popular test for evaluating B-cell function during this period (MONTANA et al. 1991; MARTIN et al. 1992). Such an approach has become more important during recent years since the first attempts were made to stop the immunological destruction of the islet B cell and obtain or prolong remission of the disease, using nicotinamide (VAGUE et al. 1989) or cyclosporine (DUPRE et al. 1991).

Finally, several therapeutic interventions have been tested during recent years for the restoration of a sufficient quantity of B cells in some IDDM subjects. It is therefore crucial to have an easy and reproducible test, such as the glucagon test, to follow the functional capacity of the new B cells, such as after pancreas islet allograft or B-cell transplantation (OSEI et al. 1990; NYBERG et al. 1990).

II. Non-Insulin-Dependent Diabetes Mellitus

One very early study comparing subjects classified on the basis of obesity or diabetes demonstrated the exaggerating influence of obesity and the limiting effect of diabetes upon the insulin response to glucagon (CROCKFORD et al. 1969). In contrast to IDDM, most NIDDM patients have residual insulin secretion as demonstrated by a positive insulin/C-peptide response to glucagon (HENDRIKSEN et al. 1977; POULSEN et al. 1985). However, insulin deficiency is not uncommon in middle-aged insulin-treated diabetic patients whose diabetes has been diagnosed after the age of 30 years (LAAKSO et al. 1989).

The use of the glucagon test in NIDDM subjects has been mainly recommended to assess the degree of defective insulin secretion and to help the physician in choosing the best treatment for a given patient (MADSBAD et al. 1981; HOEKSTRA et al. 1982; KOSKINEN et al. 1986; VAHLKAMP et al. 1990). In particular, insulin requirement in NIDDM subjects has been shown to be related to simple tests of islet B-cell function, such as a <0.600–0.700 pmol/ml plasma C-peptide value after glucagon administration (MADSBAD et al. 1981; GJESSING et al. 1988b; HOTHER-NIELSEN et al. 1988). Defective C-peptide response to glucagon has also shown to be associated with cellular and humoral autoimmunity markers in presumably NIDDM patients during the first 3 years after clinical diagnosis (GOTTSÄTER et al. 1993) or after secondary failure to oral agents (ZAVALA et al. 1992).

However, the usefulness of measuring insulin secretion in NIDDM has also been doubted (SCHEEN et al. 1996b). Indeed, post-glucagon C-peptide levels did not predict which patients could be changed successfully from insulin to tablets (KYLLASTINEN et al. 1986; VIIKARI et al. 1987). In addition, similar basal and post-glucagon C-peptide levels were observed in severely hyperglycemic newly diagnosed and secondary failure type II diabetic patients (WOLFFENBUTTEL et al. 1992). Such findings may indicate that these parameters do not provide a basis for the best choice of blood-glucose-lowering treatment.

III. Hypoglycemia

The application of glucagon to the diagnosis of hypoglycemia is based on its insulinotropic properties and the characteristic, excessive and prolonged, insulin response often recorded in patients with insulinomas. In contrast, patients with hypoglycemia caused by extrapancreatic neoplasms or of other endocrine origin generally experience an impaired plasma insulin response while those with the various types of reactive hypoglycemia usually behave normally (review in MARKS 1983).

The intravenous glucagon test proved equally, or more, reliable than most other provocative tests, i.e. tolbutamide, L-leucine and glucose tolerance tests for the diagnosis of insulinoma (MARKS 1983). However, its use for this purpose has lessened as tests of islet-cell function based upon suppression rather than stimulation of B-cell activity have become available. The classical prolonged fast and the measurement of plasma C-peptide levels during spontaneous or insulin-induced hypoglycemia are indeed safer and more reliable tests. Therefore, the use of the glucagon test as a diagnostic procedure in hypoglycemia, which was once of great importance (MARKS 1983), has considerably diminished during recent years (MARKS 1989; LEFÈBVRE 1993; LEFÈBVRE and SCHEEN 1995).

IV. Other Diseases

During the last decade, the glucagon test has been performed to evaluate insulin secretion in numerous pathological states. At present, it is regularly used to assess B-cell function in various pancreatic diseases such as chronic pancreatitis (CAVALLINI et al. 1992), cystic fibrosis (LANNG et al. 1993) and pancreatic adenocarcinoma (BASSO et al. 1994). The glucagon test is also currently used in order to better evaluate insulin secretion in subjects with impaired glucose tolerance such as individuals with polycystic ovary syndrome (WEBER et al. 1993), severe obesity (NUMATA et al. 1993) and anorexia nervosa (KOBAYASHI et al. 1992).

E. Conclusions

The glucagon test has become a classical test for the evaluation of insulin secretion in vivo. The procedure originally described, almost 20 years ago, remains the most widely used and consists of the intravenous injection of 1 mg glucagon to the subject in the fasting state with measurement of plasma C-peptide (or less frequently insulin) levels 6 min later.

Although the insulin secretory response may be influenced by the prevailing plasma glucose level, the glucagon test is widely used in clinical practice, essentially in diabetology. In type I diabetic patients, it enables the progressive decline of insulin secretion to be followed and the investigation of the possible protective effects of various interventions to delay the progression of the disease or, in some cases, to assess the endogenous insulin secretion after pancreatic or islet transplant. In type II diabetic patients, the glucagon test is mostly used to evaluate the residual insulin secretion in order to help the physician in selecting the most appropriate pharmacological treatment for a given individual.

Besides these applications in the field of diabetes, the glucagon test may be a useful alternative as a stimulatory test in several diseases causing defective insulin secretion or in pathological states characterized by significant insulin resistance and requiring appropriate compensatory insulin secretion.

References

Ahren B, Nobin A, Schersten B (1987) Insulin and C-peptide secretory responses to glucagon in man: studies on the dose-response relationships. Acta Med Scand 221:185–190

Ahren B, Schersten B, Agardh CD, Lundquist I (1981) Immunoreactive insulin and C-peptide responses to various insulin secretory stimuli in subjects with type 2 diabetes and in control subjects during continuous glucose monitoring. Acta Med Scand 210:337–348

Arnold-Larsen S, Madsbad S, Kühl C (1987) Reproducibility of the glucagon test. Diabet Med 4:288–303

Basso D, Plebani M, Fogar P, Del Favero G, Briani G, Meggiato T, Ferrara C, D'Angeli F, Burlina A (1994) Beta-cell function in pancreatic adenocarcinoma. Pancreas 9:332–335

Castillo MJ, Scheen AJ, Lefèbvre PJ (1995) Modified glucagon test allowing simultaneous estimation of insulin secretion and insulin sensitivity: application to obesity, insulin-dependent diabetes mellitus, and noninsulin-dependent diabetes mellitus. J Clin Endocrinol Metab 80:393–399

Castillo MJ, Scheen AJ, Paolisso G, Lefèbvre PJ (1996) Exhaustion of blood glucose response and enhancement of insulin response after repeated glucagon injections in type 2 diabetes: potentiation by progressive hyperglycemia (submitted)

Cavallini G, Bovo P, Zamboni M, Bosello O, Filippini M, Riela A, Brocco G, Rossi L, Pelle C, Chiavenato A, Scuro LA (1992) Exocrine and endocrine functional reserve in the course of chronic pancreatitis as studied by maximal stimulation tests. Dig Dis Sci 37:93–96

Crockford PM, Hazzard WR, Williams RH (1969) Insulin response to glucagon. The opposing effects of diabetes and obesity. Diabetes 18:216–224

Dupre J, Jenner MR, Mahon JL, Purdon C, Rodger NW, Stiller CR (1991) Endocrine-metabolic function in remission-phase IDDM during administration of cyclosporine. Diabetes 40:598–604

Escobar-Jimenez F, Herrera Pombo JL, Gomez-Villalba R, Nunez del Carril J, Aguilar M, Rovira A (1990) Standard breakfast test: an alternative to glucagon testing for C-peptide reserve? Horm Metab Res 22:339–341

Faber OK, Binder C (1977) C-peptide response to glucagon. A test for the residual β-cell function in diabetes mellitus. Diabetes 26:605–610

Ganda OP, Day JL, Soeldner JS, Connon JJ, Gleason RE (1978) Reproducibility and comparative analysis of repeated intravenous and oral glucose tolerance tests. Diabetes 27:715–725

Ganda OP, Srikanta S, Brink SJ, Morris MA, Gleason RE, Soeldner JS, Eisenbarth GS (1984) Differential sensitivity to beta cell secretagogues in "early" type 1 diabetes mellitus. Diabetes 33:516–521

Gjessing HJ, Damsgaard EM, Matzen LE, Froland A, Faber OK (1987a) Reproducibility of β-cell function estimates in non-insulin-dependent diabetes mellitus. Diabetes Care 10:558–562

Gjessing HJ, Matzen LE, Froland A, Faber OK (1987b) Correlations between fasting plasma C-peptide, glucagon-stimulated plasma C-peptide, and urinary C-peptide in insulin-treated diabetics. Diabetes Care 10:487–490

Gjessing HJ, Damsgaard EM, Matzen LE, Faber OK, Froland A (1988a) The beta-cell response to glucagon and mixed meal stimulation in non-insulin-dependent diabetes. Scand J Clin Lab Invest 48:771–777

Gjessing HJ, Matzen LE, Pedersen PC, Faber OK, Froland A (1988b) Insulin requirement in non-insulin-dependent diabetes mellitus: relation to simple tests of islet B-cell function and insulin sensitivity. Diabet Med 5:328–332

Gjessing HJ, Matzen LE, Faber OK, Froland A (1989a) Fasting plasma C-peptide, glucagon stimulated plasma C-peptide, and urinary C-peptide in relation to clinical type of diabetes. Diabetologia 32:305–311

Gjessing HJ, Reinholdt B, Pedersen O (1989b) The plasma C-peptide and insulin responses to stimulation with intravenous glucagon and a mixed meal in well controlled type 2 (non-insulin-dependent) diabetes mellitus: dependency on acutely established hyperglycaemia. Diabetologia 32:858–863

Gjessing HJ, Reinholdt B, Faber OK, Pedersen O (1991) The effect of acute hyperglycemia on the plasma C-peptide response to intravenous glucagon or to a mixed meal in insulin-dependent diabetes mellitus. Acta Endocrinol 124:556–562

Gjessing HJ, Reinhold B, Pedersen O (1992) The effect of chronic hyperglycaemia on the islet B-cell responsiveness in newly diagnosed type 2 diabetes. Diabet Med 9:601–604

Gottsäter A, Landin-Olsson M, Fernlund P, Gullberg B, Lernmark A (1992) Pancreatic beta-cell function evaluated by intravenous glucose and glucagon stimulation. A comparison between insulin and C-peptide to measure insulin secretion. Scand J Clin Lab Invest 52:631–639

Gottsäter A, Landin-Olsson M, Fernlund P, Lernmark A, Sundkvist G (1993) Beta-cell function in relation to islet cell antibodies during the first 3 yr after clinical diagnosis in type II diabetic patients. Diabetes Care 16:902–910

Heaton DA, Lazarus NR, Pyke DA, Leslie RDG (1989) B-cell responses to intravenous glucose and glucagon in non-diabetic twins of patients with Type I (insulin-dependent) diabetes mellitus. Diabetologia 32:814–817

Hendriksen C, Faber OK, Drejer J, Binder C (1977) Prevalence of residual β-cell function in insulin treated diabetics evaluated by the plasma C-peptide response to intravenous glucagon. Diabetologia 13:616–619

Hoekstra JB, Van Rijn HJM, Thijssen JHT, Erkelens DW (1982) C-peptide reactivity as a measure of insulin dependency in obese diabetic patients treated with insulin. Diabetes Care 5:585–591

Hovorka R, Jones RH (1994) How to measure insulin secretion. Diabetes Metab Rev 10:91–117

Hother-Nielsen O, Faber O, Schwartz-Sorensen N, Beck-Nielsen H (1988) Classification of newly diagnosed diabetic patients as insulin-requiring or non-insulin-requiring based on clinical and biochemical variables. Diabetes Care 11:531–537

Hsieh SD, Iwamoto Y, Matsuda A, Kuzuya T (1987) Pancreatic β-cell secretion after oral glucose and intravenous glucagon: different responses to dietary control of plasma glucose in newly diagnosed patients with NIDD. Metabolism 36:384–387

Jayyab AK, Heding LG, Czyzyk A, Malczewski B, Krolewski J (1982) Serum C-peptide and IRI levels after administration of glucagon and glucose in non-insulin-dependent diabetics. Horm Metab Res 14:112–116

Kobayashi N, Tamai H, Takii M, Matsubayashi S, Nakagawa T (1992) Pancreatic B-cell functioning after intravenous glucagon administration in anorexia nervosa. Acta Psychiatr Scand 85:6–10

Kolendorf K, Thorsteinsson B, Billebolle P, Poulsen S (1982) Correlation between the routine 6-min C-peptide values and AUC after I.V. glucagon injection in NIDDM patients. Horm Metab Res 14:675–676

Koskinen P, Viikari J, Irjala K, Kaihola HL, Seppälä P (1986) C-peptide determination in the choice of treatment in diabetes mellitus. Scand J Clin Lab Invest 46:655–663

Koskinen PJ, Viikari JSA, Irjala KMA (1988) Glucagon-stimulated and post-prandial plasma C-peptide values as measures of insulin secretory capacity. Diabetes Care 11:318–322

Kyllastinen M, Elfving S (1986) Serum C-peptide concentrations and their value in evaluating the usefulness of insulin therapy in elderly diabetics. Gerontology 32:317–326

Laakso M, Rönnemaa T, Sarlund H, Pyörälä K, Kallio V (1989) Factors associated with fasting and postglucagon plasma C-peptide levels in middle-aged insulin-treated diabetic patients. Diabetes Care 12:83–88

Lanng S, Thorsteinsson B, Roder ME, Orskov C, Holst JJ, Nerup J, Koch C (1993) Pancreas and gut hormone responses to oral glucose and intravenous glucagon in cystic fibrosis patients with normal, impaired, and diabetic glucose tolerance. Acta Endocrinol 128:207–214

Lefèbvre PJ (1993) Stratégie d'exploration des hypoglycémies de l'adulte. Ann Endocrinol (Paris) 54:409–412

Lefèbvre PJ, Scheen AJ (1995) Hypoglycemia. In: Rifkin H, Porte D Jr (eds) Ellenberg and Rifkin's diabetes mellitus: theory and practice, 5th edn. Elsevier, New York

Madsbad S, Krarup T, McNair P, Christiansen C, Faber OK, Transbol I, Binder C (1981) Practical clinical value of the C-peptide response to glucagon stimulation in the choice of treatment in diabetes mellitus. Acta Med Scand 210:153–156

Madsbad S, Sauerbrey N, Moller-Jensen B, Krarup T, Kühl C (1987) Outcome of the glucagon test depends upon the prevailing blood glucose concentration in type 1 (insulin-dependent) diabetic patients. Acta Med Scand 222:71–74

Marks V (1983) Glucagon in the diagnosis and treatment of hypoglycaemia. In: Lefèbvre PJ (ed) Glucagon II. Springer Berlin Heidelberg New York Tokyo, pp 645–666 (Handbook of experimental pharmacology, vol 66/11)

Marks V (1989) Diagnosis and differential diagnosis of hypoglycemia. Mayo Clin Proc 64:1558–1561

Martin S, Pawlowski B, Greulich B, Ziegler AG, Mandrup-Poulsen T (1992) Natural course of remission in IDDM during 1st yr after diagnosis. Diabetes Care 15:66–74

Matsuda A, Kamata I, Iwamoto Y, Sakamoto Y, Kuzuya T (1985) A comparison of serum C-peptide responses to intravenous glucagon and urine C-peptide as indexes of insulin dependence. Diabetes Res Clin Pract 1:161–167

McCulloch DK, Palmer JP (1991) The appropriate use of B-cell function testing in the preclinical period of type 1 diabetes. Diabet Med 8:800–804

Miki H, Matsuyama T, Fujii S, Komatsu R, Nishioeda Y, Omae T (1992) Glucagon-glucose (GG) test for the estimation of the insulin reserve in diabetes. Diabetes Res Clin Pract 18:99–105

Mirel RD, Ginsberg-Fellner F, Horwitz DL, Rayfield EJ (1980) C-peptide reserve in insulin-dependent diabetics: comparative responses to glucose, glucagon and tolbutamide. Diabetologia 19:183–188

Montana E, Fernandez-Castaner M, Rosel P, Gomez J, Soler J (1991) Age, sex and ICA influence on beta-cell secretion during the first year after diagnosis of type 1 diabetes mellitus. Diabete Metab 17:460–468

Nosari I, Lepore G, Maglio ML, Cortinovis F, Pagani G (1992) The effect of various blood glucose levels on post-glucagon C-peptide secretion in type 2 (non insulin-dependent) diabetes. J Endocrinol Invest 15:143–146

Numata K, Tanaka K, Saito M, Shishido T, Inoue S (1993) Very low calorie diet-induced weight loss reverses exaggerated insulin secretion in response to glucose, arginine and glucagon in obesity. Int J Obes Relat Metab Disord 17:103–108

Nyberg G, von Schenck H, Norden G, Hedman L, Frisk B (1990) Glucagon-stimulated serum C-peptide levels in the early period following pancreas transplantation. Transplant Proc 22:647–648

Oakley NW, Harrigan P, Kissebah AH, Kissin EA, Adams PW (1972) Factors affecting insulin response to glucagon in man. Metabolism 21:1001–1007

Osei K, Henry ML, O'Dorisio TM, Tesi RJ, Sommer BG, Ferguson RM (1990) Physiological and pharmacological stimulation of pancreatic islet hormone secretion in type 1 diabetic pancreas allograft recipients. Diabetes 39:1235–1242

Poulsen S, Billesbolle P, Kollendorf K, Thorsteinsson B (1985) The C-peptide response to glucagon injection in IDDM and NIDDM patients. Horm Metab Res 17:39–40

Rönnemaa T (1986) Practical aspects in performing the glucagon test in the measurement of C-peptide secretion in diabetic patients. Scand J Clin Invest 46:345–349

Samols E (1983) Glucagon and insulin secretion. In: Lefèbvre PJ (ed) Glucagon I Springer, Berlin Heidelberg New York, pp 485–518 (Handbook of experimental pharmacology, vol 66/11)

Samols E, Marri G, Marks V (1965) Promotion of insulin secretion by glucagon. Lancet II:415–416

Samols E, Marri G, Marks V (1966) Interrelationship of glucagon, insulin and glucose. The insulinogenic effect of glucagon. Diabetes 15:855–866

Sarlund H, Siitonen O, Laakso M, Pyörälä K (1987a) Repeatability of C-peptide response in glucagon stimulation test. Acta Endocrinol 114:515–518

Sarlund H, Laakso M, Pyörälä K, Penttilä (1987b) Fasting, postprandial and postprandial plus glucagon-stimulated plasma C-peptide levels in non-insulin-dependent diabetics and in control subjects. Acta Med Scand 221:377–383

Scheen AJ, Castillo MJ, Lefèbvre PJ (1996a) Assessment of residual insulin secretion in diabetic patients using the intravenous glucagon stimulatory test. Methodological aspects and clinical applications. Diab Metab 22 (in press)

Scheen AJ, Castillo MJ, Lefèbvre PJ (1996b) Glucagon-induced plasma C-peptide response in diabetic patients. Influence of body weight and relationship to insulin requirement. Diab Metab 22 (in press)

Scheen AJ, Paolisso G, Juchmes J, Lefèbvre PJ (1989) Insulin secretion rate after intravenous glucagon injection: kinetics reproducibility and potentiation by increasing initial blood glucose levels. Diabetologia 32:538 A

Scheen AJ, Paquot N, Letiexhe MR, Castillo MJ, Lefèbvre PJ (1995) Comment mesurer l'insulinosécrétion en pratique? Diabete Metab 21:458–464

Seino Y, Kurahachi H, Goto Y, Taminato T, Ikeda M, Imura H (1975) Comparative insulinogenic effects of glucose, arginine and glucagon in patients with diabetes mellitus, endocrine disorders and liver diseases. Acta Diabetol Lat 12:89–99

Small M, Cohen HN, Beastall GH, MacCuish AC (1985) Comparison of oral glucose loading and intravenous glucagon injection as stimuli to C-peptide secretion in normal men. Diabet Med 2:181–183

Snorgaard O, Hasselstrom K, Lumholtz IB, Thorsteinsson B, Siersbaek-Nielsen K (1988) Insulin/C-peptide response to intravenous glucagon. A dose-response study in normal and non-insulin-dependent diabetic subjects. Acta Endocrinol 117:109–115

Vague P, Picq R, Bernal M, Lassmann-Vague V, Vialettes B (1989) Effect of nicotinamide treatment on the residual insulin secretion in type 1 (insulin-dependent) diabetic patients. Diabetologia 32:316–321

Vahlkamp T, Lutjens A, Nauta EH (1990) The glucagon-stimulated C-peptide test: an aid in classification of patients with diabetes mellitus. Neth J Med 36:196–199

Vialettes B, Zevacq-Mattei C, Thirion X, Lassmann-Vague V, Pieron H, Mercier P, Vague P (1993) Acute insulin response to glucose and glucagon in subjects at risk of developing type 1 diabetes. Diabetes Care 16:973–977

Viikari J, Ronnemaa T, Koskinen P (1987) Glucagon C-peptide test as a measure of insulin requirement in type 2 diabetes: evaluation of stopping insulin therapy in eleven patients. Ann Clin Res 19:178–182

Weber RF, Pache TD, Jacobs ML, Docter R, Loriaux DL, Fauser BC (1993) The relation between clinical manifestations of polycystic ovary syndrome and beta-cell function. Clin Endocrinol (OXF) 38:295–300

Wolffenbuttel BH, Menheere PP, Nijst L, Rondas-Colbers GJ, Sels JP (1992) Glucagon-stimulated insulin secretion in patients with type 2 diabetes mellitus: support for the concept of glucose toxicity. Neth J Med 40:277–282

Zavala AV, Fabiano de Bruno LE, Cardoso AI, Mota AH, Capucchio M, Fainboim L, Basabe JC (1992) Cellular and humoral autoimmunity markers in type 2 (non-insulin-dependent) diabetic patients with secondary drug failure. Diabetologia 35:1159–1164

CHAPTER 14
Glucagon and the Control of Appetite

N. Geary

A. Introduction

Recent progress in behavioral neuroscience indicates that peripheral neural and endocrine feedback signals are crucial controls for the initiation, maintenance, and termination of meals, and, therefore, for total food intake and the maintenance of body weight (Campfield and Smith 1990; Scharrer and Langhans 1988; Smith and Gibbs 1992; Blundell 1991). I review here evidence that glucagon is one such signal. Results from tests of administration of glucagon or glucagon antagonists suggest that, under many conditions, glucagon released from the pancreas during meals acts in the liver to initiate a neural signal that is conveyed by vagal afferents to the brain, where it contributes to the termination of the meal.

B. Prandial Glucagon Secretion

The available data indicate that meals stimulate an immediate, transient increase in pancreatic glucagon secretion in rats (DeJong et al. 1977; Langhans et al. 1984; Berthoud and Jeanrenaud 1982) and humans (Muller et al. 1970; Day et al. 1978; Denker et al. 1975; Holst et al. 1983). DeJong et al. (1977) reported that in 2h food-deprived rats fed a 54% carbohydrate diet, peripheral plasma glucagon increased from 147 ± 24 pg/ml to 230 ± 15 pg/ml during meals. The increase began within the 1st min of feeding, and glucagon returned to the premeal values by about 10 min after meal end. The results were quite similar when the rats were fed a carbohydrate-free, 43% fat diet or when they were refed after 24h food deprivation. Rapid and transient prandial glucagon increases were demonstrated in the peripheral plasma of both normal weight men and women and obese women during each of five daily mixed-nutrient meals (except for breakfast in the obese subjects) (Holst et al. 1983). The increases were quite small (~20 pg/ml), and plasma glucagon returned to baseline within 30–60 min.

Because of the efficiency of hepatic glucagon extraction, measures of glucagon responses in the peripheral circulation underestimate the actual magnitude of prandial pancreatic glucagon secretion. Indeed, the efficiency of hepatic glucagon extraction appears greatest during rapid increases in glucagon secretion, such as apparently accompany meals. For example, isolated rat liver preparations have been reported to extract 60%–80% of the glucagon delivered during the first few minutes of infusions, with the efficiency of extraction increased by brief or pulsatile infusion patterns, again such as probably mimic the pattern of pancreatic glucagon secretion (HILDEBRANDT et al. 1991a,b, 1992; WEIGLE 1987). Unfortunately, there are few reports of prandial glucagon levels in the hepatic portal vein. The most complete study compared prandial glucagon levels in rats adapted to high-carbohydrate (77% by weight), high-fat (40%), or high-protein (88%) diets (LANGHANS et al. 1984). Rats were refed at dark onset after diurnal food deprivation. In high-carbohydrate-fed rats, hepatic portal vein plasma glucagon increased from ~140 to 290 pg/ml during meals; in high-fat-fed rats, from 120 to 230 pg/ml; and in high-protein-fed rats, from 390 to 700 pg/ml. Note that these portal vein increases occurred in a situation in which there was little or no evidence of prandial glucagon secretion in the peripheral circulation (hepatic vein plasma glucagon increases were <30 pg/ml in each group). Thus, the liver extracted nearly the entire increment in glucagon secretion during the meal. DENKER et al. (1975) reported similar results in two women and a man who were tested about 1 year after surgical treatment for colon carcinoma. Mixed meals increased portal vein glucagon concentration within 5–15 min in all three patients (by 95, 80, and 14 pg/ml, respectively), whereas peripheral glucagon concentration decreased in two of the three (changes were +56, −14, and −2 pg/ml, respectively). Thus, tests of prandial glucagonemia in the extrahepatic circulation may significantly underestimate or entirely miss the actual effect of feeding on glucagon secretion. The use of peripheral samples, often taken at times that may miss transient changes, likely explains many of the reported failures to see prandial increases in plasma glucagon (TABORSKY 1989). Rather, glucagon secretion probably generally accompanies all but pure carbohydrate meals. This phenomenon establishes the theoretical plausibility of prandial glucagon release as a feedback signal for meal termination.

Prandial glucagon secretion is probably a cephalic phase reflex because it appears during the 1st min of the meal and occurs during sham feeding (DEJONG et al. 1977; NILSSON and UVNAS-WALLENSTEIN 1977). There are numerous neural controls of the pancreatic A cells that may mediate this phenomenon (PALMER and PORTE 1983). For example, stimulation of the ventromedial hypothalamus, the dorsal motor nucleus of the vagus, or the gastric or hepatic branches of the abdominal vagus nerve all elicit glucagon secretion (BERTHOUD et al. 1990), although none of these mechanisms is known to be activated by feeding. Several hormones that are released during meals and also stimulate glucagon release may also play a role in prandial glucagon secretion.

The metabolic consequences of prandial glucagon secretion have not been extensively investigated. It is associated with hepatic glycogenolysis in rats even during low-protein, high-carbohydrate meals (LANGHANS et al. 1982a, 1984). As discussed below, whether this hepatic glucose output contributes to the satiating effect of glucagon is not clear.

C. Glucagon Administration and Food Intake

I. Animal Studies

Intravenous glucagon infusion during feeding elicits a rapid, dose-related decrease in food intake (Table 1). Behavioral observations and tests of prandial glucagon infusions during spontaneous meals demonstrate that glucagon specifically decreases meal size, with no effects on subsequent intermeal interval (LE SAUTER and GEARY 1991) (Fig. 1). This action of glucagon is typically transient, with total cumulative food intake returning to control levels within a few hours.

Subcutaneous, intramuscular, and intraperitoneal glucagon have also been reported to decrease food intake (see GEARY 1990; LE SAUTER and GEARY 1993 for reviews). Glucagon administered via these routes, however, has produced different results than intravenous glucagon: (1) the threshold dose for inhibition of feeding is 10- to 100-fold lower for hepatic portal glucagon infusion than intraperitoneal injection; (2) the inhibitory effect of intravenous glucagon is typically dose related, whereas the feeding effects of glucagon administered by other routes seldom have been; (3) the inhibitory action of intravenous glucagon is limited to the meal accompanying the infusion, whereas glucagon administered by other routes may have delayed effects on subsequent meals, sometimes including stimulatory effects; (4) depending on the length of pretest food deprivation and on circadian phase, intraperitoneal glucagon may fail to inhibit feeding, whereas intravenous glucagon's effect does not appear to depend on the context of the meal. The reason for these dramatic contrasts is unclear. In light of them and of the fact that the hepatic portal vein is the physiological route for pancreatic glucagon, studies of the feeding effects of intravenous administration of glucagon appear of the greatest physiological relevance.

It is not presently possible to say whether the inhibition of feeding by intravenous glucagon is more than a purely pharmacological effect. Hepatic-portal vein glucagon levels produced by glucagon treatments that inhibited feeding in rats have been directly compared to prandial endogenous glucagon levels only once, after intraperitoneal injections of $480\,\mu g/kg$ glucagon (LANGHANS et al. 1987). Hepatic-portal vein plasma glucagon levels 15 min into the meal were $447 \pm 112\,pg/ml$ after control injections and $13 \pm 3\,ng/ml$ after glucagon. Thus, the effect of this very large glucagon dose was clearly supraphysiological. This question can hardly be considered closed, however, since, as reviewed above, glucagon doses up to 100-fold less are sufficient to inhibit feeding when delivered via the hepatic portal vein.

Several further points are relevant to the question of the physiological status of exogenous glucagon in satiety: (1) The infusion parameters that minimize the threshold glucagon dose for inhibition of feeding have not been established. Rate of change of glucagon concentration rather than load or

Table 1. Effects of intravenous glucagon on appetite

Study	Route	Dose	Effect
A. Rat			
Martin and Novin (1977)	HP	25–65 µg/kg (bolus)	DR decrease in DI feeding. MED 35 µg/kg
Martin et al. (1978)	HP	55 µg/kg (bolus)	Decrease in DI feeding; blocked by vagotomy
Weick and Ritter (1986a)	HP	0.33–33 µg/kg/min	DR decrease in DI feeding within 15 min; MED 0.33 µg/kg/min
Weick and Ritter (1986b)	HP	1.0–10 µg/kg/min	Decrease in DI feeding during 10 µg/kg/min
Strubbe et al. (1989)	HP, IC	1.25–5 µg/kg/min	DR decreases in DI feeding within 20 min; MED 1.25 µg/kg/min
LeSauter and Geary (1991)	a) HP	1.7–13.6 µg/min (8–12 min)	DR decreases in spontaneous meal size, MED 20 µg
	b) HP	0.85–7 µg/min (2 min)	DR decreases in spontaneous meal size, MED 1.7 µg/min
Geary et al. (1993)	HP, VC	0.85–7 µg/min (2 min)	Decrease in spontaneous meal size, only after HP infusion; MED 0.85 µg/min
Geary (1995)	HP	7 µg/min (2 min)	Similar decrease in meal size after one 2-min infusion or two 1-min pulses
B. Humans			
Stunkard et al. (1955)	a) IV	125–2000 µg/min (1 min)	a), b) Decrease in sensation of hunger
	b) IV	26 µg/min (102 min)	

Table 1 (*Cont.*)

Study	Route	Dose	Effect
GEARY et al. (1992)	IV	3 ng/kg/min (10 min)	Decrease in test meal size; blocked by simultaneous 2 ng/kg/min CCK-8
C. Rabbit			
VANDERWEELE et al. (1979)	HP	4 µg/min (3 min)	Decrease in 60-min FI; blocked by vagotomy
D. Dog			
LEVINE et al. (1984)	IV	250–500 µg/kg (bolus)	Decrease scheduled meal size; blocked by vagotomy
KALOGERIS et al. (1991)	a) IV	0.35–1.4 µg/kg/min (60 min)	DR decrease in scheduled meal size; MED 0.7 µg/kg/min
	b) IV	2–290 ng/kg/min (60 min)	No effect alone; blocked effect of 1–8 ng/kg/min CCK-8
E. Sheep			
DEETZ and WANGSNESS (1981)	IV	9 ng/kg (bolus)	Infusion at each spontaneous meal, decrease in total FI
F. Chicken			
SMITH and BRIGHT-TAYLOR (1974)	IC	0.2–2 µg/kg (bolus)	Decrease in FI
HOWES and FORBES (1986)	a) HP	10–1000 µg (bolus)	Decrease in FI, MED 100 µg
	b) HP, IV	1000 µg (bolus)	Similar decreases in FI with each route
	c) HP	5–50 µg (bolus)	Decrease in FI, MED 5 µg

IV, intravenous; *HP*, hepatic portal; *IC*, intracardiac, *VC*, vena caval; *FI*, food intake; *DR*, dose-related; *DI*, deprivation induced; *MED*, minimum effective dose.

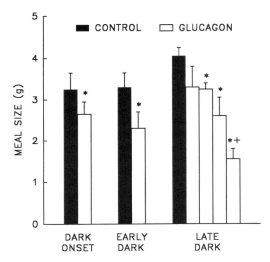

Fig. 1. Hepatic-portal glucagon infusion reduces spontaneous meal size in ad lib fed rats. Glucagon was infused for the first 2 min of either the first spontaneous meal after dark onset (*left*), the first spontaneous meal beginning at least 3 h after dark onset (*middle*), or the first spontaneous meal beginning at least 9.5 h after dark onset (*right*). *Filled bars* are meal sizes during control infusions, mean ± SEM; *open bars* are glucagon infusion data. Dark onset and early dark glucagon doses were 6.8 µg/min; late dark doses were 0.85, 1.7, 3.4, and 6.8 µg/min. *Different from control, $P < 0.05$; +different from effect of other glucagon doses, $P < 0.05$. (Data from Le Sauter and Geary 1991, by permission)

peak concentration may be crucial. The reduction in spontaneous meal size by 2-min, meal-onset glucagon infusions was a graded function of the concentration of glucagon infused (Le Sauter and Geary 1991). The total amount of glucagon infused appeared irrelevant because meal size was similarly reduced whether the infusion of one glucagon concentration was done for only the first 2 min of meals, the first 4 min of meals, or throughout entire meals, which lasted ~6–12 min. Whether this would also be true of even briefer infusions has not been tested. (2) Glucagon infusions that match the pattern of endogenous glucagon release, which may oscillate with periods of only 0.5–4 min (Hildebrandt et al. 1992; Weigle 1987), may inhibit feeding more effectively because they increase hepatic glucagon extraction. In the isolated perfused rat liver, glucagon extraction during continuous infusion fell from 63% in the 1st min to only 8% in the 2nd and 4% in the 3rd, with the result that only about 5% of the glucagon infused over 20 min was extracted, whereas when the same amount of glucagon was infused in an oscillating pattern at 1 cycle/min, about one-third of the total infused was extracted (Hildebrandt et al. 1991a, 1992). The physiological consequences of this difference and whether a similar phenomenon occurs during meals have not been established. A preliminary study failed to provide evidence that exogenous glucagon is more efficient when administered in a pulsatile pattern (Geary 1995), but this possibility needs more investigation. (3) Exogenous glucagon may be relatively ineffective if endogenous glucagon released during meals nearly saturates the mechanism by which glucagon signals satiety. If this is so, the effect of endogenous glucagon would have to be neutralized in order to evaluate the physiological status of infused glucagon. These considerations suggest that whether or not exogenous glucagon inhibits feeding when administered in a manner that matches its physiological release remains to be established.

A variety of results strongly buttress the conclusion that glucagon's inhibitory effect on meal size in animals is a specific satiety effect rather than the result of illness, toxicity, sedation, or any other nonspecific cause (see GEARY 1990; LE SAUTER and GEARY 1993 for reviews). Further, as discussed below, intravenous glucagon infusions that reduced meal size in humans failed to produce physical or subjective side effects (GEARY et al. 1992) and glucagon antagonism increases meal size in rats (LE SAUTER et al. 1991). These two results are also inconsistent with the idea that glucagon is a nonspecific anorectic agent.

II. Human Studies

Although STUNKARD et al. (1955) reported early on that intravenous glucagon reduced hunger in fasted humans, an inhibitory effect of glucagon infusion on human food intake has been only recently demonstrated (GEARY et al. 1992). A portable infusion pump began a 10-min infusion of 3 ng/kg per minute glucagon or saline into an arm vein after normal-weight males finished a soup course and 5 min before they were offered a main course of macaroni and beef in tomato sauce. Twelve volunteers were tested using a double-blind procedure. Glucagon significantly reduced macaroni intake, from 803 g to 672 g (±131 g). There were no physical signs of side effects, and the subjects did not report any adverse symptoms, nor did postmeal psychophysical ratings give any indication of side effects. Rather, the subjects were not aware when glucagon was infused and reported equal degrees of fullness and satiation after meals. Thus, glucagon appeared to reduce meal size by accelerating normal satiety processes.

Based on glucagon's plasma half-life (POLONSKY et al. 1983), this 10-min infusion of 3 ng/kg per minute glucagon probably elevated peripheral plasma glucagon level about 200–400 pg/ml during the meal. This is near enough to prandial glucagon levels reported in humans (see above) to warrant further study of the hypothesis that postprandial satiety may be a physiological function of glucagon in humans.

D. Glucagon Antagonism and Food Intake

The most direct evidence for a physiological role of glucagon in the control of meal size comes from four acute passive immunization tests in rats. LANGHANS

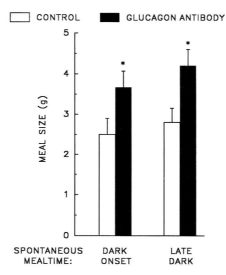

Fig. 2. Hepatic-portal infusion of glucagon antibodies increases spontaneous meal size in ad lib fed rats. Glucagon antibodies were infused for the first 2 min of either the first spontaneous meal after dark onset (*left*) or the first spontaneous meal beginning at least 9.5 h after dark onset (*right*). *Open bars* indicate meal sizes during control infusions, mean ± SEM; *filled bars* indicate meal sizes during glucagon antibody infusions. The antibody dose was sufficient to neutralize 1.5 ng glucagon in vitro. *Different from control, $P < 0.05$. (Data from Le Sauter et al. 1991, by permission)

et al. (1982b) reported that intraperitoneal injection of a highly specific polyclonal glucagon antibody increased deprivation-induced meal size by 73%. Subsequent tests of hepatic portal infusions of the same antibody during spontaneous meals revealed similarly dramatic results (Fig. 2). Finally, intraperitoneal injections of a different polyclonal glucagon antibody elicited a small but significant increase in the average meal size of lean and obese Zucker rats under some conditions (McLaughlin et al. 1986).

These increases in meal size following antagonism of glucagon by specific antibodies provide compelling evidence for the necessary participation of glucagon in normal meal-ending satiety in the rat. Future work should determine whether: (1) Whether neutralization of the prandial increment in glucagon secretion, without affecting basal glucagon level, is sufficient to increase meal size. If not, it would suggest that glucagon is only a permissive factor in satiety. (2) Whether different doses of glucagon antibody can elicit a graded stimulatory effect on feeding, which would be expected if glucagon contributed to the graded intensity of postprandial satiety with respect to the amount of food ingested (McHugh et al. 1975). In Langhans et al.'s (1982b) work, smaller intraperitoneal glucagon antibody doses produced a smaller effect. Hepatic portal infusion of larger antibody amounts, however, resulted in a dose reversal (Geary et al. 1993). The reason for this is unclear. Interestingly, however, similar reversals also occurred in tests of antagonism of the putative satiety peptide cholecystokinin (DOURISH et al. 1989; MIESNER et al. 1992).

E. Mechanism of Glucagon Satiety

I. Site of Action

Glucagon's satiating effect appears to originate in the liver, because in dual-catheterized rats 2-min infusions of 0.85–7 µg/min glucagon that reduced rats' spontaneous meal size when infused into the hepatic-portal vein did not significantly affect meal size when infused into the inferior vena cava near the junction of the hepatic vein (GEARY et al. 1993). Due to the efficiency of first-pass hepatic glucagon extraction (ISHIDA and FIELD 1983; LANGHANS et al. 1984), such a difference is explicable only if glucagon acted in the liver to inhibit feeding. Infusion parameters may have been crucial to this effect, in that 20-min, 1.25–5-µg/kg per minute hepatic portal and intracardiac glucagon infusions appeared equipotent (STRUBBE et al. 1989). Further evidence that the satiating effect of glucagon originates in the liver comes from a report that glucagon failed to reduce test meal size in rats that received hepatic-portal injections of an alloxan dose that appeared to produce cytotoxic effects only in the liver (RITTER et al. 1986).

II. Transduction

Little progress has been made in identifying which of glucagon's many hepatic actions initiates its satiety effect (for a review, see GEARY 1990). For obvious reasons, attention has focussed on glucagon's stimulatory effect on hepatic glucose production. This effort has not produced any compelling evidence for the idea that glucose mediates glucagon satiety. Perhaps the strongest support is the finding that hepatic portal administration of alloxan, which is toxic to some glucoreceptive cells, blocked exogenous glucagon's satiating action (RITTER et al. 1986). On the other hand, there are several dissociations between glucagon satiety and indices of glucose metabolism (GEARY 1990). For example, reducing the duration of pretest food deprivation increased the stimulation of hepatic glucose production rats produced by intraperitoneal glucagon, but decreased glucagon's satiating potency (GEARY et al. 1987). Glucagon's failure to inhibit sham feeding (GEARY and SMITH 1982) also mitigates against the hypothesis that increased glucose utilization mediates glucagon's satiating action, because glucagon still increases hepatic glucose production in sham feeding rats (GEARY and SMITH 1982). Finally, if insulin-dependent glucose utilization mediates glucagon satiety, then co-infusion of insulin and glucagon should increase glucagon's satiating potency, but it does not (GEARY et al. 1995). Further, chronic subcutaneous glucagon administration decreased food intake as much in streptozotocin-diabetic as in healthy rats (DeCastro et al. 1978).

Alternative hepatic mechanisms for glucagon satiety have not been extensively tested. One possibility is that glucagon satiety originates from increased hepatic β-fatty acid oxidation. Control of fatty acid oxidation appears to be a physiological function of glucagon (Foster 1984; Seifter and Englard 1993), and a number of recent studies link fatty acid oxidation to the control of food intake (Langhans and Scharrer 1992). Glucagon's effect on hepatocyte membrane potential provides another possibility. Ouabain, which depolarizes the membrane by inactivating Na^+/K^+-ATPase, stimulated feeding (Langhans and Scharrer 1987) and antagonized the inhibitory effect of intraperitoneal glucagon (Duss 1986). A nonmetabolic effect could also mediate glucagon satiety. Exogenous glucagon relaxes visceral and vascular smooth muscle in the liver and elsewhere in the gut (Blackmore et al. 1991; D'Almeida and Lautt 1991), and the majority of hepatic vagal afferents innervate the biliary tree and the hepatic vasculature (Berthoud et al. 1992; Sawchenko and Friedman 1979). Two attempts to relate glucagon satiety to gastrointestinal smooth muscle effects, however, failed (Geary 1987; Stockinger and Geary 1989). The role of vascular smooth muscle has not been tested. It is relevant in this connection that prandial glucagon secretion does not appear to be involved in the prandial increase in splanchnic blood flow (Braatvedt et al. 1993).

III. Hepatic Vagal Afferents

Selective lesion of the hepatic branch of the abdominal vagus has been reported to block the satiating effects of intraperitoneal glucagon injection on scheduled meals (Geary and Smith 1983; MacIssac and Geary 1985; Weatherford and Ritter 1986) and of hepatic portal glucagon infusion during spontaneous meals (Geary et al. 1993). Hepatic vagotomy also blocked the stimulatory effect of hepatic portal infusions of glucagon antibodies on spontaneous feeding (Geary et al. 1993). The hepatic branch appeared sufficient to mediate glucagon satiety because abdominal vagotomies that spared only the hepatic branch did not affect it (Geary and Smith 1983). Surgical resection of the hepatic artery combined with phenol denervation of the surfaces of the portal vein and bile duct also failed to affect glucagon satiety (MacIssac and Geary 1985), indicating: (1) that glucagon satiety does not depend on the many hepatic vagal fibers that pass through the liver to more distal sites (Prechtl and Powley 1987; Berthoud et al. 1991, 1992) and (2) that spinal visceral afferents that reach the liver along the portal vein and bile ducts are not necessary for glucagon satiety. The hepatic vagal contribution to this satiety effect appears to be sensory. Antagonism of peripheral postganglionic muscarinic receptors with atropine methyl nitrate did not block glucagon satiety (Geary and Smith 1983), whereas both selective lesion of unmyelinated peripheral afferents with capsaicin (Ritter and Weatherford 1986) and lesion of the central hepatic vagal afferent terminals in the nucleus of the solitary tract did (WEATHERFORD and RITTER 1988). These data point to the conclusion that information about glucagon's hepatic satiety action reaches the central nervous system via the vagus nerve.

The extent of direct vagal innervation of the hepatocytes is controversial. One authoritative study indicated that the rat hepatic vagal afferents terminate in the bile ducts, the paraganglia of the nerve plexus, and the portal vein, but not the liver parenchyma (BERTHOUD et al. 1992). On the other hand, immunocytochemical studies of nerve-specific proteins have produced evidence for direct innervation of the parenchyma (LEE et al. 1992; FRIEDMAN et al. 1994).

F. Clinical Aspects

I. Pathophysiology of Glucagon Satiety

No dysfunction in glucagon physiology has been established in human obesity, anorexia nervosa, or bulimia nervosa (HOLST 1983; JOHNSON et al. 1991). To what extent this reflects the disorders' heterogeneity, however, is not clear. Similarly, there are only scattered reports of alterations in glucagon's feeding effects in animal models of obesity. For example, intraperitoneal glucagon's effect on feeding was reduced in a test of genetically obese fa/fa Zucker rats (McLAUGHLIN et al. 1986), although glucagon metabolism appears relatively normal in these animals (HOLST 1983).

Disorders of glucagon satiety may occur in some diseases. Glucagonomas associated with very elevated plasma glucagon levels and severe anorexia have been reported in humans and rats (WYNICK et al. 1993; MADSEN et al. 1995). Increased postprandial glucagon occurs in NIDDM and may be an important cause of hyperglycemia in both NIDDM and IDDM (BRAND et al. 1994; DINNEEN et al. 1995; GUTNIAK et al. 1986). A missense mutation in the glucagon receptor gene that results in the expression of a glucagon receptor with greatly reduced affinity has recently been reported in late-onset NIDDM (HAGER et al. 1995). If this defect reduces satiating potency of endogenous glucagon, it may account for some of the increased food intake and obesity associated with NIDDM.

II. Therapeutic Potential

Tests of chronic glucagon administration in humans question the safety and the efficacy of the potential of this treatment as an appetite suppressant. In each of three trials, intramuscular injections of 1–2 mg glucagon before meals decreased food intake and body weight, but also produced side effects such as nausea and glycosuria (CLAYTON and LIBRIK 1963; PENICK and HINKLE 1961; SCHULMAN et al. 1957). In view of the effects of route of administration on glucagon's actions and the potency of intravenous glucagon doses in the microgram range to reduce meal size acutely in humans (GEARY et al. 1992), however, these results should not discourage attempts to develop safe and effective procedures to administer glucagon chronically. Novel developments in drug delivery may be helpful (CHIOU and CHUANG 1988; WALLACE and LASKER 1993).

G. Conclusions

Recent work has greatly strengthened the hypothesis that glucagon signals postprandial satiety. In rats, glucagon infusion produces a dose-related, behaviorally specific reduction in meal size and glucagon antibody infusion increases meal size. In humans, intravenous infusion of a moderate glucagon dose produced a subjectively specific decrease in meal size. Key questions for the next decade are: (1) What are the physiological mechanisms of glucagon's satiety effect? (2) What is glucagon's role in normal and disordered human appetite? (3) What is glucagon's therapeutic potential in disordered human food intake?

References

Berthoud H-R, Jeanrenaud B (1982) Sham feeding-induced cephalic phase insulin release in the rat. Am J Physiol 242:E280–E285
Berthoud H-R, Fox EA, Powley TL (1990) Localization of vagal preganglionics that stimulate insulin and glucagon secretion. Am J Physiol 258:R160–R168
Berthoud H-R, Carlson NR, Powley TL (1991) Topography of efferent vagal innervation of the rat gastrointestinal tract. Am J Physiol 260:R200–R207
Berthoud H-R, Kressel M, Neuhuber WL (1992) An anterograde tracing study of the vagal innervation of rat liver, portal vein and biliary system. Anat Embryol 186:431–442
Blackmore PF, Mojsov S, Exton JH, Habener JF (1991) Absence of insulinotropic glucagon-like peptide-1 (7-37) receptors on isolated rat liver hepatocytes. FEBS Lett 283:7–10
Blundell JE (1991) Pharmacological approaches to appetite suppression. Trends Pharmacol Sci 12:147–157
Braatvedt GD, Stanners A, Newrick PG, Halliwell M, Corrall RJM (1993) Evidence against a putative role for glucagon as a physiological splanchnic vasodilator in man. Clin Sci 84:193–199
Brand CL, Rolin B, Jorgensen PN, Svendsen I, Kristensen JS, Holst JJ (1994) Immunoneutralization of endogenous glucagon with monoclonal glucagon antibody normalizes hyperglycaemia in moderately streptozotocin-diabetic rats. Diabetologia 37:985–993
Campfield LA, Smith FJ (1990) Systemic factors in the control of food intake. In: Stricker EM (ed) Neurobiology of food and fluid intake. Plenum, New York, pp 183–206 (Handbook of behavioral neurobiology, vol 10)
Chiou GCY, Chuang CY (1988) Treatment of hypoglycemia with glucagon eye drops. J Ocular Pharm 4:179–186
Clayton GW, Librik L (1963) Therapy of exogenous obesity in childhood and adolescence. Pediatr Clin North Am 10:99–107

D'Almeida MS, Lautt WW (1991) Glucagon pharmacodynamics and modulation of sympathetic nerve and norepinephrine-induced constrictor responses in the superior mesenteric artery of the cat. J Pharmacol Exp Ther 259:118–123

Day J, Johansen LK, Ganda OP, Soeldner JS, Gleason RE, Medgley W (1978) Factors governing insulin and glucagon response during normal meals. Clin Endocrinol (Oxf) 9:443–454

DeCastro JM, Paullin SK, DeLugas GM (1978) Insulin and glucagon as determinants of body weight set point and microregulation in rats. J Comp Physiol Psychol 92:571–579

Deetz LE, Wangsness PJ (1981) Influence of intrajugular administration of insulin, glucagon and propionate on voluntary feed intake of sheep. J Anim Sci 53:427–433

DeJong A, Strubbe JH, Steffens AB (1977) Hypothalamic influence on insulin and glucagon release in the rat. Am J Physiol 233:E380–E388

Denker H, Hedner P, Holst J, Tranberg KG (1975) Pancreatic glucagon response to an ordinary meal. Scand J Gastroenterol 10:471–474

Dinneen S, Alzaid A, Turk D, Rizza R (1995) Failure of glucagon suppression contributes to postprandial hyperglycaemia in IDDM. Diabetologia 38:337–343

Dourish CT, Rycroft W, Iversen SD (1989) Postponement of satiety by blockade of brain cholecystokinin (CCK-8) receptors. Science 245:1509–1511

Duss M (1986) Untersuchungen zur Sättingungswirkung von Glucagon. University of Zurich, Inaugural Dissertation. Maus, Konstanz

Foster D (1984) From glycogen to ketones – and back. Diabetes 33:1188–1199

Friedman MI, Ketchum M, Ulrich P, Breslin PAS, DellaCorte C (1994) Evidence for innervation of the hepatic parenchyma in the rat. Soc Neurosci Abstr 20:1226 (abstract)

Geary N (1987) Glucagon 1–21 fails to inhibit feeding. Peptides 8:943–945

Geary N (1990) Pancreatic glucagon signals postprandial satiety. Neurosci Biobehav Rev 14:323–338

Geary N (1995) A failure of pulsatile infusion to increase glucagon's satiating potency. Physiol Behav 59:613–616

Geary N, Smith GP (1982) Pancreatic glucagon fails to inhibit sham feeding in the rat. Peptides 3:163–166

Geary N, Smith GP (1983) Selective hepatic vagotomy blocks pancreatic glucagon's satiety effect. Physiol Behav 31:391–394

Geary N, Farhoody N, Gersony A (1987) Food deprivation dissociates pancreatic glucagon's effects on satiety and hepatic glucose production at dark onset. Physiol Behav 39:507–511

Geary N, Kissileff HR, Pi-Sunyer FX, Hinton V (1992) Individual, but not simultaneous, glucagon and cholecystokinin infusions inhibit feeding in men. Am J Physiol 262:R975–R980

Geary N, Le Sauter J, Noh N (1993) Glucagon acts in the liver to control spontaneous meal size in rats. Am J Physiol 264:R116–R122

Geary N, Asarian L, Langhans W (1995) Insulin does not increase glucagon's potency to reduce meal size in spontaneously feeding rats. Soc Neurosci Abstr 21: 8 (abstract)

Gutniak M, Grill V, Efendic S (1986) Effect of composition of mixed meals–low- versus high-carbohydrate content – on insulin, glucagon, and somatostatin release in healthy humans and in patients with NIDDM. Diabetes Care 9(3):244–249

Hager J, Hansen L, Vaisse C, Vionnet N, Philippi A, Poller W, Velho G, Carcassi C, Contu L, Julier C, Cambien F, Passa P, Lathrop M, Kindsvogel W, Demenais F, Nishimura E, Froguel P (1995) A missense mutation in the glucagon receptor gene is associated with non-insulin-dependent diabetes mellitus. Nature Genet 9:299–304

Hildebrandt W, Blech W, Kohnert K-D (1991a) Kinetic studies on hepatic handling of glucagon using the model of non-recirculating perfused rat livers. Horm Metab Res 23:410–413

Hildebrandt W, Blech W, Kohnert K-D (1991b) Hepatische Reaktion auf pulsatile Anderungen der Pankreasglukagonkonzentration. Wiss Z Univ Halle 40:137–149

Hildebrandt W, Blech W, Kohnert K-D (1992) Amplitude smoothing and phase shifting effect of the isolated perfused rat liver on oscillating insulin and glucagon. Horm Metab Res 24:345–346

Holst JJ (1983) Glucagon in obesity. In: Lefèbvre PJ (ed) Glucagon II. Springer, Berlin Heidelberg New York, pp 507–521 (Handbook of experimental pharmacology, vol 66/II)

Holst JJ, Schwartz TW, Lovgreen NA, Pedersen O, Beck-Nielsen H (1983) Diurnal profile of pancreatic polypeptide, pancreatic glucagon, gut glucagon and insulin in human morbid obesity. Int J Obes 7:529–538

Howes GA, Forbes JM (1987) A role for the liver in the effects of glucagon on food intake in domestic fowl. Physiol Behav 39:587–592

Ishida R, Field JB (1983) Hepatic handling of glucagon. In: Lefèbvre PJ (ed) Glucagon II. Springer, Berlin Heidelberg New York, pp 361–388 (Handbook of experimental pharmacology, vol 66/II)

Johnson PR, Greenwood MRC, Horwitz BA, Stern JS (1991) Animal models of obesity: genetic aspects. Annu Rev Nutr 11:325–353

Kalogeris TJ, Reidelberger RD, Mendel VE, Solomon TE (1991) Interaction of cholecystokinin-8 and pancreatic glucagon in control of food intake in dogs. Am J Physiol 260:R688–R692

Langhans W, Scharrer E (1987) Evidence for a role of the sodium pump of hepatocytes in the control of food intake. J Auton Nerv Syst 20:199–205

Langhans W, Scharrer E (1992) Metabolic control of eating, energy expenditure and the bioenergetics of obesity. In: Simopoulos AP (ed) World review of nutrition and dietetics. Karger, Basel, pp 1–67

Langhans W, Geary N, Scharrer N (1982a) Liver glycogen content decreases during meals in rats. Am J Physiol 243:R450–R453

Langhans W, Zieger U, Scharrer E, Geary N (1982b) Stimulation of feeding in rats by intraperitoneal injection of antibodies to glucagon. Science 218:894–896

Langhans W, Pantel K, Muller-Schell W, Eggengerger E, Scharrer E (1984) Hepatic handling of pancreatic glucagon and glucose during meals in rats. Am J Physiol 247:R827–R832

Langhans W, Duss M, Scharrer E (1987) Decreased feeding and supraphysiological plasma levels of glucagon after glucagon injection in rats. Physiol Behav 41:31–35

Le Sauter J, Geary N (1991) Hepatic portal glucagon infusion decreases spontaneous meal size in rats. Am J physiol 261:R154–R161

Le Sauter J, Geary N (1993) Le glucagon pancréatique: signal physiologique de la satiété post prandiale. Ann Endocrinol 54:149–161

Le Sauter J, Noh U, Geary N (1991) Hepatic portal infusion of glucagon antibodies increases spontaneous meal size in rats. Am J Physiol 261:R162–R165

Lee JA, Ahmed Q, Burt AD (1992) Disappearance of hepatic parenchymal nerves in human liver cirrhosis. Gut 33:87–91

Levine S, Sievert CE, Morley JE, Gosnell BA, Silvis SE (1984) Peptidergic regulation of feeding in the dog (Canis familiaris). Peptides 5:675–679

MacIssac L, Geary N (1985) Partial liver denervations dissociate the inhibitory effects of pancreatic glucagon and epinephrine on feeding. Physiol Behav 35:233–237

Madsen OD, Karlsen C, Blume N, Jensen HI, Larsson L-I, Holst JJ (1995) Transplantable glucagonomas derived from pluripotent rat islet tumor tissue cause severe anorexia and adipsia. Scand J Clin Lab Invest 55 [Suppl 220]:27–36

Martin JR, Novin D (1977) Decreased feeding in rats following hepatic-portal infusion of glucagon. Physiol Behav 19:461–466

Martin JR, Novin D, VanderWeele DA (1978) Loss of glucagon suppression of feeding after vagotomy in rats. Am J Physiol 234:E314–E318

McHugh P, Moran T, Barton GN (1975) A graded behavioral phenomenon regulating caloric intake. Science 190:167–169

McLaughlin CL, Gingerich RL, Baile CA (1986) Role of glucagon in the control of food intake in Zucker obese and lean rats. Brain Res Bull 17:419–426

Miesner J, Smith GP, Gibbs J, Tyrka A (1992) Intravenous infusion of CCKA-receptor antagonist increases food intake in rats. Am J Physiol 262:R216–R219

Muller WA, Faloona GR, Aguilar-Parada E, Unger RH (1970) Abnormal alpha-cell function in diabetes. N Engl J Med 283:109–115

Nilsson G, Uvnas-Wallenstein K (1977) Effect of teasing and sham feeding on plasma glucagon concentration in dogs. Acta Physiol Scand 100:298–302

Palmer JP, Porte D Jr (1983) Neural control of glucagon secretion. In: Lefèbvre PJ (ed) Glucagon II. Springer, Berlin Heidelberg New York, pp 115–132 (Handbook of experimental pharmacology, vol 66/II)

Penick SB, Hinkle LE Jr (1961) Depression of food intake induced in healthy subjects by glucagon. N Engl J Med 264:893–897

Polonsky KS, Jaspan JB, Rubenstein AH (1983) The metabolic clearance rate of glucagon. In: Lefèbvre PJ (ed) Glucagon II. Springer, Berlin Heidelberg New York, pp 353–359 (Handbook of experimental pharmacology, vol 66/II)

Prechtl JC, Powley TL (1987) A light and electron microscopic examination of the vagal hepatic branch of the rat. Anat Embryol (Berl) 176:115–126

Ritter S, Weatherford SC (1986) Capsaicin pretreatment blocks glucagon-induced suppression of food intake. Appetite 7:291

Ritter S, Weatherford SC, Stone SL (1986) Glucagon-induced inhibition of feeding is impaired by hepatic portal alloxan injection. Am J Physiol 250:R682–R690

Sawchenko PE, Friedman MI (1979) Sensory functions of the liver – a review. Am J Physiol 236:R5–R20

Scharrer E, Langhans W (1988) Metabolic and hormonal factors controlling food intake. Int J Vit Nutr Res 58:249–261

Schulman JL, Carleton JL, Whitney G, Whitehorn JC (1957) Effect of glucagon on food intake and body weight in man. J Appl Physiol 11:419–421

Seifter S, Englard S (1993) Energy metabolism. In: Arias IM, Jakoby WB, Popper H, Schachter D, Shafritz DA (eds) The liver: biology and pathobiology. Raven, New York, pp 279–315

Smith CJV, Bright-Taylor B (1974) Does a glucostatic mechanism for food intake control exist in chickens? Poultry Sci 53:1720–1724

Smith GP, Gibbs J (1992) The development and proof of the CCK hypothesis of satiety. In: Dourish CT, Cooper SJ, Iversen SD, Iversen LL (eds) Multiple cholecystokinin receptors in the CNS. Oxford University Press, Oxford, pp 166–182

Stockinger Z, Geary N (1989) Pancreatic glucagon does not alter intrameal gastric emptying of milk in the rat. Physiol Behav 45:1259–1261

Strubbe JH, Wolsink JG, Schutte AM, Prins AJA (1989) Hepatic-portal and cardiac infusion of CCK-8 and glucagon induce different effects on feeding. Physiol Behav 46:643–646

Stunkard AJ, Van Itallie TB, Reis BB (1955) The mechanism of satiety: effect of glucagon on gastric hunger contractions in man. Proc Soc Exp Biol Med 89:258–261

Taborsky GJ Jr (1989) The endocrine pancreas: control of secretion. In: Patton HD, Fuchs AF, Hille B, Scher AM, Steiner R (eds) Textbook of physiology, 21st edn. Saunders, Philadelphia, pp 1522–1543

VanderWeele DA, Geiselman PJ, Novin D (1979) Pancreatic glucagon, food deprivation and feeding in intact and vagotomized rabbits. Physiol Behav 23:155–158

Wallace B, Lasker J (1993) Stand and deliver: getting peptide drugs into the body. Science 260:912–915

Weatherford SC, Ritter S (1986) Glucagon satiety: diurnal variation after hepatic branch vagotomy or intraportal alloxan. Br Res Bull 17:545–549

Weatherford SC, Ritter S (1988) Lesion of vagal afferent terminals impairs glucagon-induced suppression of food intake. Physiol Behav 43:645–650

Weick BG, Ritter S (1986a) Dose-related suppression of feeding by intraportal glucagon infusion in the rat. Am J Physiol 250:R676–R681
Weick BG, Ritter S (1986b) Stimulation of insulin release and suppression of feeding by hepatic portal glucagon. Physiol Behav 38:531–536
Weigle DS (1987) Pulsatile secretion of fuel-regulatory hormones. Diabetes 36:764–775
Wynick D, Hammond P, Bloom S (1993) The glucagonoma syndrome. Clin Dermatol 11:93–97

CHAPTER 15
Glucagonoma and Its Management

A.J. SCHEEN and P.J. LEFÈBVRE

A. Introduction

Glucagonomas are considered to be among the rarest of the islet-cell tumors, with approximately 200 cases reported to date, even though they may be underdiagnosed clinical entities (EDNEY et al. 1990). Because of their uncommon occurrence, recognition is often delayed until metastases have developed. This occurs despite the fact that glucagonomas elicit a well-defined clinical syndrome that comprises distinctive dermatosis ("necrolytic migratory erythema"), diabetes, diarrhea, weight loss, anaemia and, more rarely, dementia. A detailed description of the glucagonoma syndrome has been previously reported (MALLINSON et al. 1974; STACPOOLE 1981; MONTENEGRO-RODAS and SAMAAN 1981; LUYCKX and LEFEBVRE 1981; GUILLAUSSEAU et al. 1982) and particularly in Chap. 43 of the previous volume in this series, *Glucagon II* (WOOD et al. 1983). Since then, several reviews have been devoted to this syndrome (PARKER et al. 1984; HOLST 1985; BLOOM and POLAK 1987; HASHIZIUME et al. 1988; SOMERS and DE VROEDE 1988; BODEN 1989; JOCKENHÖVEL and REINWEIN 1992).

Glucagonoma tumors are generally bulky and localization, in contrast to what is frequently observed with insulinomas, generally poses few problems. It is estimated that approximately 70% of patients with this syndrome have malignant tumors. When glucagonomas are malignant, about 50% of patients already have hepatic metastases at the time of diagnosis. Glucagonomas may exert clinical effects in several different ways, including mechanical compression around the primary tumor, the effects of metastases, particularly in the liver, and, as with other functioning neuroendocrine carcinomas, profound systemic effects through the biologically active agents (in the present case, glucagon) the tumors secrete. Because the glucagonoma syndrome can be debilitating, patients may benefit from relief of the syndrome even if the tumor itself does not regress.

We will focus the present review on glucagonoma on two specific topics (MIGNON and JENSEN 1995; JENSEN and NORTON 1995); (1) the new imaging techniques allowing accurate localization of the primary tumor and possible metastases and (2) the various therapeutic modalities which have been tested during the last 10 years in patients with glucagonoma syndrome.

B. Diagnosis and Localization of the Tumor

Diagnosis of the glucagonoma syndrome is made on the basis of the presenting clinical features, supported by an elevated fasting plasma glucagon concentration (Wood et al. 1983). Localization of the tumor is then important to assess resectability and the presence on metastases, on which the choice of treatment will depend. Because of the large size of glucagonomas, intraoperative palpation can be routinely used to localize the tumors, and it may be questioned whether preoperative localization is indeed warranted. However, preoperative localization is useful in planning surgical strategy, and one might hope that with increased clinical awareness of the syndrome more tumors will be diagnosed earlier, at smaller sizes that could escape intraoperative palpation. The techniques used for tumor localization and the rationale for using various localization methods are similar for all the pancreatic endocrine tumors (Rossi et al. 1989).

Ultrasonography of the abdomen is the established first-line procedure for evaluation of patients presenting with gastrointestinal symptoms. Consequently, a considerable number of gastroenteropancreatic tumors are readily detected by the demonstration of liver metastases or large primary tumors such as is often the case for glucagonomas. Endoscopic ultrasound is reported to detect from 80% to 100% of primary pancreatic endocrine tumors and to be much more effective than conventional transabdominal ultrasonography (Jensen and Norton 1995). Only two cases of glucagonoma studied using this technique have been reported in two large series of islet-cell tumors (Rösch et al. 1992; Glover et al. 1992). However, the sensitivity of this new technique for small tumors as well as extrapancreatic tumors and hepatic metastases has not been established (Jensen and Norton 1995).

The challenge is the patient with an overt clinical syndrome related to hormonal secretion, in whom ultrasound does not show the presence of metastatic spread. A great number of other radiological techniques have been studied and compared (Hercot et al. 1989; Arnaud et al. 1992; Lamberts et al. 1993). Computed tomography scanning is the technique of choice both to localize the primary glucagonoma and to demonstrate metastatic disease (Krudy et al. 1984; Breatnach et al. 1985). Since most glucagonoma tumors are usually 3cm in diameter or larger at the time of diagnosis, their identification is rarely difficult (Lax et al. 1986). The value of nuclear magnetic resonance imaging is at present unclear. Nevertheless, for detection of metastatic disease to the liver of gastrinomas, one study (Pisegna et al. 1993) demonstrated that magnetic resonance imaging is more sensitive than ultrasound, computer tomography or angiography, so that it is now considered as the imaging procedure of choice for malignant pancreatic endocrine tumors (Jensen and Norton 1995). Selective angiography is also effective in demonstrating glucagonomas, which are characteristically highly vascular and typically show a pronounced tumor "blush" (Wawrukiewicz et al. 1982). Percutaneous transhepatic portal venous sampling (Ingemansson et al. 1977), which is particularly useful for diagnosis and localization of small insulinomas, is rarely used for glucagonomas (EDNEY et al. 1990; LAMBERTS et al. 1993). When all imaging studies fail to identify a tumor but the biochemical diagnosis is certain (which is rarely the case for glucagonomas), the final step is exploration of the pancreas (and the liver) by a surgeon experienced in skillful organ palpation and intraoperative ultrasonography.

The most exciting progress in the field is the recent increasing experience of peptide receptor scintigraphy, which has greatly increased the success of accurate localization of gastroenteropancreatic endocrine tumors and metastases (LAMBERTS et al. 1990, 1991, 1993). The presence of large numbers of high-affinity somatostatin receptors on many of these tumors led LAMBERTS and colleagues to explore whether it was possible to detect receptor-positive tumor in vivo using a radioactive iodine-labeled somatostatin analogue (LAMBERTS et al. 1990). They demonstrated a close relationship between the in vitro detection of somatostatin receptors using autoradiography after surgical removal of the tumor and the gamma camera pictures obtained in vivo after injection of ^{123}I-Tyr^3-octreotide. Furthermore, there was a close parallel between the presence of somatostatin receptors and the in vivo and in vitro effects of octreotide on hormonal secretion by these tumors (LAMBERTS et al. 1991, 1993). This means that a positive scan predicts a beneficial effect of octreotide therapy on hormonal hypersecretion and vice versa (NOCAUDIE-CALZADA et al. 1994) (see below). In contrast, the presence of somatostatin receptors does not predict the success of octreotide therapy on tumor growth (ARNOLD et al. 1993).

The ^{123}I-labeled Tyr^3-octreotide scanning technique has been shown to be a rapid and safe procedure for the visualization of some endocrine tumors as small as 1 cm in diameter that have somatostatin receptors (LAMBERTS et al. 1990, 1991). This technique appears to be valuable in locating not only primary but also secondary deposits of these tumors, which may be more difficult to localize with current diagnostic techniques. One problem, important especially in the visualization of islet cell tumors, is that ^{123}I-Tyr^3-octreotide is excreted via the liver, gallbladder and bile ducts, and makes this part of the body so "hot" with radioactivity that small tumors in these regions can be missed. In order to circumvent this drawback, alternative peptides, such as ^{111}Ind-DTPA-octreotide (LAMBERTS et al. 1993; VAN EYCK et al. 1993) or ^{111}In-pentetreotide (PAUWELS et al. 1994), have been designed. Such peptides, with longer half-life and excretion via the kidneys, allow visualization of gastroenteropancreatic somatostatin receptor-positive tumors even better than ^{123}I-Tyr^3-octreotide. However, because of the relative rarity of glucagonomas, only one individual with glucagonoma has been mentioned in the list of patients with endocrine tumors studied with this technique until now (LAMBERTS et al. 1993; VAN EYCK et al. 1993). Although the preliminary results have been encouraging, the sensitivity and specificity of this scanning procedure in the localization of glucagonomas must be confirmed by studies of larger numbers of patients.

C. Management of the Glucagonoma Syndrome

As is the case with most other endocrine tumors, a dual therapeutic strategy is required to optimally manage glucagonomas: eradication of as much tumor as is clinically feasible and alleviation of the systemic effects of secreted peptides (FRIESEN 1987; VINIK and MOATTARI 1989). While removal of all tumor on operation should resolve both problems, this may not be possible (McENTEE et al. 1990).

The treatment of choice and, at present, the only chance for cure, is surgical extirpation. However, many other therapeutic choices exist for patients who are not suitable for surgery or remain symptomatic after operation. Conservative therapy with minimum side-effects should be recommended first. More aggressive therapy should be proposed to patients in whom compromise of vital organ function either exists or is anticipated within a few months. Systemic (chemo- or biological) therapy or regional (vascular occlusion or radiation) therapy provide useful options for patients with advanced symptomatic islet-cell carcinomas, among which are glucagonomas.

I. Surgical Treatment

The role of surgical treatment includes possible cure, prolonged survival and potential control of symptoms (FRAKER and NORTON 1989). In those patients fortunate enough to have an early diagnosis and a localized small lesion, enucleation alone has been suggested as adequate therapy (FRIESEN 1987). However, most glucagonomas are large and malignant, and a few are multicentric. The latter two instances require pancreatic resection. Fortunately, most glucagonomas are located in the pancreatic body and tail and, thus, distal pancreatectomy is more frequently possible. The chance of sugical cure of malignant endocrine pancreatic tumors is low, however, comprising only 30% in one series (GRAMA et al. 1992). The presence of a large or metastatic tumor should not preclude an attempt at radical resection or debulking (McENTEE et al. 1990; CARTY et al. 1992; GRANT 1993). These tumors are very slow growing, and most patients are debilitated by the effects of excessive glucagon for many years rather than by tumor bulk. Debulking may result in a prolonged remission, with several years of improved well-being for the patient, and may also augment the success of consecutive medical treatment, such as antisecretory therapy with somatostatin analogues (see below).

Prior to surgery, the patient should be stabilized. Hyperalimentation will help reverse the catabolic state, especially in combination with octreotide treatment. Insulin should be given if necessary. Perioperative heparin may be useful to counteract the thrombotic tendencies frequently seen in patients with glucagonomas.

II. Vascular Occlusion

Liver metastases from islet cell tumors, particularly glucagonomas, are highly vascular and thus lend themselves to vascular occlusion therapy (ALLISON et al. 1985). Most of the blood supply to hepatic metastases is derived from the hepatic artery while the normal liver parenchyma receives the majority of its vascular supply from the portal vein. So, after occlusion of the hepatic artery, the viability of the normal liver parenchyma can be maintained by the portal vein. Tumors have been embolized with either gel foam pledgets (JIAN et al. 1984; LOKICH et al. 1987) or polyvinyl alcohol foam particles (ASSAAD et al. 1987; FREIMANN and PAZDUR 1990) via percutaneous femoral arterial catheters directed into the hepatic artery. Approximately two-thirds of patients with endocrine tumors have responded to hepatic arterial embolization (AJANI et al. 1988). Rapid symptomatic response is usually observed, with a median response duration of about 12–15 months. Repeated embolizations may be performed to maintain a clinical response (AJANI et al. 1988).

A review of the literature published in 1990 (FREIMANN and PAZDUR 1990) revealed 12 cases of glucagonoma with liver metastases treated with hepatic arterial embolization. Detailed reports were available on eight cases only. Most patients demonstrated rapid resolution of the characteristic rash accompanied by decreased glucagon plasma levels and improvement of plasma glucose levels. Five of these eight patients had some radiological evidence of antitumor response after vascular occlusion (FREIMAN and PAZDUR 1990). Four additional glucagonoma patients were included in a subgroup of individuals with metastatic endocrine tumors in a large series of patients treated with hepatic arterial embolization, but the results in those patients with glucagonoma were not specified (ALLISON et al. 1985). Hepatic artery embolization provides the potential for both relief of symptoms from excess hormone production and reduction in the size of hepatic metastases. Vascular occlusion, if it could be coupled with more effective systemic therapy, may thus offer great benefit to patients with glucagonoma syndrome.

Hepatic artery chemoembolization may also be performed by placing an arterial catheter selectively in the artery supplying the dominant tumor mass(es) in patients with liver metastases of endocrine tumors (RUSZNIEWSKI et al. 1993). Streptozocin, for instance, has been selectively administered suspended in a powdered Gelfoam (FREIMANN and PAZDUR 1990) or a Lipiodol emulsion (NESOVIC et al. 1992) in patients with malignant glucagonomas (see below).

III. Radiation Therapy

Experience with radiation therapy in patients with islet-cell carcinoma has been scarce. Only a few reports are available for review (AJANI et al. 1989). RICH (1985) and TORRISI et al. (1987) have each reported effective palliation by radiotherapy in three patients with locally advanced islet-cell carcinoma. This experience, in addition to that from other published reports (see review in TORRISI et al. 1987), suggests that radiotherapy is a useful mode for treating malignant islet-cell carcinoma. However, to our knowledge, none of the reported cases concerned documented glucagonomas.

Several radionuclides have been proposed and investigated for radiotherapeutic applications. KRENNING et al. (1994) recently reported a patient with an inoperable, metastasized glucagonoma in whom peptide receptor radiotherapy with electrons emitting (^{111}In-DTPA-D-Phe^1)-octreotide affected the growth of the tumor and the circulating glucagon levels. This positive effect contrasted with the previous lack of effect of octreotide alone and even of a combination of octreotide and interferon-α. Future nuclear radiotherapy is being directed to use of α- or β-particles emitting radionuclide-labeled peptides, because of their more appropriate physical characteristics. It is to be expected that the radiotherapeutic use of radionuclides with these higher energies, coupled to small peptides, leads to higher radiation doses and more appropriate particle ranges. Peptide receptor radiotherapy with radiolabeled peptides (-derivatives), such as hormones and growth factors, is at the moment in its infancy but preliminary results are promising (KRENNING et al. 1994).

IV. Chemotherapy

A number of different chemotherapeutic regimens have been evaluated in patients with metastatic pancreatic endocrine tumors. Because of the rare occurrence of pancreatic endocrine tumors, patients in most studies on chemotherapy have been considered together as a group, including only a few cases of glucagonomas, and sometimes included in series with metastatic carcinoid tumors. It has, however, not been established that each type of pancreatic endocrine tumor responds equally to chemotherapy (JENSEN and NORTON 1995). Only a few therapeutic agents have been adequately investigated, and very few patients with glucagonoma were indeed involved in such trials (FRIESEN 1987; AJANI et al. 1989, 1991; MODLIN et al. 1993). The most frequently used chemotherapeutic agents in the treatment of glucagonoma include streptozotocin and dacarbazine in single-agent or combination with 5-fluorouracil or doxorubicin regimens. Local chemotherapy, following hepatic artery catheterization, may improve efficacy and reduce the side-effects of some agents when compared to systemic chemotherapy (KVOLS and BUCK 1987).

1. Streptozotocin

a) Single-Agent Chemotherapy

Most studies using systemic administration of streptozotocin alone for the treatment of islet-cell tumors were performed in the 1970s (BRODER and

CARTER 1973). Streptozotocin was, however, rather toxic and frequently not very effective. The intraoperative placement of a catheter in the hepatic artery via the gastroduodenal artery allowed for localized intra-arterial administration and reduction of systemic toxicity. A beneficial response could be expected in approximately 50% of patients with metastatic endocrine pancreatic tumors (FRIESEN 1987).

b) Combination Chemotherapy

Streptozotocin has been more effective in combination with 5-fluorouracil or with doxorubicin than as a single agent (OBERG and ERIKSSON 1989; AJANI et al. 1989, 1991). In a large multicenter trial on 105 patients with advanced islet-cell carcinoma (but only two patients with high plasma glucagon levels), the combination of streptozotocin and doxorubicin was superior to the current standard regimen of streptozotocin plus 5-fluorouracil (MOERTEL et al. 1992). Only partial remission of symptoms has been reported in a patient with malignant glucagonoma and liver metastases using selective hepatic artery chemoembolization of Lipiodol emulsified with streptozotocin and 5-fluorouracil (NESOVIC et al. 1992). In the latter study, the results were more impressive in two other cases with malignant insulinomas treated with the same procedure.

2. Dacarbazine

a) Single-Agent Chemotherapy

Since the demonstration that dimethyltriazenoimidazole carboxamide (DTIC or dacarbazine) was an effective drug in the treatment of malignant glucagonoma resistant to streptozotocin, several studies have reported at least 15 glucagonoma cases with an objective, clinical and biological, response to the drug lasting from 2 to 48 months (PRINZ et al. 1981; review in ALTIMARI et al. 1987 and KVOLS and BUCK 1987). In particular, glucagonoma-related necrolytic migratory erythema may disappear with dacarbazine (van der LOOS et al. 1987). Only one paper reported failure to respond to dacarbazine (HALLENGREN et al. 1983). In some cases of malignant glucagonomas, a marked reduction in size of hepatic metastases (KUROSE et al. 1984; FUJITA et al. 1986) or even a long-term remission up to 4–7 years (KESSINGER et al. 1983; ALTIMARI et al. 1987; JEANMOUGIN et al. 1988), has been described. This high level of response possibly represents some reporting bias. Nevertheless, this activity has prompted some authors to recommend dacarbazine as the drug of choice for malignant glucagonoma after surgery (PRINZ et al. 1981; KESSINGER et al. 1983; KVOLS and BUCK 1987; EDNEY et al. 1990; MODLIN et al. 1993).

b) Combination Chemotherapy

In a detailed report on seven glucagonomas initially treated by aggressive cytoreductive surgery, dacarbazine has been given in five patients with extensive disease at the time of operation. Darcarbazine was used in combination with various other chemotherapeutic compounds, such as 5-fluorouracil, streptozotocin, doxorubicin or somatostatin (EDNEY et al. 1990).

3. Others

Surprisingly, *5-fluorouracil* as single agent has not been extensively investigated as therapy in patients with islet-cell tumors. Following the use of this agent, occasional responses have been observed (but no reports are available in clearly documented glucagonomas) (AJANI et al. 1989, 1991). However, 5-fluorouracil has been tested in combination with streptozotocin (KHANDEKAR and SRIRATANA 1986; MOERTEL et al. 1992), dacarbazine (EDNEY et al. 1990), chlorozotocin (BUKOWSKI et al. 1992) or α-interferon (JONES et al. 1992) in some patients with glucagonoma syndrome.

Chlorozotocin alone has been shown to be similar in efficacy to streptozotocin plus fluorouracil in a multicenter study on patients with advanced islet-cell carcinoma (among whom only two had glucagonomas) (MOERTEL et al. 1992). Since it produced fewer gastrointestinal and renal side-effects than streptozotocin, it merits study as a constituent of combination drug regimens in islet-cell carcinomas, particularly glucagonomas (BUKOWSKI et al. 1992).

Lomustine has been shown to be effective on the clinical symptoms, plasma glucagon levels and metastatic masses in a patient with recidivant malignant glucagonoma, not responding to a second cure combining 5-fluorouracil and streptozotocin (KHANDEKAR and SRIRATANA 1986).

The responsiveness to combined *etoposide and cisplatin* therapy was poor in 14 patients with islet-cell carcinomas (elevation of plasma glucagon levels in 5 cases) in contrast to the good response observed in 18 patients with anaplastic neuroendocrine carcinomas (MOERTEL et al. 1991).

V. Biological Therapy

On the basis of observations in patients with carcinoid syndrome who showed improvement in biochemical and subjective parameters when treated with leukocyte interferon (review in OBERG and ERIKSSON 1989, AJANI et al. 1991), human leukocyte interferon therapy has also been tested in patients with advanced islet-cell tumors, mostly patients with Zollinger-Ellison syndrome (ERIKSSON et al. 1986). Partial responses were noted in 16 of 20 assessable patients, but apparently none of these patients had glucagonomas.

A considerable fall in plasma glucagon, resolution of the associated rash and a reduction in hepatic tumor mass demonstrated by computed tomography have been reported after treatment with human lymphoblastoid interferon given subcutaneously for 10 weeks over a 4-month period in a patient with metastatic glucagonoma (SHEEHAN-DARE et al. 1988). JONES et al. (1992) reported the case of a patient with a glucagonoma who experienced a dramatic clinical improvement when treated with a combination of 5-fluorouracil and α-interferon. Further evaluation is required to determine what role 5-fluorouracil with α-interferon may play in the treatment of such neuroendocrine malignancies.

VI. Antisecretory Peptide Therapy

Most complaints of patients with glucagonomas are related to excessive glucagon secretion, especially dermatosis, diabetes and debilitation. Consequently, well-being of the patients may be improved by inhibiting glucagon secretion even if no effect on tumor growth is obtained. Glucagon secretion can be blocked by somatostatin and its long-acting analogues.

In 1986 were reported the first four cases of inoperable glucagonoma where treatment with the long-acting somatostatin analogue SMS 201–995 (octreotide or Sandostatine) had sustained and beneficial effects on hyperglucagonemia, hypoalaninemia, weight loss and skin changes (BODEN et al. 1986; ALTIMARI et al. 1986; CH'NG et al. 1986). Since these original observations, at least 20 patients with glucagonoma have been treated with octreotide (O'DORISIO 1987; ROSENBAUM et al. 1989; MATON et al. 1989; LAMBERTS et al. 1991; BLANCHIN et al. 1992; MATON 1993; DEBAS and GITTES 1993; JENSEN and NORTON 1995). All patients had metastatic disease. Some patients had prior surgery to debulk their tumors. Other had been treated with chemotherapy or by arterial embolization. Daily doses of octreotide generally ranged between 100 and 500 μg (in exceptional cases up to 2250 μg) and were administered subcutaneously, usually in two or three injections. The characteristic migratory necrolytic dermatitis resolved over several days in 90% of these patients after octreotide. Weight loss, diarrhea and pain improved in more than 80% of the patients with these symptoms. The diabetes, however, responded in less than 10% of patients despite a fall in plasma glucagon in 75% of the patients. This can be explained by the simultaneous inhibitory effect of octreotide on insulin secretion. Most patients continued with octreotide for several months and the drug continued to be effective on the rash despite the fact that in some patients plasma concentrations of glucagon have returned to pretreatment levels (MATON et al. 1989). In one well-studied patient given octreotide, the rash resolved with no change in plasma concentrations of glucagon, amino acid or zinc, suggesting that octreotide may have direct action on the skin (SANTANGELO et al. 1986). Spectacular reversal with octreotide of a neurologic paraneoplastic syndrome in a patient with glucagonoma has also been reported (HOLMES et al. 1991).

In metastatic glucagonomas, a dissociation between alleviation of symptoms and reduction of plasma glucagon levels on the one hand, and persistence of tumor growth on the other hand, has generally been observed (MOATIARI et al. 1990; JOCKENHÖVEL et al. 1994). This is in agreement with the findings reported in most patients with other metastatic endocrine gastropancreatic tumors (ARNOLD et al. 1993). Therefore, final resistance of metastatic glucagonomas after long-term treatment with octreotide may frequently occur (WYNICK et al. 1989).

Thus, octreotide seems to be effective in controlling the rash and possibly other symptoms in patients with glucagonoma. It has a place in the treatment of glucagonomas and may prove to be a good adjunct to debulking procedures such as surgery and possibly chemotherapy (O'DORISIO 1987; ROSENBAUM et al. 1989; LAMBERTS et al. 1991; MATON 1993; DEBAS and GITTES 1993). Octreotide may be helpful prior to surgery, by preparing the patients to the intervention through the suppression of the debilitating effects of severe hyperglucagonemia, as well as after surgery, by improving the comfort of the patients remaining symptomatic because of residual glucagonoma metastases. The potential interest of longer-acting somatostatin analogues such as lanreotide remains to be evaluated.

VII. Symptomatic Treatment

The initial priority in the management of patients with tumor hyperglucagonemia is a vigorous attempt to correct the effect of catabolism. Total parenteral nutrition with intravenous supplementation of proteins, lipids and glucose is most beneficial. Insulin administration may be necessary, depending on the degree of hyperglycemia and glycosuria. If phlebitis is present or if there is a history of pulmonary embolism, subcutaneous heparin anticoagulation is advisable.

The etiology of the dermatitis in glucagonoma syndrome has been thought to be related to hypoaminoacidemia mediated via glucagon (MALLINSON et al. 1974). However, opinions about the efficacy of intravenous amino acids in treating necrolytic migratory erythema associated with glucagonoma syndrome are mixed. Although a few early investigators found it to be of benefit, subsequent studies have either disagreed or suggested other measures be implemented (WOOD 1983; SHEPHERD et al. 1991). However, it has been shown that patients with severe cutaneous manifestations can benefit from intravenous administration of amino acids before surgical or chemotherapeutic intervention (FUJITA et al. 1986; SHEPHERD et al. 1991).

D. Prognosis

Glucagonomas, as other islet-cell tumors, are slow growing, and patients present with significant clinical features often many years after the start of the disease (WOOD et al. 1983). For those with benign tumors surgical cure is possible. The prognosis for those patients with metastases at diagnosis is obviously less favorable, the median survival being approximately 3 years (WOOD et al. 1983). In a series of 41 patients with malignant islet-cell tumors (among 85 patients with endocrine pancreatic tumors associated with clinical syndromes of hormone excess and 7 glucagonomas), the overall 5-year and 10-

year survivals were 54% and 28%, respectively (GRAMA et al. 1992). Absence of liver metastases at time of operation/diagnosis, smaller size of the primary tumor, grossly radical tumor resection as well as response to medical therapy predicted the more favorable survival. In another 20-year experience on 58 patients with islet-cell carcinomas (of which four were malignant glucagonomas), the absence of hepatic metastases was also reported to be a major predictor of survival at 3 years (82% vs. 56%) (THOMPSON et al. 1988). Unfortunately, the effects of hepatic artery embolization, chemotherapy, biological therapy and antisecretory peptide therapy on the prognosis of glucagonomas remain unknown as they have not been assessed in a sufficient number of patients (MIGNON and JENSEN 1995; JENSEN and NORTON 1995). Even if octreotide clearly improves the quality of life of the patients with malignant glucagonomas (O'DORISIO 1987; ROSENBAUM et al. 1989; LAMBERTS et al. 1991; MATON 1993; DEBAS and GITTES 1993), its effect on tumor growth is doubtful (ARNOLD et al. 1993) and its efficacy on survival is presently unknown. It has been reported in a small series that all patients died within a period of 5 months once the clinical and biological resistance phase of their illness to maximum dosages of octreotide had been reached (WYNICK et al. 1989).

E. Conclusions

Glucagonomas are rare endocrine tumors which are usually slow growing, with a long life expectancy from the time of diagnosis despite frequent malignancy. Death results from local growth, metastatic disease and the sequelae of uncontrolled hormonal production progressively debilitating the patients with sustained and severe hyperglucagonemia. Most experts agree that chemotherapy or biological therapy is indicated only in patients with extensive metastatic disease (usually to the liver), and is not indicated for treatment of the primary tumor when surgery is not curative. There is no agreement, however, about when drug therapy should be started. Some groups recommend treatment only when symptoms develop due to the pancreatic endocrine tumor (nevertheless, antisecretory peptide therapy with octreotide seems more rapidly effective and should be now considered as the first choice for treating symptomatic patients), whereas others recommend treatment only when the tumor is demonstrated to be increasing in size (JENSEN and NORTON 1995).

Many advances have been made in the recognition, diagnosis and management of patients with functional islet-cell tumors during the past 3 decades. Recently developed therapeutic agents, especially somatostatin analogues, and treatment methods for these tumors are exciting and give hope to investigators and patients alike (MIGNON and JENSEN 1995). Future directions in the management of glucagonomas include development of somatostatin-receptor-labeled antibodies for diagnosis, staging and therapy, treatment of liver metastases with vascular embolization with chemotherapeutic agents or selective vascular occlusion, combinations of biological and cytotoxic agents, and development of longer-acting somatostatin analogues.

References

Ajani JA, Carrasco CH, Charnsangavej C, Samaan NA, Levin B, Wallace S (1988) Islet cell tumors metastatic to the liver: effective palliation by sequential hepatic artery embolization. Ann Intern Med 108:340–344

Ajani JA, Levin B, Wallace S (1989) Systemic and regional therapy of advanced islet cell tumors. Gastroenterol Clin North Am 18:923–930

Ajani JA, Carrasco CH, Samaan NA, Wallace S (1991) Therapeutic options for patients with advanced islet cell and carcinoid tumors. Reg Cancer Treat 3:235–242

Allison DJ, Jordan H, Hennessy O (1985) Therapeutic embolization of the hepatic artery. A review of 75 procedures. Lancet I:595–599

Altimari AF, Bhoopalam N, O'Dorsio T, Lange CL, Sandberg L, Prinz RA (1986) Use of a somatostatin analog (SMS 201–995) in the glucagonoma syndrome. Surgery 100:989–996

Altimari AF, Badrinath K, Reisel HJ, Prinz RA (1987) DTIC therapy in patients with malignant intra-abdominal neuroendocrine tumors. Surgery 102:1009–1017

Arnaud A, Fetissof F, Lorette G, Reigner J, Bertrand G, Lecomte P (1992) Le syndrome du glucagonome: trois nouveaux cas. Rev Med Interne 13:103–108

Arnold R, Neuhaus C, Benning R, Schwerk WB, Trautmann ME, Joseph K, Bruns C (1993) Somatostatin analog sandostatin and inhibition of tumor growth in patients with metastatic endocrine gastroenteropancreatic tumors. World J Surg 17:511–519

Assaad SN, Carrasco CH, Vassilopoulou-Sellin R, Samaan NA (1987) Glucagonoma syndrome: rapid response following arterial embolization of glucagonoma metastatic to the liver. Am J Med 82:533–535

Blanchin M, James-Deidier A, Chaumet-Riffaud PD, Chayvialle J-A (1992) Utilisation de l'octréotide dans les tumeurs endocrines digestives. Etude française multicentrique. Presse Med 21:697–702

Bloom SR, Polak JR (1987) Glucagonoma syndrome. Am J Med 82 (Suppl 5B): 25–36

Boden G (1989) Glucagonomas and insulinomas. Gastroenterol Clin North Am 18:831–845

Boden G, Ryan IG, Eisenschmid BL, Shelmet JJ, Owen OE (1986) Treatment of inoperable glucagonoma with the long acting somatostatin analogue SMS 201–995. N Engl J Med 314:1686–1689

Breatnach ES, Han SY, Rahatzad MT, Stanley RJ (1985) CT evaluation of glucagonomas. J Comput Assist Tomogr 9:25–29

Broder LE, Carter SK (1973) Pancreatic islet cell carcinoma. Results of therapy with streptozotocin in 52 patients. Ann Intern Med 79:108–118

Bukowski RM, Tangen C, Lee R, Macdonald JS, Einstein AB Jr, Peterson R, Fleming TR (1992) Phase II trial of chlorozotocin and fluorouracil in islet cell carcinoma: a southwest oncology group study. J Clin Oncol 10:1914–1918

Carty SE, Jensen RT, Norton JA (1992) Prospective study of aggressive resection of metastatic pancreatic endocrine tumors. Surgery 112:1024–1032

Ch'ng JLC, Anderson JV, Williams SJ, Carr DH, Bloom SR (1986) Remission of symptoms during long term treatment of metastatic pancreatic endocrine tumours with long acting somatostatin analogue. Br Med J 292:981–982

Debas HT, Gittes G (1993) Somatostatin analogue therapy in functioning neuroendocrine gut tumors. Digestion 54 [Suppl] 1:68–71

Edney JA, Hofmann S, Thompson JS, Kessinger A (1990) Glucagonoma syndrome is an underdiagnosed clinical entity. Am J Surg 160:625–629

Eriksson B, Öberg K, Alm G, Karlsson A, Lundqvist G, Andersson T, Wilander E, Wide L (1986) Treatment of malignant endocrine pancreatic tumours with human leucocyte interferon. Lancet II:1307–1309

Fraker DL, Norton JA (1989) The role of surgery in the management of islet cell tumors. Gastroenterol Clin North Am 18:805–830

Freimann J, Pazdur R (1990) Hepatic arterial embolization for treatment of glucagonomas: antineoplastic and palliative benefits. Am J Clin Oncol 13:271–275

Friesen SR (1987) Update on the diagnosis and treatment of rare neuroendocrine tumors. Surg Clin North Am 67:379–393

Fujita J, Seino Y, Ishida H, Taminato T, Matsukura S, Horio T, Imamura S, Naito A, Tobe T, Takahashi K, Midorikawa O, Imura H (1986) A functional study of a case of glucagonoma exhibiting typical glucagonoma syndrome. Cancer 57:860–865

Glover JR, Shorvon PJ, Lees WR (1992) Endoscopic ultrasound for localisation of islet cell tumours. Gut 33:108–110

Grama D, Eriksson B, Martensson H, Cedermark B, Ahren B, Kristoffersson A, Rastad J, Oberg K, Akerström G (1992) Clinical characteristics, treatment and survival in patients with pancreatic tumors causing hormonal syndromes. World J Surg 16:632–639

Grant CS (1993) Surgical management of malignant islet cell tumors. World J Surg 17:498–503

Guillausseau P-J, Guillausseau C, Villet R, Kaloustian E, Valleur P, Hautefeuille P, Lubetzki J (1982) Les glucagonomes. Aspects cliniques, biologiques, anatomo-pathologiques et thérapeutiques (revue générale de 130 cas). Gastroenterol Clin Biol 6:1029–1041

Hallengren B, Dymling JF, Manhem P, Tennvall L, Tibblin S (1983) Unsuccessful DTIC treatment of a patient with glucagonoma syndrome. Acta Med Scand 213:317–318

Hashizume T, Kiryu H, Noda K, Kano T, Nakano R (1988) Glucagonoma syndrome. J Am Acad Dermatol 19:377–383

Hercot O, Legmann P, Humbert M, Sibert A, Somveille E, Picard C, Mignon M, Benacerraf R (1989) Diagnostic du glucagonome. Intérêt du scanner, de l'échographie et de l'artériographie. A propos de deux observations et revue de la littérature. J Radiol 70:309–316

Holmes A, Kilpatrick C, Proietto J, Green MD (1991) Reversal of a neurologic paraneoplastic syndrome with octreotide (Sandostatin) in a patient with glucagonoma. Am J Med 91:434–436

Holst JJ (1985) Glucagon-producing tumors. In: Cohen S, Soloway RD (eds) Hormone-producing tumors of the gastrointestinal tract. Contemporary issues in gastroenteroloyg, vol 5. Churchill Livingstone, New York, pp 57–84

Ingemansson S, Holst J, Larsson LI, Lunderquist A (1977) Localization of glucagonomas by catheterization of the pancreatic veins and with glucagon assay. Surg Gynecol Obstet 145:509–516

Jeanmougin M, Civatte J, Bonvalet D, Passa P, Verola O, Zylberait D (1988) Glucagonome métastatique. Rémission complète par la dacarbazine (Recul de 5 ans). Ann Dermatol Venereol 115:833–838

Jensen RT, Norton JA (1995) Endocrine neoplasms of the pancreas. In: Yamada T (ed) Textbook of gastroenterology. Lippincott, Philadelphia, chap. 93, pp 2131–2160

Jian R, Seyrig JA, Roche A, Modigliani R, Lenormand Y, Hautefeuille M (1984) Improvement of metastatic glucagonoma by hepatic artery embolization. Gastroenterology 87:481–482

Jockenhövel F, Reinwein D (1992) Das Glukagonom: klinische Aspekte, Diagnostik und Therapie. Aktuel Endokrinol 13:66–80

Jockenhövel F, Lederbogen S, Olbricht T, Schmidt-Gayk H, Krenning EP, Lamberts SWJ, Reinwein D (1994) The long-acting somatostatin analogue octreotide alleviates symptoms by reducing posttranslational conversion or prepro-glucagon to glucagon in a patient with malignant glucagonoma, but does not prevent tumor growth. Clin Invest Med 72:127–133

Jones DV, Samaan NA, Sellin RV, Ajani JA (1992) Metastatic glucagonoma: clinical response to a combination of 5-fluorouracil and α-interferon. Am J Med 93:348–349

Kessinger A, Foley JF, Lemon HM (1983) Therapy of malignant APUD cell tumors. Effectiveness of DTIC. Cancer 51:790–794

Khandekar JD, Sriratana P (1986) Response of glucagonoma syndrome to lomustine. Cancer Treat Rep 70:433–434

Krenning EP, Kooij PPM, Bakker WH, Breeman WAP, Postema PTE, Kwekkeboom DJ, Oei HY, de Jong M, Visser TJ, Reijs AEM, Lamberts SWJ (1994) Radiotherapy with a radiolabeled somatostatin analogue, (^{111}In-DTPA-D-Phe^1)-octreotide. A case history. Ann NY Acad Sci 733:496–506

Krudy AG, Doppman LJ, Jensen RT, Norton JA, Collen MJ, Shawker TH, Gardner JD, McArthur K, Gorden P (1984) Localization of islet tumors by dynamic CT: comparison with plain CT, arteriography, sonography and venous sampling. Am J Roentgenology 143:585–589

Kurose T, Seino Y, Ishida H, Fujita J, Taminato T, Matsukura M, Imura H (1984) Successful treatment of metastatic glucagonoma with dacarbazine. Lancet I:621–622

Kvols LK, Buck M (1987) Chemotherapy of endocrine malignancies: a review. Semin Oncol 14:343–353

Lamberts SWJ, Bakker WH, Reubi J-C, Krenning EP (1990) Somatostatin-receptor imaging in the localization of endocrine tumors. N Engl J Med 323:1246–1249

Lamberts SWJ, Krenning EP, Reubi J-C (1991) The role of somatostatin and its analogs in the diagnosis and treatment of tumors. Endocr Rev 12:450–482

Lamberts SWJ, Chayvialle J-A, Krenning EP (1993) The visualization of gastroenteropancreatic endocrine tumors. Digestion 54 [Suppl 1]:92–97

Lax E, Leibovici V, Fields SI, Gordon RL (1986) Neglected radiologic signs of the glucagonoma syndrome. Diag Imag Clin Med 55:321–326

Lokich J, Bothe A, O'Hara C, Federman M (1987) Metastatic islet cell tumor with ACTH, gastrin, and glucagon secretion. Clinical and pathologic studies with multiple therapies. Cancer 59:2053–2058

Luyckx AS, Lefèbvre PJ (1981) Les glucagonomes. Diabete Metab 7:289–300

Mallinson CN, Bloom SR, Warin AP, Salmon PR, Cox B (1974) A glucagonoma syndrome. Lancet II:1–5

Maton PN (1993) Use of octreotide acetate for control of symptoms in patients with islet cell tumors. World J Surg 17:504–510

Maton PN, Gardner JD, Jensen RT (1989) Use of long-acting somatostatin analog SMS 201–995 in patients with pancreatic islet cell tumors. Dig Dis Sci 34 [Suppl]:28S–39S

McEntee GP, Nagorney DM, Kvols LK, Moertel CG, Grant CS (1990) Cytoreductive hepatic surgery for neuroendocrine tumors. Surgery 108:1091–1096

Mignon M, Jensen RT (eds) (1995) Endocrine tumors of the pancreas: recent advances in research and management. Karger, Basel

Moattari AR, Cho K, Vinik AI (1990) Somatostatin analogue in treatment of coexisting glucagonoma and pancreatic pseudocyst: dissociation of responses. Surgery 108:581–587

Modlin IM, Lewis JJ, Ahlman H, Bilchik AJ, Kumar RR (1993) Management of unresectable malignant endocrine tumors of the pancreas. Surg Gynecol Obstet 176:507–518

Moertel CG, Kvols LK, O'Connell MJ, Rubin J (1991) Treatment of neuroendocrine carcinomas with combined etoposide and cisplatin. Cancer 68:227–232

Moertel CG, Lefkopoulo M, Lipsitz S, Hahn RG, Klaassen D (1992) Streptozotocin-doxorubicin, streptozotocin-fluorouracil, or chlorozotocin in the treatment of advanced islet-cell carcinoma. N Engl J Med 326:519–523

Montenegro-Rodas F, Samaan NA (1981) Glucagonoma tumors and syndrome. Curr Probl Cancer 6:1–54

Nesovic M, Ciric J, Radojkovic S, Zarkovic M, Durovic M (1992) Improvement of metastatic endocrine tumors of the pancreas by hepatic artery chemoembolization. J Endocrinol Invest 15:543–547

Nocaudie-Calzada M, Huglo D, Deveaux M, Carnaille B, Proye Ch, Marchandise X (1994) Iodine-123-Tyr-3-octreotide uptake in pancreatic endocrine tumors and in carcinoids in relation to hormonal inhibition by octreotide. J Nucl Med 35:57–62

Obeg K, Eriksson B (1989) Medical treatment of neuroendocrine gut and pancreatic tumors. Acta Oncol 28:425–431

O'Dorisio TM (1987) Sandostatin in the treatment of gastroenteropancreatic endocrine tumors. Springer, Berlin Heidelberg New York

Parker CM, Hanke CW, Madura JA, Lisse EC (1984) Glucagonoma syndrome: case report and literature review. J Dermatol Surg Oncol 10:884–889

Pauwels S, Leners N, Fiasse R, Jamar F (1994) Localization of gastroenteropancreatic neuroendocrine tumors with [111]-Indium pentetreotide scintigraphy. Semin Oncol 21 [Suppl 13]:15–20

Pisegna JR, Doppman JL, Norton JA, Metz DC, Jensen RT (1993) Prospective comparative study of ability of MR imaging and other imaging modalities to localize tumors in patients with Zollinger-Ellison syndrome. Dig Dis Sci 38:1318–1328

Prinz RA, Badrinath K, Banerji M, Sparagana M, Dorsch TR, Lawrence AM (1981) Operative and chemotherapeutic management of malignant glucagon producing tumors. Surgery 90:713–719

Rich TA (1985) Radiation therapy for pancreatic cancer: eleven year experience at the JCRT. Int J Radiat Oncol Biol Phys 11:759–763

Rösch T, Lightdale CJ, Botet JF, Boyce GA, Sivak MV, Yasuda K, Heyder N, Palazzo L, Dancygier H, Schusdziarra V, Classen M (1992) Localization of pancreatic endocrine tumors by endoscopic ultrasonography. N Engl J Med 326:1721–1726

Rosenbaum A, Flourie B, Chagnon S, Blery M, Modigliani R (1989) Octreotide (SMS 201–995) in the treatment of metastatic glucagonoma: report of one case and review of the literature. Digestion 43:116–120

Rossi P, Allison DJ, Bezzi M (1989) Radiology of the pancreas. Endocrine tumors of the pancreas. Radiol Clin North Am 27:129–161

Ruszniewski P, Rougier P, Roche A, Legmann P, Sibert A, Hochlaf S, Ychou M, Mignon M (1993) Hepatic arterial chemoembolization in patients with liver metastases of endocrine tumors. A prospective phase II study in 24 patients. Cancer 71:2624–2630

Santangelo WC, Unger RH, Orci, L, Dueno MI, Popma JJ, Krejs GJ (1986) Somatostatin analog-induced remission of necrolytic migratory erythema without changes in plasma glucagon concentration. Pancreas 1:464–469

Sheehan-Dare RA, Simmons AV, Cotterill JA, Janke PG (1988) Hepatic tumors with hyperglucagonemia. Response to treatment with human lymphoblastoid interferon. Cancer 62:912–914

Shepherd ME, Raimer SS, Tyring SK, Smith EB (1991) Treatment of necrolytic migratory erythema in glucagonoma syndrome. J Am Acad Dermatol 25:925–928

Somers G, De Vroede M (1988) Islet cell tumors and diabetes mellitus. In: Lefèbvre PJ, Pipeleers DG (eds) The pathology of the endocrine pancreas in diabetes. Springer, Berlin Heidelberg New York, pp 171–190

Stacpoole PW (1981) The glucagonoma syndrome: clinical features, diagnosis, and treatment. Endocr Rev 2:347–361

Thompson GB, van Heerden JA, Grant CS, Carney JA, Ilstrup DM (1988) Islet cell carcinomas of the pancreas: a twenty-year experience. Surgery 104:1011–1017

Torrisi JR, Treat J, Zeman R, Dritschilo A (1987) Radiotherapy in the management of pancreatic islet cell tumors. Cancer 60:1226–1231 van der Loos TLJM, Lambrecht ER, Lambers JCCA (1987) Successful treatment of glucagonoma-related necrolytic erythema with dacarbazine. J Am Acad Dermatol 16:468–472

van Eyck CHJ, Bruining HA, Reubi J-C, Bakker WH, Oei HY, Krenning EP, Lamberts SWJ (1993) Use of isotope-labeled somatostatin analogs for visualization of islet cell tumors. World J Surg 17:444–447

Vinik AI, Moattari AR (1989) Treatment of endocrine tumors of the pancreas. Endocrinol Metab Clin North Am 18:483–518

Wawrukiewicz AS, Rösch J, Keller FS, Lieberman DA (1982) Glucagonoma and its angiographic diagnosis. Cardiovasc Intervent Radiol 5:318–324

Wynick D, Anderson JV, Williams SJ, Bloom SR (1989) Resistance of metastatic pancreatic endocrine tumours after long-term treatment with the somatostatin analogue octreotide (SMS 201-995). Clin Endocrinol (Oxf) 30:385–388

Wood SM, Polak JM, Bloom SR (1983) The glucagonoma syndrome. In: Lefèbvre PJ (ed) Glucagon II. Springer, Berlin Heidelberg New York, pp 411–430 (Handbook of experimental pharmacology, vol 66/11)

CHAPTER 16
Structure and Function of the Glucagon-Like Peptide-1 Receptor

B. THORENS and C. WIDMANN

A. Introduction

The postprandial increase in plasma insulin level is the result not only of a rise in blood glucose concentration but of a combined effect of glucose and hormones secreted by gut endocrine cells (EBERT and CREUTZFELD 1987; DUPRE 1991). The contribution of intestinal factors in the stimulation of pancreatic endocrine secretions has been recognized for a long time. At the beginning of this century studies suggested that gut "secretins" were able to stimulate pancreatic endocrine and exocrine secretions (MOORE et al. 1906). The identification of neuronal and hormonal factors originating from the intestine which control either positively or negatively the function of the endocrine pancreas led to the notion of an entero-insular axis (UNGER and EISENTRAUT 1969). With the development of peptide chemistry and the more recent progress in the molecular biological characterization of hormone genes, the most important factors involved in the stimulation of glucose-induced insulin secretion have now been identified. These are GIP (glucose-dependent insulinotropic polypeptide or gastric inhibitory polypeptide) and GLP-1 (glucagon-like peptide-1).

GIP is a 42-amino-acid polypeptide synthesized by K cells of the duodenum and secreted in response to fat or carbohydrate ingestion (BROWN 1982). GLP-1 is derived by a specific proteolytic processing of the preproglucagon molecule in L cells of the jejunum and colon (MOJSOV et al. 1986; ØRSKOV 1992). Different forms of this peptide are generated: GLP-1-(1-37), which is biologically inactive, and GLP-1-(7-37) or GLP-1-(7-36)amide, which are the biologically active forms (DRUCKER et al. 1987; MOJSOV et al. 1990; HOLST et al. 1994). GLP-1 will be used in this text to refer to the short, active forms of the molecule.

The best-described effect of GIP and GLP-1 is their stimulatory action on insulin secretion (HOLST et al. 1987; WEIR et al. 1989). Both peptides are found in the circulation shortly after absorption of a meal and they reach their highest plasma concentration after about 30 min (ELLIOTT et al. 1993). The stimulation of insulin secretion by both peptides is glucose dependent, i.e., they require glucose to be present at normal or elevated concentration and do not stimulate insulin secretion when the glucose concentration falls below the normal value of 5 mM. The stimulatory effect of both peptides is observed at physiologically relevant concentrations. For GIP these concentrations may vary between 50 and 300–500 pM and between 5 and 30 pM for GLP-1 (ELLIOTT et al. 1993; ØRSKOV et al. 1991; CREUTZFELD and NAUCK 1992). At these concentrations both hormones strongly potentiate glucose-induced insulin secretion. An additional important effect of GLP-1 in B cells of the islets of Langerhans is the stimulation of insulin gene expression and thus the replenishment of insulin stores (DRUCKER et al. 1987). Extrapancreatic effects of GLP-1 and GIP have been described, in particular on the in vivo inhibition of hydrochloric acid secretion by gastric parietal cells (SCHJOLDAGER et al. 1989), the inhibition of gastric emptying (WETTERGREN et al. 1993) and possibly also on the stimulation of fatty acid synthesis by adipocytes (OBEN et al. 1991).

One interesting feature of GLP-1 is that in type II diabetic patients its insulinotropic action is still observed although at higher than physiological concentrations. Administration of this peptide to diabetic patients can prevent the development of postprandial hyperglycemia (GUTNIAK et al. 1992; NATHAN et al. 1992; NAUCK et al. 1993b), an effect not observed with GIP (NAUCK et al. 1993a). Because of this preserved stimulatory effect in diabetes and the glucose dependence of its action, GLP-1 or agonists of its B-cell receptor may be valuable new drugs for the treatment of type II diabetes.

B. GLP-1 Receptor

I. Structure

Initial characterization of the GLP-1 receptor was achieved by an expression cloning strategy. In this experimental approach a rat islet cDNA library was constructed in a vector which, when transfected in Cos cells, drives the expression of the cDNA it carries. Expression of the GLP-1 receptor at the cell surface was then detected by its ability to bind radioiodinated GLP-1 using an autoradiographic detection system (THORENS 1992). This very powerful technique allows identification of surface receptors even if they are expressed by a rare cell population, such as pancreatic B cells, which would not permit the use of standard biochemical characterization. Following the initial cloning of the rat islet GLP-1 receptor, corresponding receptor cDNAs were cloned from human islets (DILLON et al. 1993; THORENS et al. 1993) and from rat lung (LANKAT-BUTTGEREIT et al. 1994) and hypothalamus libraries (ALVAREZ et al. 1995). The human GLP-1 receptor was localized to the long arm of chromosome 6 by STOFFEL et al. (1993).

Analysis of the deduced primary amino acid sequence revealed that the GLP-1 receptor contained seven transmembrane domains, preceded by a long hydrophilic, extracellular segment and a short leader sequence required for receptor translocation across the endoplasmic reticulum at the time of biosynthesis (THORENS 1992) (Fig. 1). The presence of the seven transmembrane segments is the signature for heterotrimeric G protein-coupled receptors

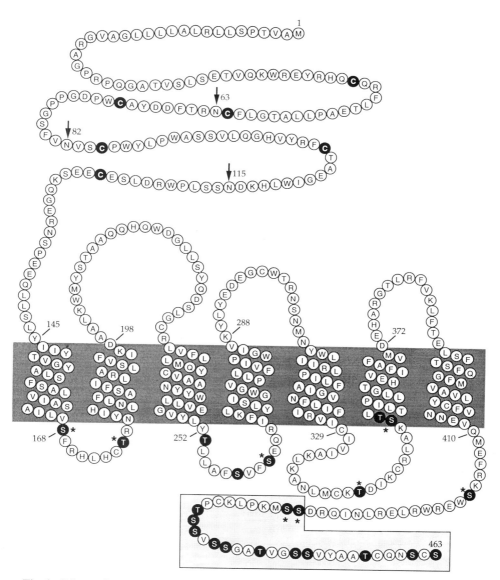

Fig. 1. Schematic representation of the GLP-1 receptor. The receptor consists of seven transmembrane domains linked by hydrophilic intracellular and extracellular loops. The amino-terminal extracellular domain is shown here with the leader sequence, which probably extends to residue 16 and which is absent from the mature molecule. The extracellular region contains six cysteines which are conserved in number and location in all the receptors of this subfamily. Three N-glycosylation sites are indicated by *arrows* in the extracellular region of the receptor. *Asterisks* indicate potential protein kinase C phosphorylation sites. The last 32 amino acids of the receptor cytoplasmic tail contain four serine doublets, some of which are phosphorylated in GLP-1- or protein kinase C-desensitized receptor. Removal of these 32 amino acids leads to inhibition of receptor internalization, loss of receptor phosphorylation and absence of homologous and heterologous desensitization but no change in ligand-binding affinity or dose-response curve of GLP-1-induced cAMP production

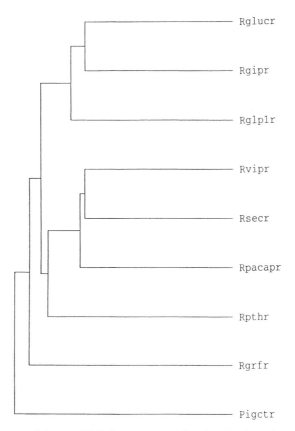

Fig. 2. Dendogram of the rat GLP-1 receptor subfamily. The length of the connecting line between two given receptors is a measure of their sequence divergence. The receptors are those for: *Rglucr*, glucagon; *Rgipr*, gastric inhibitory polypeptide; *Rglp1r*, GLP-1; *Rvipr*, vasoactive intestinal polypeptide; *Rsecr*, secretin; *Rpacapr*, pituitary adenylyl cyclase activating polypeptide; *Rpthr*, parathyroid hormone; *Rgrfr*, growth hormone releasing factor; *Pigctr*, pig calcitonin

which form a very large family. However, the GLP-1 receptor does not show any significant homology with these other receptors, except with a subfamily of peptide receptors which include those for glucagon, secretin and others (Fig. 2, Table 1). Interestingly, these receptors bind peptides which are also structurally related.

II. Tissue Distribution

Northern blot analysis of GLP-1 receptor in rat islet mRNA preparations revealed the presence of two main transcripts with sizes of 2.7 and 3.6kb (THORENS 1992). The same mRNAs are also detected in lung and stomach

Table 1. Percentage similarity between receptors of the GLP-1/glucagon/GIP receptor subfamily

	Receptors							
	GLU	GIP	VIP	SEC	PACAP	PTH	GRF	CT
GLP-1	83	74	69	70	63	63	55	55
GLU		85	68	69	62	61	55	55
GIP			65	68	61	63	56	54
VIP				68	85	69	67	54
SEC					85	56	67	56
PACAP						81	65	51
PTH							56	55
GRF								47

GLP-1, glucagon-like peptide-1; GLU, glucagon; GIP, glucose-dependent insulinotropic polypeptide; VIP, vasoactive intestinal polypeptide; SEC, secretin; PACAP, pituitary adenylate cyclase activating polypeptide; PTH, parathyroid hormone; GRF, growth hormone releasing factor; CT, calcitonin. The sequences compared are for the rat receptors except for the pig calcitonin receptor.

(THORENS 1992), in hypothalamus (ALVAREZ et al. 1996) and in a somatostatin cell line (GROS et al. 1993). This distribution is consistent with the identification of the receptor in these tissues assessed by cross-linking studies (RICHTER et al. 1990b; UTTENTHAL et al. 1992, and CALVO et al. 1995). In stomach, the receptor mRNA was detected in highly purified parietal cells (GROS et al. 1995; SCHMIDTLER et al. 1994). In human islets, a major 2.7-kb mRNA was detected together with lower abundance 4.1-, 5.0- and 6.0-kb mRNAs (THORENS et al. 1993; DILLON et al. 1993). By amplification by polymerase chain reaction, GLP-1 receptor mRNAs were also detected in the adipocyte cell line 3T3-L1 (EGAN et al. 1994) and in different segments of the mouse intestine, at various stages of development (CAMPOS et al. 1994). GLP-1 receptors were also suggested to be present in adipocytes as detected by binding assays on solubilized membrane proteins (VALVERDE et al. 1993). These data, however, await further confirmation.

The GLP-1 receptor was also characterized at the protein level using specific antibodies raised against the amino-terminal (extracellular) or carboxy-terminal (intracellular) domains of the receptor by WIDMANN et al. (1995a). As shown in Fig. 3, the B-cell receptor is detected as ~64 and ~44-kDa forms. The low molecular weight form is core glycosylated as demonstrated by its sensitivity to endo-β-glycosaminidase H. The high molecular weight form contains complex N-linked oligosaccharides and can be converted to the same ~43-kDa form as the core-glycosylated receptor by treatment with endoglycosidase F, an enzyme which eliminates all the N-linked oligosaccharides. Only the high molecular weight form is able to bind GLP-1 and is thus present on the cell surface. Why a sizeable fraction of total receptor is present inside islet B cells in a core-glycosylated form, and whether it plays a physi-

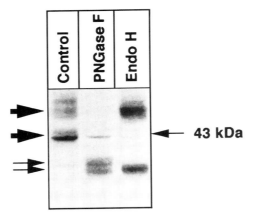

Fig. 3. Western blot analysis of the GLP-1 receptor of pancreatic islets. In total islet lysate the receptor is detected as two bands of 44 and 64 kDa *(lower and upper thick arrows)*. The low molecular weight form can be converted to a ~35-kDa form *(lower thin arrow)* by endo-β-glycosaminidase H, which indicates that it is core-glycosylated. The 64-kDa form can be converted to 35/37-kDa species *(double thin arrows)* by treatment with PNGaseF, which cleaves all the N-linked oligosaccharides. The upper form of this doublet may be due to additional modifications of the receptor present on the plasma membrane and which are not present in the core-glycosylated, intracellular form of the receptor. Only the high molecular weight form can be cross-linked to ^{125}I-GLP-1 and is thus expressed at the cell surface

ological role, is still not known. Characterization of the receptor from insulinoma by cross-linking of radioiodinated GLP-1 and gel electrophoresis analysis also indicated that the receptor was a ~65-kDa protein (GÖKE et al. 1989). In lung the receptor appears as a 55-kDa protein (RICHTER et al. 1991) although this different size compared to the B-cell receptor does not appear to be due to a different amino acid sequence (LANKAT-BUTTGEREIT et al. 1994).

Cellular expression of the GLP-1, GIP and glucagon receptors in pancreatic islet cells has been determined following autofluorescence-activated cell sorting of B and non-B cells from dispersed islet cells (PIPELEERS et al. 1985). Using a combination of Northern and Western blot analysis, it was determined that the GLP-1 and glucagon receptors were present only in B cells while the GIP receptor was present in both B and non-B cells, which are mostly A cells (MOENS et al. 1996). This is consistent with the effect of these peptides on insulin and glucagon secretion. Indeed, GLP-1 induces a strong stimulation of insulin and a decrease in glucagon secretion (ØRSKOV et al. 1993). In contrast, while GIP stimulates insulin secretion, it also stimulates glucagon secretion (SUZUKI et al. 1990). The presence of the GLP-1 receptor only in B cells thus indicates that the inhibitory effect of GLP-1 on glucagon secretion is most likely indirect, and results from the inhibitory action of insulin on A cells. GIP may, however, stimulate glucagon secretion by a direct effect on islet A cells.

III. Binding Characteristics

1. GLP-1 and Related Peptides

Binding of GLP-1 to the endogenous receptor of insulinoma (Göke and Conlon 1988) and lung (Richter et al., 1990b) membranes and to gastric glands (Uttenthal and Blazquez 1990) is with a K_d of 0.2–1 nM. Binding to the cloned receptor expressed in different transformed cell lines shows a very similar affinity (0.6 nM) (Fig. 4) (Thorens 1992; Widmann et al. 1993). Furthermore, this receptor is very specific for GLP-1 since it does not bind other peptidic hormones such as VIP, secretin or GIP (Thorens 1992) (Fig. 5). Glucagon, however, binds to the receptor but with a 1000-fold lower affinity, and oxyntomodulin, which is also derived from preproglucagon and consists of the glucagon sequence plus six amino acids on the carboxy-terminal side (see Chap. 19, this volume), binds the GLP-1 receptor with an affinity about 100-fold lower than GLP-1 (Gros et al. 1993). Previous studies have demonstrated an effect of oxyntomodulin on the stimulation of insulin secretion and on the inhibition of gastric acid secretion (Jarousse et al. 1984, 1986). These effects were, however, observed at relatively high peptide concentrations and were almost certainly due to binding of this peptide to the GLP-1 receptor.

2. Exendins

Two peptides isolated from the venom of *Heloderma suspectum*, a lizard from New Mexico, bind the GLP-1 receptor with high affinity. These peptides are called exendin-4 and exendin-(9-39), the latter being an amino-terminal truncation of the former (Eng et al. 1992). Exendin-4 shows a high homology to GLP-1: eight out of the first nine amino acids are identical between both peptides. In the rest of the molecule, only nine amino acids are conserved between exendin-4, exendin-(9-39) and GLP-1. Binding of both exendins to the GLP-1 receptor is with an affinity similar to that of GLP-1, with exendin-4 even having a slightly higher affinity than GLP-1 itself (Göke et al. 1993; Thorens et al. 1993). Interestingly, however, while exendin-4 is an agonist of the receptor, exendin-(9-39) is an antagonist (see below).

Because of their structure, the binding of exendins to the GLP-1 receptor indicates that, while the carboxy-terminal region of the peptide may be sufficient for binding, the first amino-terminal residues are required for agonist activity. Studies with GLP-1/glucagon chimeric peptides also suggest that the carboxy-terminal end of GLP-1 is essential for binding of the peptide to the receptor (Hjorth et al. 1994). Mutated GLP-1 peptides in which each amino acid was replaced one at a time by alanine residues indicated that several amino-terminal residues were important for binding and coupling. In the N-terminal regions some residues essential for GLP-1 peptide activity were located, in particular histidine 1, glycine 4, phenylalanine 6, threonine 7 and isoleucine 9. In the carboxy-terminal region, phenylalanine 22 and isoleucine 23 were shown to be crucial for binding (Adelhorst et al. 1994).

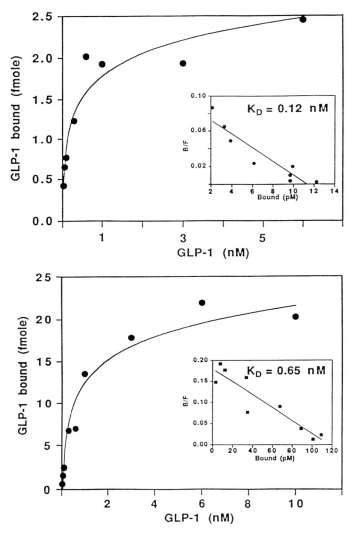

Fig. 4. Binding of ^{125}I-GLP-1 to the endogenous receptor of the INS-1 insulinoma cells and to the cloned receptor transiently transfected in Cos cells. The binding is saturable and Scatchard transformation of the binding data indicates dissociation constants (K_d) of 0.12 nM and 0.65 nM for the insulinoma and transfected receptors, respectively (THORENS 1992)

IV. Coupling to Intracellular Second Messengers

As indicated by its structure with seven transmembrane domains, the GLP-1 receptor is coupled to heterotrimeric G-proteins. Receptors from this family can activate several different intracellular signaling pathways, depending on

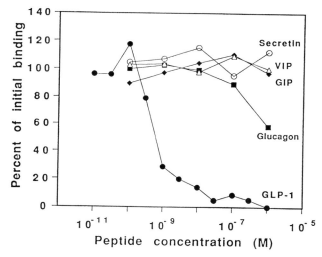

Fig. 5. Binding of GLP-1 to the receptor is specific. GLP-1 but not GIP or secretin can displace the binding of radioiodinated GLP-1 from the cloned receptor transiently transfected in Cos cells. As discussed in the text, oxyntomodulin has a ~100-fold lower affinity for the receptor and glucagon a ~1000-fold lower affinity (THORENS 1992)

the G-protein they are associated with. Binding of GLP-1 to the endogenous receptor of insulinoma cells activates the adenylyl cyclase pathway (DRUCKER et al. 1987). Studies with the cloned receptor expressed in different cell lines confirmed that activation of the adenylyl cyclase and production of cAMP was the principal second messenger pathway activated (Fig. 6) (THORENS 1992; WIDMANN et al. 1993). The dose-response curves for GLP-1-activated cAMP production with the endogenous receptor of insulinomas and the cloned receptor transfected in fibroblasts (Fig. 6) are indistinguishable and have EC_{50}s of ~0.5 nM. This value is thus very close to the K_d for peptide binding. A similar production of cAMP could be observed with exendin-4 while exendin-(9-39), which also binds to the receptor with high affinity, does not induce cAMP formation. In contrast, it inhibits GLP-1-induced activation of adenylyl cyclase and is thus an antagonist of the receptor (GÖKE et al. 1993; THORENS et al. 1993).

In some transfected cells, however, especially in Cos cells, the receptor may also activate at a low level the phospholipase C pathway with production of inositol phosphate and mobilization of cytosolic calcium (WIDMANN et al. 1993; WHEELER et al. 1993). This pathway is, however, unlikely to be of significance in B cells for the stimulation of insulin secretion since no direct coupling to inositol phosphate production or mobilization of intracellular calcium stores has been demonstrated (FRIDOLF and AHRÉN 1991). Thus, in islet B cells, cAMP is the only mediator of the GLP-1-dependent amplification of the insulin secretory response.

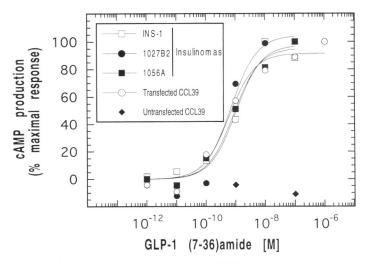

Fig. 6. GLP-1 dose-dependent production of cAMP in insulinomas and in receptor-transfected or not transfected fibroblasts. With both the endogenous receptor of insulinoma cells and the transfected receptor, similar EC_{50} values (0.5–1 nM) for cAMP production are measured (WIDMANN et al. 1994)

V. Cross-talk Between the GLP-1 and Glucose Signaling Pathways

A key feature of the gluco-incretin action is its glucose dependence. That means that GLP-1 and GIP are not secretagogues but amplify the glucose signal which induces insulin secretion. Since the GLP-1 receptor (as well as the GIP receptor) is coupled to the production of cAMP, the amplification signal will be mediated by activation of cAMP-dependent protein kinase (PKA) and phosphorylation of key proteins participating in the B-cell glucose sensor.

Glucose-induced insulin secretion first requires glucose to be taken up by the low-affinity glucose transporter GLUT2. Glucose is then phosphorylated by glucokinase and metabolized. This results in an increased intracellular ATP/ADP ratio which induces the closure of an ATP-dependent K$^+$ channel. As a result of the closure of this channel there is a depolarization of the plasma membrane which leads to the opening of voltage-gated calcium channels. The influx of Ca^{2+} then induces the exocytosis of insulin granules (see Fig. 7) (HENQUIN et al. 1992; ASHCROFT et al. 1992; HELLMAN et al. 1992). Which protein in this pathway is phosphorylated by PKA?

Addition of cAMP to permeabilized cells in which intracellular Ca^{2+} is maintained to stimulatory concentrations can still potently stimulate insulin secretion (VALLAR et al. 1987; ÄMMÄLÄ et al. 1993). This indicates that protein(s) required for the exocytosis of insulin granules may be activated by phosphorylation by PKA. These proteins have, however, not yet been

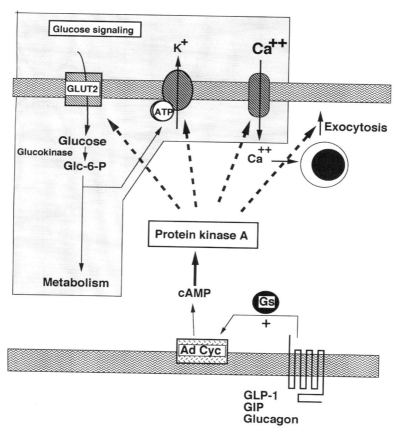

Fig. 7. Cross-talk between the glucose and gluco-incretin signaling pathways. Glucose signaling in islet B cells requires its uptake by GLUT2. This is followed by phosphorylation by glucokinase and glycolysis. The consequent rise in the intracellular ATP/ADP ratio induces the closure of KATP channels, depolarization of the plasma membrane and opening of voltage-gated Ca^{2+} channels. The resulting increase in intracellular Ca^{2+} concentration triggers the exocytosis of insulin granules. Activation of the GLP-1 receptor by ligand binding activates adenylyl cyclase and production of cAMP. The effects of cAMP are mediated by stimulation of protein kinase A. It is thought that the phosphorylation of key elements of the glucose-signaling pathway by this kinase are responsible for the amplification of the secretory response. *Broken arrows* suggest possible targets for PKA phosphorylation. See text for discussion (THORENS and WAEBER 1993)

identified. Phosphorylation of L-type Ca^{2+} channels may increase channel activity although this may be of marginal importance for the glucose-dependent secretory response (SMITH et al. 1990; ÄMMÄLÄ et al. 1993). Whether the K_{ATP} channel acivity can be modified by PKA has been suggested recently but it is not known whether this is by direct phosphorylation of the channel itself (HOLZ et al. 1993). The glucose transporter GLUT2 is rapidly and strongly phosphorylated by PKA in B cells following GLP-1 binding to its receptor. The phosphorylation takes place both on serine and threonine residues of the carboxy-terminal cytoplasmic tail and this phosphorylation decreases the transporter intrinsic activity by ~40% (THORENS et al. 1996). Whether phosphorylation of GLUT2 participates in the stimulatory effect of GLP-1 on insulin secretion is not known. The rate of insulin secretion is directed by the rate of glucose metabolism and the rate-limiting step in this process is the phosphorylation of glucose by glucokinase (MEGLASSON and MATSCHINSKY 1984, 1986). Uptake of glucose is usually 50- to 100-fold in excess over the rate of phosphorylation (HEIMBERG et al. 1993), and a ~40% decrease in transporter intrinsic activity may thus be of no effect on glucose metabolism. This is also supported by experiments performed on purified B cells which showed that the rate of glucose oxidation was not modified by addition of forskolin (GORUS et al. 1984), which, in these conditions, induces a strong phosphorylation of GLUT2. This therefore suggests that if GLUT2 phosphorylation plays a role in the potentiation effect of gluco-incretins, it may be by a mechanism not involving a change in transport activity but, for instance, by promoting interaction of the transporter with other proteins participating in the control insulin secretion. GLUT2 may thus have a signaling role distinct from its transport function, as also suggested by others (HUGHES et al. 1993). This role, however, if it exists, needs to be characterized.

Finally, another potential target for PKA may be the ryanodine receptor, a calcium-activated Ca^{2+} release channel present in intracellular calcium storage compartments (GROMADA et al. 1995). Phosphorylation may sensitize this channel to glucose-induced Ca^{2+} entry in the cells and trigger Ca^{2+} release from internal stores. This phosphorylation may thus participate in a feed-forward mechanism sustaining and amplifying the glucose-induced secretory response.

VI. GLP-1 Versus GIP in the Stimulation of Insulin Secretion

The characterization of exendin-(9-39) as a strong antagonist of the GLP-1 receptor has permitted the role of GLP-1 in the postprandial elevation of plasma insulin to be addressed. When this peptide was injected subcutaneously into rats before a meal, the resulting postprandial glycemia was increased and the plasma insulin concentration decreased as compared to control rats (WANG et al. 1995). In similar experiments, the elevation of plasma insulin levels was greatly reduced by intravenous injection of exendin-(9-39) in rats receiving an intraduodenal glucose bolus (KOLLIGS et al. 1995). The concentrations of exendin-(9-39) given in these experiments were not sufficient to antagonize the activation of the GIP receptor. The extent of plasma insulin reduction by the GLP-1 receptor antagonist was surprisingly high in the latter study and it was suggested that a greater part of the incretin effect may be due to GLP-1 rather than GIP.

VII. Regulation of Receptor Function

1. Regulated Expression

The B-cell secretory response to GLP-1 can be modulated by different mechanisms. First, a regulation of total receptor expression could change the sensitivity of B cells to elevations in GLP-1 levels. A small decrease in GLP-1 receptor mRNA expression in pancreatic islets maintained in tissue culture was induced by elevated glucose concentrations. A more significant reduction (~50%) was achieved by exposure of the islets to dexamethasone (ABRAHAMSEN and NISHIMURA 1995). The effect of dexamethasone on decreasing cell surface receptor expression by RINm5F insulinoma was also reported by RICHTER et al. (1990a). However, in contrast to the B-cell glucagon receptor, no regulation of GLP-1 receptor expression could be induced by changes in cAMP levels (ABRAHAMSEN and NISHIMURA 1995).

2. Desensitization

Another level of regulation of the GLP-1 response is by desensitization of the receptor. Fehmann and Habener have reported that the insulin secretory response of the HIT insulinoma cells was decreased following GLP-1 pretreatment of the cells (FEHMANN and HABENER 1991). We have studied desensitization of the receptor by measuring the production of cAMP induced by GLP-1 following a first preexposure of the cells to the peptide (WIDMANN et al. 1996). A strong desensitization of the cAMP response was observed which affected both the maximal production of second messengers and the EC_{50} of the dose-response curve. Beside this homolgous desensitization process, activation of protein kinase C by phorbol esters also desensitized the GLP-1-dependent cAMP production. Therefore, production of cAMP by the GLP-1 receptor, and thus potentiation of the insulin secretory response, is tightly regulated either by a previous activation of the receptor by GLP-1 or by activation of protein kinase C (WIDMANN et al. 1996). In islet B cells, several receptors are linked to the phospholipase C pathway such as the muscarinic cholinergic receptors or the receptors for vasopressin or cholecystokinin. Activation of these receptors, which also potentiate insulin secretion, may thus reduce the stimulation of the secretory activity induced by increasing GLP-1 levels.

The molecular basis for both homologous and heterologous desensitization processes involves phosphorylation of the receptor on the carboxy-terminal tail. Since desensitization and phosphorylation of the receptor induced by both processes are additive, the phosphorylation sites might be on different amino acids of the cytoplasmic tail (WIDMANN et al. 1996).

3. Internalization

GLP-1 binding also induces a fast internalization of the receptor, probably via coated pits. Following endocytosis, the ligand is degraded in lysosomes while the receptor recycles back to the plasma membrane. The kinetics for receptor internalization is much faster than that for re-expression to the cell surface. As a result, prolonged exposure of the receptor-expressing cells to GLP-1 leads to a time- and dose-dependent decrease in cell surface binding sites (WIDMANN et al. 1995). Interestingly, internalization of the receptor is not induced by binding of the antagonist exendin-(9-39). This thus suggests that a conformational change of the receptor induced by agonist binding may also be required to induce internalization.

The role of internalization in receptor physiology is not yet understood. Internalization does not participate in heterologous desensitization since activation of protein kinase C does not decrease the number of cell surface binding sites. For homologous desensitization, it has not yet been possible to exclude participation of internalization in this regulatory phenomenon. It is, however, possible that, as described for the β-adrenergic receptor, internalization is required for receptor resensitization, by allowing dephosphorylation of the receptor at a stage along the endocytic/exocytic route.

VIII. GLP-1 in Non-Insulin-Dependent Diabetes

As discussed in Chaps. 17 and 18, this volume, in type II diabetic patients, GLP-1 can stimulate insulin secretion. This potentiation effect is, however, observed at pharmacological doses of the peptide (GUTNIAK et al. 1992; NATHAN et al. 1992; NAUCK et al. 1993b). This therefore suggests that signaling by the receptor can still occur, although less efficiently. It is therefore likely that in diabetes the islet B-cell GLP-1 receptors are desensitized. This would be compatible with the observed elevated levels of GLP-1 observed in some diabetic patients which could induce homologous desensitization (ØRSKOV et al. 1991). In addition, high glucose levels may activate protein kinase C in B cells (CALLE et al. 1992). It is thus possible that hyperglycemia induces receptor desensitization in diabetes and thus progressively decreases the insulinotropic effect of GLP-1. It is not yet known whether GIP receptor is desensitized in diabetes or whether its expression is reduced. However, even given at pharmacological levels, GIP cannot correct the hyperglycemia of type II diabetic patients (NAUCK et al. 1993a). This thus indicates that a dysfunction of the GIP receptor may be an important pathogenic event in the development of NIDDM. Mutations in the genes for these receptors could potentially be involved in the pathogenesis of diabetes. No such mutations have, however, been detected in diabetic patients (VAXILLAIRE et al. 1994; ZHANG et al. 1994; TANIZAWA et al. 1994).

C. Conclusions

Molecular biology techniques have permitted the structural characterization of the GLP-1 receptor which led to new insights in one important aspect of the control of glucose homeostasis: the modulation of insulin secretion by gluco-incretin hormones. We have learned that specific B-cell receptors linked to the adenylyl cyclase pathway could be activated specifically by either GLP-1 or GIP. The activation of protein kinase A thus leads to phosphorylation of proteins participating in the glucose sensor of B cells. The identification of these proteins represents an important goal of future research to better understand the regulation of insulin secretion. The characterization of exendin-(9-39) as a strong antagonist of the GLP-1 receptor and its use in in vivo experiments has revealed that GLP-1 may be a more important incretin in the postprandial stimulation of insulin secretion than GIP.

Interestingly, studies on receptor desensitization have demonstrated that both homologous and heterologous desensitization processes could tightly control the production of cAMP by GLP-1-activated receptor. These mechanisms may be important in preventing oversecretion of insulin which could otherwise lead to the development of hypoglycemic episodes.

Finally, the use of the cloned receptor expressed stably in different cell lines is being used to analyze the peptide-binding sites and the structure of GLP-1, which is important for receptor-binding and agonist activities. These studies will hopefully lead to the development of receptor agonists. Such agonists, especially if they were small non-peptidic molecules that may be given orally, could be useful in the control of postprandial glycemia in non-insulino-dependent diabetic subjects. These could represent major new drugs for the management of NIDDM.

References

Abrahamsen N, Nishimura E (1995) Regulation of glucagon and glucagon-like peptide-1 receptor messenger ribonucleic acid expression in cultured rat pancreatic islets by glucose, cyclic adenosine 3',5'-monophosphate, and glucocorticoids. Endocrinology 136:1572–1578

Adelhorst K, Hedegaard BB, Knudsen LB, Kirk O (1994) Structure-activity studies of glucagon-like peptide-1. J Biol Chem 269:6275–6278

Alvarez E, Roncero I, Chowen JA, Thorens B, Blazquez E (1996) Gene expression of the rat glucagon-like peptide-1 (GLP-1) receptor in rat brain. J Neurochem 66:920–927

Ämmälä C, Ashcroft FM, Rorsman P (1993) Calcium-independent potentiation of insulin release by cyclic AMP in single β cells. Nature 363:356–358

Ashcroft FM, Williams B, Smith PA, Fewtrell CMS (1992) Ion channels involved in the regulation of nutrient-stimulated insulin secretion. In: Flatt PR (ed) Nutrient regulation of insulin secretion. Portland, London, p 193

Brown JC (1982) Gastric inhibitory polypeptide. Berlin, Heidelberg, New York (Monographs on endocrinology, vol 24)

Calle R, Ganesan S, Smallwood JI, Rasmussen H (1992) Glucose-induced phosphorylation of myristoylated alanine-rich C kinase substrate (MARCKS) in isolated rat pancreatic islets. J Biol Chem 267:18723–18727

Calvo JC, Yusta B, Mora F, Blazquez E (1995) Structural characterization by affinity cross-linking of glucagon-like peptide-1(7-36) amide receptor in rat brain. J Neurochem 64:299–306

Campos RV, Lee YC, Drucker DJ (1994) Divergent tissue-specific and developmental expression of receptors for glucagon and glucagon-like peptide-1 in the mouse. Endocrinology 134:2156–2164

Creutzfeld W, Nauck M (1992) Gut hormones and diabetes mellitus. Diabetes Metab Rev 8:149–177

Dillon JS, Tanizawa Y, Wheeler MB, Leng X-H, Ligon BB, Rabin DU, Yoo-Warren H, Permutt MA, Boyd III AE (1993) Cloning and functional expression of the human glucagon-like peptide-1 (GLP-1) receptor. Endocrinology 133:1907–1910

Drucker DJ, Philippe J, Mojsov S, Chick WL, Habener JF (1987) Glucagon-like peptide I stimulates insulin gene expression and increases cyclic AMP levels in rat islet cell line. Proc Natl Acad Sci USA 84:3434–3438

Dupre J (1991) Influences of the gut on the endocrine pancreas. An overview of established and potential physiological mechanisms. In: Samols E (ed) The endocrine pancreas. Raven, New York, p 253

Ebert R, Creutzfeld W (1987) Gastrointestinal peptides and insulin secretion. Diabetes Metab Rev 3:1–16

Egan JM, Montrose-Rafizadeh C, Wang Y, Bernier M, Roth J (1994) Glucagon-like peptide-1(7-36)amide (GLP-1) enhances insulin-stimulated glucose metabolism in 3T3-L1 adipocytes: one of several potential extrapancreatic sites of GLP-1 action. Endocrinology 135:2070–2075

Elliott RM, Morgan LM, Tredger JA, Deacon S, Wright J, Marks V (1993) Glucagon-like peptide-1(7-36)amide and glucose-dependent insulinotropic polypeptide secretion in response to nutrient ingestion in man: acute post prandial and 24-h secretion pattern. J Endocrinol 138:159–166

Eng J, Kleinman WA, Singh L, Singh G, Raufman J-P (1992) Isolation and characterization of exendin-4, an exendin-3 analogue, from Heloderma suspectum venom. Further evidence for an exendin receptor on dispersed acini from guinea pig pancreas. J Biol Chem 267:7402–7405

Fehmann H-C, Habener JF (1991) Homologous desensitization of the insulinotropic glucagon-like peptide-I(7-37) receptor on insulinoma (HIT-T15) cells. Endocrinology 126:2880–2888

Fridolf T, Ahrèn B (1991) GLP-1(7-36)amide stimulates insulin secretion in rat islets: studies on the mode of action. Diabetes Res 16:185–191

Gorus FK, Malaisse WJ, Pipeleers D (1984) Differences in glucose handling by pancreatic A and B cells. J Biol Chem 259:1196–1200

Gromada J, Dissing S, Bokvist K, Renström E, Frqkjaer-Jensen J, Wulff BS, Rorsman P (1995) Glucagon-like peptide-1 increases cytoplasmic calcium in insulin-secreting βTC3 cells by enhancement of intracellular calcium mobilization. Diabetes 44:767–774

Gros L, Thorens B, Bataille D, Kervran A (1993) Glucagon-like peptide-1-(7-36)amide, oxyntomodulin, and glucagon interact with a common receptor in a somatostatin-secreting cell line. Endocrinology 133:631–638

Gutniak M, Ørskov C, Holst JJ, Ahrèn B, Efendic S (1992) Antidiabetogenic effect of glucagon-like peptide-1 (7-36)amide in normal subjects and patients with diabetes mellitus. Ne Engl J Med 326:1316–1322

Göke R, Conlon JM (1988) Receptors for glucagon-like peptide-1(7-36)amide on rat insulinoma-derived cells. J Endocrincol 116:357–362

Göke R, Cole T, Conlon JM (1989) Characterization of the receptor for glucagon-like peptide-1 (7-36)amide on plasma membranes from rat insulinoma derived cells by covalent cross-linking. J Mol Endocrinol 2:93–98

Göke R, Fehmann H-C, Linn T, Schmidt H, Krause M, Eng J, Göke B (1993) Exendin-4 is a high potency agonist and truncated exendin-(9-39)-amide an antagonist at the glucagon-like peptide 1-(7-36)-amide receptor of insulin-secreting B-cells. J Biol Chem 268:19650–19655

Gros L, Hollande F, Thorens B, Kervran A, Bataille D (1995) Comparative effects of GLP-1(7-36)amide, oxyntomodulin, and glucagon on rabbit gastric parietal cell function. Eur J Pharmacol 288:319–327

Heimberg H, De Vos A, Vandercammen A, Van Schaftingen E, Pipeleers D, Schuit F (1993) Heterogeneity in glucose sensitivity among pancreatic B-cells is correlated to differences in glucose phosphorylation rather than glucose transport. EMBO J 12:2873–2879

Hellman B, Gylfe E, Grapengiesser E, Lund P-E, Marcström A (1992) Cytoplasmic calcium and insulin secretion. In: Flatt PR (ed) Nutrient regulation of insulin secretion. Portland, London, p 213

Henquin JC, Debuyser A, Drews G Plant TD (1992) Regulation of K^+ permeability and membrane potential in insulin-secreting cells. In: Flatt PR (ed) Nutrient regulation of insulin secretion. Portland, London, p 173

Hjorth S, Adelhorst K, Brogaard Pedersen B, Kirk O, Schwartz TW (1994) Glucagon and glucagon-like peptide 1: selective receptor recognition via distinct peptide epitopes. J Biol Chem 269:30121–30124

Holst JJ, Ørskov C, Vagn Nielsen O, Schwartz TW (1987) Truncated glucagon-like peptide I, an insulin-releasing hormone from the distal gut. FEBS Lett 211:169–174

Holst JJ, Bersani M, Johnsen AH, Kofod H, Hartmann B, Ørskov C (1994) Proglucagon processing in porcine and human pancreas. J Biol Chem 269:18827–18833

Holz GG, IV Kühtreiber WM, Habener JF (1993) Pancreatic beta-cells are rendered glucose-competent by the insulinotropic hormone glucagon-like peptide-1(7-37). Nature 361:362–365

Hughes SD, Quaade C, Johnson JH, Ferber S, Newgard CB (1993) Transfection of AtT-20ins cells with GLUT-2 but not GLUT-1 confers glucose-stimulated insulin secretion. Relationship to glucose metabolism. J Biol Chem 268:15205–15212

Jarousse C, Bataille D, Jeanrenaud B (1984) A pure enteroglucagon, oxyntomodulin (glucagon 37), stimulates insulin release in perfused rat pancreas. Endocrinology 115:102–105

Jarousse C, Niel H, Audousset-Puech M-P, Martinez J, Bataille D (1986) Oxyntomodulin and its C-terminal octapeptide inhibit liquid meal-stimulated acid secretion. Peptides 7:253–256

Kolligs F, Fehmann H-C, Göke R, Göke B (1995) Reduction of the incretin effect in rats by the glucagon-like peptide 1 receptor antagonist exendin (9-39) amide. Diabetes 44:16–19

Lankat-Buttgereit B, Fehamann HC, Richter G, Göke B (1994) Molecular cloning of a cDNA encoding for the GLP-1 receptor expressed in rat lung. Exp Clin Endocrinol 102:341–347

Meglasson MD, Matschinsky FM (1984) New perspectives on pancreatic islet glucokinase. Am J Physiol 246:E1–E13

Meglasson MD, Matschinsky FM (1986) Pancreatic islet gucose metabolism and regulation of insulin secretion. Diabetes Metab Rev 2:163–214

Moens K, Heimberg H, Flamez D, Huypens P, Quartier E, Ling Z, Pipeleers D, Gremlich S, Thorens B, Schuit F (1996) Expression and functional activity of glucagon-, GLP-1, and GLP-receptors in rat pancreatic islet cells. Diabetes 45:257–261

Mojsov S, Heinrich G, Wilson IB, Ravazzola M, Orci L (1986) Preproglucagon gene expression in pancreas and intestine diversifies at the level of post-translational processing. J Biol Chem 261:11880–11889

Mojsov S, Kopzynski MG, Habener JF (1990) Both amidated and nonamidated forms of glucagon-like peptide-1 are synthesized in the rat intestine and the pancreas. J Biol Chem 265:8001–8008

Moore B, Edie ES, Abram JH (1906) On the treatment of diabetes mellitus by acid extract of duodenal mucous membrane. Biochem J 1:28–38

Nathan DM, Schreiber E, Fogel H, Mojsov S, Habener JF (1992) Insulinotropic action of glucagonlike peptide-1-(7-37) in diabetic and nondiabetic subjects. Diabetes Care 15:270–276

Nauck MA, Heimesaat MM, Ørskov C, Holst JJ, Ebert R, Creutzfeld W (1993a) Preserved incretin activity of glucagon-like peptide 1 (7-36)amide but not of synthetic human gastric inhibitory polypeptide in patients with type-2 diabetes mellitus. J Clin Invest 91:301–307

Nauck MA, Kleine N, Ørskov C, Holst JJ, Willms B, Creutzfeld W (1993b) Normalization of fasting hyperglycaemia by exogenous glucagon-like peptide 1 (7-36 amide) in type 2 (non-insulin-dependent) diabetic patients. Diabetologia 36:741–744

Oben J, Morgan L, Fletcher J, Marks V (1991) Effect of entero-pancreatic hormones, gastric inhibitory polypeptide and glucagon-like polypeptide-1 (7-36)amide, on fatty acid synthesis in explants of rat adipose tissue. J Endocrinol 130:267–272

Ørskov C (1992) Glucagon-like peptide-1, a new hormone of the entero-insular axis. Diabetologia 35:701–711

Ørskov C, Jeppesen J, Madsbad S, Holst JJ (1991) Proglucagon products in plasma of noninsulin-dependent diabetics and nondiabetic controls in the fasting state and after oral glucose and intravenous arginine. J Clin Invest 87:415–423

Ørskov C, Wettergren A, Holst JJ (1993) Biological effects and metabolic rates of glucagonlike peptide-1 7-36 amide and glucagonlike peptide-1 7-37 in healthy subjects are indistinguishable. Diabetes 42:658–661

Pipeleers DG, In't Veld PA, Van De Winkel M, Maes E, Schuit FC, Gepts W (1985) A new in vitro model for the study of pancreatic A and B cells. Endocrinology 117:806–816

Richter G, Göke R, Göke B, Arnold R (1990a) Dexamethasone pretreatment of rat insulinoma cells decreases binding of glucagon-like peptide-1(7-36)amide. J Endocrinol 126:445–450

Richter G, Göke R, Höke B, Arnold R (1990b) Characterization of receptors for glucagonlike peptide-1 (7-36)amide on rat lung membranes. FEBS Lett 267:78–80

Schjoldager BTG, Mortensen PE, Christiansen J, Ørskov C, Holst JJ (1989) GLP-1 (glucagon-like peptide 1) and truncated GLP-1, fragments of human proglucagon, inhibit gastric acid secretion in humans. Dig Dis Sci 34:703–708

Schmidtler J, Dehne K, Allescher H-D, Schusdziarra V, Classen M, Holst JJ, Polack A, Schepp W (1994) Rat parietal cell receptors for GLP-1-(7-36)amide: northern blot, cross-linking, and radioligand binding. Am J Physiol 267:G423–G432

Smith PA, Fewtrell CMS, Ashcroft FM (1990) Cyclic AMP potentiates L-type Ca-channel activity in murine pancreatic β cells. Diabetologia 33:A104

Stoffel M, Espinosa III R, Le Beau MM, Bell GI (1993) Human glucagon-like peptide-1 receptor gene. Localization to chromosome band 6p21 by fluorescence in situ hybridization and linkage of a highly polymorphic simple tandem repeat DNA polymorphism to other markers on chromosome 6. Diabetes 42:1215–1218

Suzuki S, Kawai K, Ohashi S, Mukai H, Murayama Y, Yamashita K (1990) Reduced insulinotropic effects of glucagonlike peptide I-(7-36)-amide and gastric inhibitory polypeptide in isolated perfused diabetic rat pancreas. Diabetes 39:1320–1325

Thorens B (1992) Expression cloning of the pancreatic beta cell receptor for the gluco-incretin hormone glucagon-like peptide I. Proc Natl Acad Sci USA 89:8641–8645

Thorens B, Waeber G (1993) Glucagon-like peptide-I and the control of insulin secretion in the normal state and in NIDDM. Diabetes 42:1219–1225

Thorens B, Porret A, Bühler L, Deng S-P, Morel P, Widmann C (1993) Cloning and functional expression of the human islet GLP-1 receptor: demonstration that exendin-4 is an agonist and exendin-(9-39) an antagonist of the receptor. Diabetes 42:1678–1682

Thorens B, Dériaz N, Bosco D, De Vos A, Pipeleers D, Schuit F, Meda P, Porret A (1996) Protein kinase A dependent phosphorylation of GLUT2 in pancreatic β cells. J Biol Chem 271:8075–8081

Unger RH, Eisentraut AM (1969) Entero-insular axis. Arch Intern Med 123:261–266

Uttenthal LO, Blazquez E (1990) Characterization of high-affinity receptors for truncated glucagon-like peptide-1 in rat gastric glands. FEBS Lett 262:139–141

Uttenthal LO, Toledano A, Blazquez E (1992) Autoradiographic localization of receptors for glucagon-like peptide-1(7-36)amide in the rat brain. Neuropeptides 21:143–146

Tanizawa Y, Riggs AC, Elbein SC, Whelan A, Donis-Keller H, Permutt A (1994) Human glucagon-like peptide-1 receptor gene in NIDDM. Identification and use of simple sequence repeat polymorphisms in genetic analysis. Diabetes 43:752–757

Vallar L, Biden TJ, Wollheim CB (1987) Guanine nucleotide induces Ca^{2+}-independent insulin secretion from permeabilized RINm5F cells. J Biol Chem 262:504–5056

Valverde I, Merida E, Delgado E, Trapote MA, Villanueva-Peñacarillo ML (1993) Presence and characterization of glucagon-like peptide-1-(7-36)amide receptors in solubilized membranes of rat adipose tissue. Endocrinology 132:75–79

Vaxillaire M, Vionnet N, Vigouroux C, Sun F, Espinosa R, Lebeau MM, Stoffel M, Lehto M, Beckmann JS, Detheux M, Passa P, Cohen D, Van Schaftingen E, Velho G, Bell GI, Froguel P (1994) Search for a third susceptibility gene for maturity-onset diabetes of the young. Studies with eleven candidate genes. Diabetes 43:389–395

Wang Z, Wang RM, Owji AA, Smith DM, Ghatei MA, Bloom SR (1995) Glucagon-like peptide-1 is a physiological incretin in rat. J Clin Invest 95:417–421

Weir GC, Mojsov S, Hendrick GK, Habener JF (1989) Glucagon-like peptide 1 (7-37) actions on endocrine pancreas. Diabetes 38:338–342

Wettergren A, Schjoldager B, Mortensen PE, Myhre J, Christiansen J, Holst JJ (1993) Truncated GLP-1 (proglucagon 78-107-amide) inhibits gastric and pancreaic functions in man. Dig Dis Sci 38:665–673

Wheeler MB, Lu M, Dillon JS, Leng XH, Chen C, Boyd III AE (1993) Functional expression of the rat glucagon-like peptide-1 receptor, evidence for coupling to both adenylyl cyclase and phospholipase C. Endocrinology 133:57–62

Widmann C, Bürki E, Dolci W, Thorens B (1993) Signal transduction by the cloned glucagon-like peptide-1 receptor. Comparison with signalling by the endogenous receptors of β cell lines. Mol Pharmacol 45:1029–1035

Widmann C, Dolci W, Thorens B (1995) Agonist-induced internalization and recycling of the glucagon-like peptide-1 receptor in transfected fibroblast and in insulinomas. Biochem J 310:203–214

Widmann C, Dolci W, Thorens B (1996) Desensitization and phosphorylation of the glucagon-like peptide-1 (GLP-1) receptor by GLP-1 and 4-phorbol 12-myristate 13-acetate. Mol Endocrinol (in press)

Zhang Y, Cook JTE, Hattersley AT, Firth R, Saker PJ, Warren-Perry M, Stoffel M, Turner RC (1994) Non-linkage of the glucagon-like peptide 1 receptor gene with maturity onset diabetes of the young. Diabetologia 37:721–724

CHAPTER 17
Physiology and Pathophysiology of GLP-1

B. GÖKE, R. GÖKE, H.-C. FEHMANN, and H.-P. BODE

A. The Incretin Concept

The presence of intestinal factors regulating the function of the endocrine secretion from the pancreas was first described in 1906 by MOORE and colleagues. The physiologic functional connection between gastrointestinal tract and endocrine pancreas was actually proven in the 1960s when insulin became measurable in plasma. In these classic studies, the insulin response to oral or intravenous glucose – resulting in nearly identical plasma glucose levels – was compared (MCINTYRE et al. 1964; PERLEY and KIPNIS 1967). A much higher insulin secretion was found after oral glucose. It was calculated that up to 50% of the insulin release after oral glucose was triggered by the "entero-insular axis," a term introduced in the literature in 1969 (UNGER and EISENTRAUT 1969).

An important breakthrough in incretin research was the isolation of GIP (gastric inhibitory polypeptide or glucose-dependent insulin-releasing polypeptide), which possesses strong insulin-releasing actions in vitro as well as in vivo (see reviews in BROWN 1982 and PEDERSON 1994). On the other hand, it was clearly documented, employing different experimental approaches, that GIP alone does not represent the "incretin." In vivo immunoneutralization of GIP reduced the insulin secretion in response to an oral glucose load by only approximately 20%–50%. Significant insulin-releasing activities remained after removal of GIP from intestinal mucosa preparations as well as from venous perfusate from isolated perfused rat intestine (LEVIN et al. 1979; EBERT and CREUTZFELDT 1982; EBERT et al. 1983; LAURITSEN et al. 1981). Patients after intestinal resection with only a small part remaining or with a total loss of the ileum showed a smaller incretin effect than patients with larger amounts of residual ileum, although GIP levels were identical in all groups (LAURITSEN et al. 1980). From these findings it was concluded that GIP alone cannot account for the full "incretin effect" (CREUTZFELDT 1979; CREUTZFELDT and EBERT 1985). Several other gut hormones such as cholecystokinin were discussed as incretin hormones in man but had to be discounted for various reasons (CREUTZFELDT 1979; CREUTZFELDT and EBERT 1985; ENSINCK and D'ALESSIO 1992). Today, we have learned that glucagon-like peptide (GLP-1)(7-37)/(7-36)amide must additionally be considered as an incretin hormone with powerful insulinotropic effects (GÖKE and CONLON 1988; GÖKE et al. 1991; GÖKE

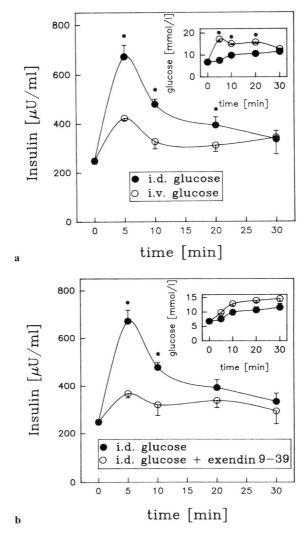

Fig. 1. a Incretin effect in rat after administration of glucose intraduodenally into the gut lumen or intravenously into the jugular vein. Values are means ± SE. *$P < 0.05$, difference between intraduodenal and intravenous glucose administration ($n = 5$, respectively). **b** Plasma insulin and glucose concentrations after an intraduodenal glucose challenge and preinjection of 5.94 nmol exendin (9-39)amide (GLP-1 receptor antagonist) in rat. *$P < 0.05$, differences between insulin levels after intraduodenal glucose alone versus intraduodenal glucose plus preinjection of exendin (9-39) amide. Glucose levels after exendin (9-39) amide were significantly ($P < 0.05$) higher than those in control subjects at 10, 20 and 30 min. (From KOLLIGS et al. 1995)

1993; FEHMANN et al. 1992, 1995c). Injection of the GLP-1 receptor antagonist exendin (9-39)amide into rats resulted in a suppression of the incretin effect by approximately 70% (KOLLIGS et al. 1995; Fig. 1). These studies indicate that besides GIP and GLP-1 no additional incretins are necessary to explain the full incretin effect in rat, especially since both incretins might additively interact (FEHMANN et al. 1989). So far, no immunoneutralization/antagonist studies in man with combinations of antisera/antagonists directed against incretin hormone action have been published. Therefore, it is still unknown whether other mechanisms contribute to the incretin effect in man. Clearly, the species differences, the individual roles and the mode of interaction of physiologically circulating concentrations of GIP and GLP-1 need to be characterized in more detail.

During the last few years, several important findings have extended our knowledge about the physiology and pathophysiology of the entero-insular axis, including the isolation and sequencing of the genes that encode the hormones with incretin activity. Specific receptors for these hormones were discovered and characterized on islet cells as well as the respective intracellular second messengers. Interactions of these hormones with each other, with nutrients and with neural mediators involved in the regulation of the endocrine pancreatic function were characterized. Furthermore, in addition to the insulin-releasing effects incretin hormones were described as potent regulators of proinsulin gene transcription and biosynthesis, suggesting that they influence pancreatic B-cell function on several levels. Furthermore, extra-pancreatic effects were suggested to contribute to the whole wealth of actions induced by these gut hormones.

B. Origin, Processing, Secretion and Fate of GLP-1

I. GLP-1 as Post-translational Product of Proglucagon Processing in Gut, Postsecretory Fate

In man and pig, the cell number of GLP-1-positive cells increases from the duodenum to the distal jejunum and ileum and, again, from the proximal to the distal colon, reaching a maximum cell density in the rectum (EISSELE et al. 1992). In rats, the L-cell number follows a similar pattern from the proximal small bowel to the ileum, where the maximal cell number is found (EISSELE et al. 1992). L cells are also present in the rat colon with increasing abundance from the proximal to the distal parts. GLP-1 is colocalized in the L cells with other proglucagon-derived peptides and, at least partly, with peptide YY (EISSELE et al. 1992; BÖTTCHER et al. 1984). Interestingly, GLP-1-producing cells were also found in the human anal canal, where the vast majority of PYY-immunoreactive cells labeled also for GLP-1 (HÖRSCH et al. 1994).

As reviewed in Chap. 2, this volume, molecular cloning techniques have allowed the determination of the structure and sequence of the gene-encoding proglucagon in several species. A very similar structure of the proglucagon gene was found in several mammals, including rat, hamster, guinea pig and man. The amino acid sequences of GLP-1 are identical in all mammals studied so far and highly homologous in many lower vertebrates (BELL et al. 1983; HEINRICH et al. 1984a,b). GLP-1 was first described in anglerfish based on the cloning of the respective cDNAs. Of all peptides so far described from the glucagon/vasoactive intestinal polypeptide (VIP)/secretin family, the non-mammalian exendin-4, isolated from Helodermatidae venom, shares the highest degree of sequence homology with GLP-1 (ENG et al. 1992).

After translation of the mRNA, hormone precursor protein is subject to extensive post-translational modifications as discussed in detail in Chap. 3, this volume. The mechanisms involved in the regulation of the post-translational processing of prohormones have begun to emerge. The intestinal L cells produce glicentin (proglucagon 1-69), oxyntomodulin (proglucagon 33-69), GLP-1 and GLP-2. GLP-1 is cleaved after position 6, resulting in the bioactive molecule GLP-1(7-37), which is, at least partly, further C-terminal truncated and amidated to yield GLP-1(7-36)amide (ORSKOV et al. 1987, 1989; UTTENTHAL et al. 1985; KREYMANN et al. 1988b; MOJSOV et al. 1990). In man, approximately 80% of the GLP-1 occurs and circulates in the truncated and amidated form (ORSKOV et al. 1994), which will be termed mature GLP-1. The reason for this last step of post-translational modification, which includes the cleavage of the amino acid at position 37 and the amidation of an arginine at position 36, is still unknown. Interestingly, the N-terminal part is responsible for the receptor binding and activation, and both peptides, amidated or not, possess identical insulin-releasing potencies in equimolar concentrations (ORSKOV et al. 1993). There are well-characterized examples that some peptide hormones are more stable after amidation than the non-amidated isoforms (BRADBURY and SMYTH 1991). Nevertheless, it was demonstrated that this is not true for the amidated truncated GLP-1: experiments in healthy human subjects showed that the plasma half-lives and metabolic clearance rates of GLP-1(7-37) and GLP-1(7-36)amide are not significantly different (ORSKOV et al. 1993). In studies with isolated rat islets, it has been shown that pancreatic proglucagon is O-glycosylated, an unusual post-translational modification for a prohormone (PATZELT and WEBER 1986). Whether this is a pancreas-specific modification or also occurs in the intestinal L cells is not known; nor is the biological significance of this phenomenon known.

Multiple forms of glucagon-like immunoreactive material in plasma were first described more than 20 years ago. Later, gel-chromatography studies in conjunction with specific radioimmunoassays allowed the determination of the circulating molecules' nature containing the GLP-1 sequence (ORSKOV et al. 1991). This was of interest since the GLP-1 sequence is contained in mature GLP-1(7-37)/(7-36)amide, the major end product of the intestinal processing of proglucagon and in proglucagon 72-158 (MPGF), an end product of the pancreatic processing of proglucagon that is secreted into circulation. After an oral glucose load, nearly all of the circulating GLP-1 immunoreactivity was found as mature GLP-1. Infusion of arginine, a direct potent secretagogue of the pancreatic A cell, selectively elevated a high molecular protein reacting with a GLP-1 antiserum, probably representing proglucagon 72-158 (ORSKOV et al. 1991). Therefore, after a meal nearly all GLP-1 plasma immunoreactivity derives from the intestine and represents the bioactive molecule (proglucagon 78-108/78-107 amide; mature GLP-1). The postsecretory fate of GLP-1 is probably influenced by one or more of a small group of membrane-bound ectopeptidases. A recent study demonstrated that GLP-1(7-36)amide appears to be a good substrate for human neutral endopeptidase in vitro (NEP) 24.11 (HUPE-SODMANN et al. 1995). However, whether NEP 24.11 is important for the metabolic clearance of GLP-1 in vivo needs to be proven. Another recent study has suggested that dipeptidyl peptidase IV may be the primary mechanism for GLP-1 degradation in human plasma (DEACON et al. 1995).

Removal of the N-terminal histidine-residue of GLP-1(7-37) resulted in a nearly completely loss of biological activity and GLP-1(8-37) stimulates insulin secretion only at concentrations above 10nmol/l (GEFEL et al. 1990). In good agreement with these data, substitution of the N-terminal histidine residue of GLP-1(7-37) with L-alanine (ala$^{7(1)}$-GLP-1) caused a dramatic loss of affinity to the GLP-1 receptor (ADELHORST et al. 1994). Interestingly, GLP-1(6-37) was also completely biologically inactive (SUZUKI et al. 1989). GLP-1(7-34), generated by removal of C-terminal amino acids, was 1000-fold less potent than GLP-1(7-37) at the perfused rat pancreas. GLP-1(7-35) is almost equipotent with GLP-1(7-37) (GEFEL et al. 1990). Both GLP-1(7-25) and GLP-1(7-20) are also completely inactive. In one study, further analysis of the GLP-1 molecule by substitution of each amino acid by alanine demonstrated that in addition to His$^{7(1)}$ also Gly$^{10(4)}$, Phe$^{12(6)}$, Thr$^{13(7)}$, Asp$^{15(9)}$, Phe$^{28(16)}$ and Ile$^{29(17)}$ are important for receptor binding and its activation. [First superscript number refers to amino acid position relative to nontruncated GLP-1(1-37); number in parenthesis indicates position of consecutively numbered amino acids of truncated GLP-1(7-37)]. These data were largely confirmed by another recent study (GALLWITZ et al. 1994). Interestingly, all these residues are homologous in exendin-4, a potent GLP-1 receptor agonist. Furthermore, His$^{7(1)}$, Gly$^{10(4)}$, Phe$^{12(6)}$, Thr$^{13(7)}$ and Asp$^{15(9)}$ of the GLP-1 molecule are identical in glucagon and peptide histidine isoleucine (PHI), both known as weak GLP-1 receptor agonists. This stresses the importance of Phe$^{28(16)}$ and Ile$^{29(17)}$ in the GLP-1 molecule. In a very recent study, two-dimensional magnetic resonance imaging (MRI) technology was employed to characterize the structure of GLP-1 in a membrane-like environment (a dodecylphosphocholine micelle) (THORNTON and GORENSTEIN 1994).

Here, GLP-1 showed a striking similarity to glucagon. It consisted of an N-terminal random coil segment (residues 1–7 of truncated GLP-1), two helical segments (residues 7–14 and 18–29) and a linker region (residues 15–17). Such structural data combined with information obtained from the molecular characterization of the recombinant GLP-1 receptor will eventually facilitate the design of new GLP-1 analogs for clinical purposes. However, first attempts to design GLP-1 agonists with a higher potency have failed (WATANABE et al. 1994; GALLWITZ et al. 1990).

Only few studies have been published which have addressed the intestinal regulation of genes encoding incretin hormones in general. This is surprising since the nutrient-dependent adaptation of gut hormones could be of pathophysiological interest. In primary cultures of fetal rat intestinal mucosa, a cAMP analog increased proglucagon mRNA (DRUCKER and BRUBAKER 1989). At the present time, no definitive information is available which bears directly on the mechanisms of nutrient-dependent glucagon gene expression in the gut L cell.

Higher proglucagon mRNA levels were found after massive small bowel resection, resulting in a marked increase in synthetic activity of the remaining L cells (FULLER et al. 1993; TAYLOR et al. 1992). This effect does not depend on the presence of nutrients, but seems to be secondary to increased polyamine synthesis, since it is dependent on ornithine decarboxylase activity (TAYLOR et al. 1992; ROUNTREE et al. 1992). The mRNA levels of proglucagon were not different between diabetic and nondiabetic NOD mice (BERGHÖFER et al. 1994). Furthermore, the distribution pattern of expression of these genes along the gastrointestinal tract remained unchanged (EFRAT et al. 1988). These findings argue against an effect of insulin on proglucagon expression.

II. Secretion of GLP-1

So far, only few studies have addressed the question of which stimuli trigger GLP-1 secretion in man. We have learned that the other incretin hormone GIP is mainly an upper small intestinal hormone found in highest concentrations in the duodenum and released following nutrient intake depending on the rate of absorption and not the mere presence of nutrients in the gut (CREUTZFELDT 1979; CREUTZFELDT and EBERT 1985; BROWN 1982). The mechanisms which control GLP-1 secretion must differ from those controlling GIP release. This is because the GLP-1-producing cells of the intestine are mainly positioned in the distal parts of the gut. This would only allow GLP-1-secreting cells to respond to direct mucosal contact of nutrients which have escaped the absorptive surface of the upper small intestine. Since nutrients rarely reach the ileum before postprandial insulin secretion occurs, the role of GLP-1 as a physiological incretin has been questioned. However, in man GLP-1 is promptly released into the circulation after oral ingestion of nutrients (MOJSOV et al. 1990; CREUTZFELDT and NAUCK 1992; D'ALESSIO et al. 1993; HERRMANN et al. 1995; RICHTER et al. 1995; Fig. 2).

It has been known for some time that after meals plasma levels of gut glucagon-like immunoreactivity increase, especially when nutrients reach the lower small intestine (HAYAKAWA et al. 1989). This circumstance very likely accounts for a prolonged and increased GLP-1 secretion under glucosidase inhibitor treatment (FUKASE et al. 1992b; GÖKE et al. 1995). There is evidence in the literature that GLP-1-like immunoreactivity in plasma increases after

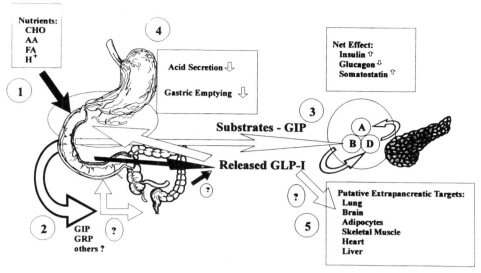

Fig. 2. Overview of the physiology of GLP-1 secretion and effects. *1*, Nutrient ingestion (*CHO*, carbohydrates; *AA*, amino acids; *FA* fat) induces a prompt release of GLP-1. *2*, Since GLP-1-producing cells reside in the lower intestines, it is likely that mediators released from the upper gastrointestinal tract and/or reflex mechanisms stimulate GLP-1 secretion from ileum and colon. *3*, Substrates derived from nutrient digestion in concert with GIP (released from the upper small intestine) and GLP-1 impact on the endocrine pancreatic islet. Insulin secretion and synthesis is stimulated and glucagon release inhibited, the latter probably by an increased somatostatin secretion. *4*, GLP-1 delays gastric emptying (physiologically) and inhibits gastric acid secretion (only phamacologically?), both possibly by central nervous effects. *5*, Circulating GLP-1 may also have an effect on other extrapancreatic tissues. Whereas the effects on adipocytes, skeletal muscle and liver are subject to controversy, the action in lung is rather mediated by peptidergic nerves of the non-adrenergic non-cholinergic pathway (NANC)

oral glucose or mixed meals (ORSKOV et al. 1991; CREUTZFELDT and NAUCK 1992; ELLIOTT et al. 1993; FUKASE et al. 1992a,b; HIROTA et al. 1990; UNGER et al. 1968). However, a wide variability in basal and stimulated values has been reported in the literature and most likely reflects differences in the specificity of the antisera used and, also of importance, in extraction procedures of plasma. After an oral glucose load (HERRMANN et al. 1995) in humans, basal plasma GLP-1 levels increase approximately six- to eightfold in a biphasic manner. Intravenous infusion of glucose does not release GLP-1. Oral galactose and a mixture of amino acids are also good stimulators of GLP-1 release. GLP-1 is also released by oral, but not intravenous, arginine loads. After a mixed meal, GLP-1 plasma levels increase rapidly, reaching basal values after 90 min (HERRMANN et al. 1995). The exact connection between gastric emptying and GLP-1 secretion after glucose ingestion has been only recently addressed. It was found that a threshold rate of gastric glucose emptying (>1.4 kal/min) is necessary to release GLP-1, whereas the secretion of the other incretin, GIP, is not under control by gastric emptying (KATSCHINSKI et al. 1995; SCHIRRA et al. 1996).

In the isolated perfused rat ileum, both intraluminal glucose and a polydiet induced sodium dependently a biphasic release of GLP-1. Intraluminal infusion of a mixture of amino acids or fat did not induce GLP-1 secretion (HERRMANN et al. 1995).

There is evidence that GIP might represent a GLP-1 secretagogue in rats. This was demonstrated in an in vitro culture system employing fetal rat intestinal mucosa cells as well as in the perfused rat intestine (BRUBAKER 1991; HERRMANN-RINKE et al. 1995). Furthermore, cholinergic agonists were good stimulators of GLP-1 secretion in vitro (BRUBAKER 1991; HERRMANN-RINKE et al. 1995). Therefore, it seems possible that GLP-1 release is controlled by several mechanisms including nutrients, the autonomic nervous system and humoral mediators. One can hypothesize the existence of an intestinal neuroneuroendocrine loop between proximal and distal gut to control incretin hormone release after a meal (ROBERGE and BRUBAKER 1993). Recent in vitro data obtained from studies utilizing a murine intestinal cell line STC1 indicated that secretion of GLP-1 induced by cholinergic agonist depends on muscarinic M3-subtype receptors (ABELLO et al. 1994).

In contrast to findings from animal studies, in humans infusion of exogenous GIP does not lead to increments in plasma GLP-1 (NAUCK et al. 1993a). Furthermore, preliminary experiments in our laboratory revealed that infusion of atropine (5 µg/kg per hour) blocking exocrine pancreatic secretion is not able to alter GLP-1 plasma levels in response to an oral load of glucose (100 g) in healthy volunteers (T. Schilling and B. Göke, unpublished results). Therefore, the fine-tuning of postprandial GLP-1 secretion in man still needs to be evaluated.

C. Tissue Distribution of GLP-1 Receptors and Biological Actions

I. General

Specific receptors for GLP-1 were first detected on RINm5F cells (GÖKE and CONLON 1988) and, consecutively, also on several other insulinoma-cell lines (GALLWITZ et al. 1993; ORSKOV and NIELSEN 1988; FEHMANN and HABENER 1991a,b, 1992) and on human insulinoma cell membranes (LANKAT-BUTTGEREIT et al. 1994b). Furthermore, GLP-1 receptors were found on somatostatin-secreting cells (FEHMANN and HABENER 1991a; GROS et al. 1992). Data obtained with proglucagon-producing cells are controversial. In contrast to previous data (MATSUMARA et al. 1992), others failed to demonstrate GLP-1-

binding sites and GLP-1 receptor mRNA in such cells (FEHMANN and HABENER 1991a). However, one should keep in mind that different cell lines might express widely different patterns of peptide receptors and, furthermore, that the same cell line can express different patterns upon repeated passages. This could also explain reported discrepancies of receptor affinities.

Analysis of binding data obtained from experiments with RINm5F cells showed a single class of binding sites for GLP-1 (GÖKE and CONLON 1988). This agrees with the finding that covalent-cross-linking of ^{125}I-GLP-1 to these cells labeled a single ligand-binding protein (GÖKE et al. 1989b), thereby allowing the calculation of an apparent molecular weight of about 63 kDa (GÖKE et al. 1989b, 1992). With RINm5F, HIT-TI5, RIN 1027B2 cells, glucagon was an only weak agonist at the GLP-1 receptor (GÖKE and CONLON 1988; FEHMANN and HABENER 1991a,b), similar to oxyntomodulin (GROS et al. 1992, 1993; FEHMANN et al. 1994a) and PHI (FEHMANN et al. 1994a). GIP and several other members of the glucagon-VIP-secretin family including (VIP), pituitary adenylate cyclase activating polypeptide (PACAP-38), helodermin and helospectin I and II up to 1 μmol/l, did not influence GLP-1 receptor binding, indicating a high specificity of ligand binding to the GLP-1 receptor (GÖKE and CONLON 1988; FEHMANN et al. 1994a,b).

Exendin-4 is an interesting new tool for the study of the ligand-GLP-1 receptor interaction. It is a 39-amino-acid-containing peptide isolated from the venom of the Gila monster *Heloderma suspectum* and has been found to be a potent agonist for the GLP-1 receptor expressed in B cells and lung (GÖKE et al. 1993c), and for the recombinant human B-cell GLP-1 receptor (THORENS et al. 1993). The N-terminal truncated exendin (9-39) amide is a potent GLP-1 receptor antagonist at B cells (FEHMANN et al. 1994b; GÖKE et al. 1993c; THORENS et al. 1993).

After binding to its receptor, GLP-1 is internalized into B cells (GÖKE et al. 1989d). So far, it is unknown whether the receptor protein itself is internalized after ligand binding. The recent cloning of the human and rat B-cell GLP-1 receptor cDNAs revealed the primary structure of these proteins (THORENS et al. 1993; THORENS 1992; DILLON et al. 1993; VAN EYLL et al. 1994; GRAZIANO et al. 1993). The receptor consists of 463 amino acids and contains seven hydrophobic regions, each most probably representing a transmembrane domain. The sequence homology at the amino acid level between the human and the rat GLP-1 receptor is ~90%. The human GLP-1 receptor gene is localized on the long arm of chromosome 6 (STOFFEL et al. 1993). Northern blot analysis with rat RNA showed a strong expression of the GLP-1 receptor in pancreatic islets and lung, and, after long exposure of films, putative transcripts in brain, liver, skeletal muscle and kidney (WHEELER et al. 1993). Another recent report describes the expression of the GLP-1 receptor on a rat medullary carcinoma of thyroid cell line 6/23 (VERTONGEN et al. 1994).

The GLP-1 receptor expressed in rat lung has an apparent molecular weight of 55 kDa (RICHTER et al. 1990a, 1991). The reason for the difference of molecular weights between the lung and the B-cell receptor has not yet been elucidated since molecular cloning of a lung GLP-1 receptor cDNA revealed a nearly identical sequence to that of the B-cell receptor cDNA (LANKAT-BUTTGEREIT et al. 1994a). It is possible that both receptors are subject to tissue-specific post-transcriptional modifications, e.g., glycosylation. It has recently been shown that glycosylation is a prerequisite for regular GLP-1 receptor function (GÖKE et al. 1994). Also, tissue-specific splicing of the receptor mRNA is possible if the receptor gene is not intronless. In fact, the human glucagon receptor gene contains several introns (LOK et al. 1994). This might also be the case for the GLP-1 receptor gene. A third possibility is the existence of several genes encoding GLP-1 receptors. However, this seems rather unlikely at present.

Northern blot analysis of RNA isolated from a human islet preparation showed multiple specifically labeled bands of 2.4, 4.1, 5.0 and 6.0kb (LANKAT-BUTTGEREIT et al. 1994a). In human insulinoma we found a major GLP-1 receptor transcript of 7.0kb (LANKAT-BUTTGEREIT et al. 1994a). Cloning of a GLP-1 receptor from a human insulinoma cDNA library showed several amino acid exchanges compared to the cDNAs from human islet cell libraries (VAN EYLL et al. 1994). Whether these mutations are of functional importance is currently under investigation. In any case, comparison of the deduced amino acid sequences of the various published sequences from human GLP-1 receptors reveals a high degree of homology between receptors in normal and tumor tissue, which corresponds to the very similar biological function of recombinantly expressed receptors.

Only few studies have been performed to characterize the regulation of expression of the incretin hormone receptors. A recent study showed a downregulation of the GLP-1 receptor number after treatment of RINm5F cells with dexamethasone, while the affinity of the receptor to its ligand remained unchanged (RICHTER et al. 1990b). A similar mechanism might contribute to the pathophysiology of steroid-induced diabetes. In studies with other G-protein-coupled receptors it was shown that steroid hormones regulate the receptor gene expression, at least in part, at the mRNA level (BENOVIC et al. 1988). Our studies showed that the steady-state levels of rat GLP-1 receptor mRNA in RINm5F cells are downregulated by dexamethasone and agents that increase intracellular cAMP levels including GLP-1 itself (JIANG et al. 1994; FEHMANN et al., in press). Treatment of RINm5F cells with forskolin resulted in a 40% reduction of GLP-1 receptor mRNA levels. GLP-1-binding sites on these cells were downregulated by 35%. The protein kinase C activator PMA (phorbol 12-myristate 13-acetate) decreased GLP-1 receptor mRNA concentrations by 20%. Thus, in insulin-secreting cells the expression of the GLP-1 receptor is negatively regulated after the activation of several second messenger systems. These data from RIN cells are in contrast to those from a recent study which reports that under all conditions that alter intracellular cAMP levels the GLP-1 receptor mRNA levels remained unchanged in cultured rat islets (ABRAHAMSEN and NISHIMURA 1995).

II. Endocrine Pancreas

GLP-1 has true insulinotropic actions on the B cells, i.e., it stimulates insulin secretion in the presence of elevated glucose levels and enhances proinsulin gene expression. Its impact on the other cells of the pancreatic islet is less well defined.

The potent insulin-releasing activities of GLP-1 have been documented in numerous species in vitro (MOJSOV et al. 1987; HOLST et al. 1987; CLARK et al. 1990; WEIR et al. 1989; GÖKE et al. 1989c, 1993b; ORSKOV et al. 1988; SHIMA et al. 1988; SUZUKI et al. 1990). Concentrations as low as 50 pmol/l were shown to release insulin from the perfused pancreas. The insulin-releasing action of GLP-1 is dependent upon the glucose concentrations (GÖKE et al. 1993b; NAUCK et al. 1993a). In man, GLP-1 lowers basal plasma glucose levels by approximately 20% before it loses its insulinotropic effect (NAUCK et al. 1993a). GLP-1 also augments the amino-acid-induced insulin secretion in man (FIESELER et al. 1995).

Both forms of bioactive GLP-1, GLP-1(7-36) amide and (7-37), are equally potent enhancers of insulin secretion. Most interestingly, GLP-1 but not GIP, was shown to release insulin and to lower plasma glucagon levels in patients with diabetes mellitus type II (NAUCK et al. 1993b; GUTNIAK et al. 1993). Furthermore, GLP-1 infusion in healthy man strongly increased the glucose clearance, which suggests that GLP-1 may have a great impact on glucose turnover in man (HVIDBERG et al. 1994).

GLP-1 very likely has a direct effect on somatostatin secretion from the pancreatic D cells. GLP-1 in low concentrations releases somatostatin from pig and rat pancreata and possesses functionally active receptors on a somatostatin-secreting cell line (FEHMANN and HABENER 1991a; GROS et al. 1992; SCHMID et al. 1990; ORSKOV and NIELSEN 1988). Furthermore, GLP-1 potently stimulates somatostatin secretion from isolated human islets. This effect was not dependent upon elevated glucose concentration (FEHMANN et al. 1995a).

GLP-1 at physiological levels inhibited glucagon secretion in various animal species (HOLST and ORSKOV 1994), this inhibition being more pronounced at low than at high blood glucose levels. An inhibitory effect on glucagon secretion was also demonstrated in man (KREYMANN et al. 1987). Although GLP-1 induced an inhibition of glucagon release from isolated human islets (FEHMANN et al. 1995a), it did not inhibit glucagon secretion in a pancreatic islet cell culture (D'ALESSIO et al. 1989) or in glucagonoma INR1 G9 cells (FEHMANN and HABENER 1991a). It was suggested that the effect of GLP-1 on insulin and somatostatin cells is direct, while the effect on glucagon-producing cells may be indirect, possibly mediated by a paracrine effect on the somatostatin cells (HOLST and ORSKOV 1994; HELLER and APONTE 1995). As discussed above, GLP-1 releases somatostatin secretion from isolated human islets and the inhibitory effect of GLP-1 on glucagon secretion negatively parallels the somatostatin release. Therefore, glucagon secretion might be suppressed by the stimulation of B and D cells (HELLER and APONTE 1995). This agrees with the concept of intra-islet interactions in which both insulin and somatostatin are known as inhibitors of glucagon secretion (SAMOLS et al. 1986). Furthermore, glucagonoma-derived A cells in culture express both insulin and somatostatin receptors (FEHMANN et al. 1994d, 1995b), which, however, still only hints at but does not prove their existence in normal cells.

So far, the effects of GLP-1 on PP cells have not been described. However, miscellaneous effects of GLP-1 have been reported on their release from other secretory products of the endocrine pancreas. Amylin is costored and coreleased with insulin (FEHMANN et al. 1990b) and, therefore, it is easy to understand that GLP-1 stimulates amylin release from isolated perfused rat pancreas (INOUE et al. 1991).

Infusion of GLP-1 in man significantly elevated basal insulin levels and decreased glucagon concentrations; the insulinotropic, but not the glucagonostatic, effect was more pronounced during co-infusion with glucose (CREUTZFELDT and NAUCK 1992; NAUCK et al. 1993a; HVIDBERG et al. 1994; KREYMANN et al. 1987; GUTNIAK et al. 1992; NATHAN et al. 1992). These data correlate with the glucose-dependent insulin-releasing action of GLP-1 found in experimental studies (GÖKE et al. 1993b).

Pre-exposure of the B cells to GLP-1 enhances the insulin secretory response during subsequent stimulation with glucose or other secretagogues ("priming effect") (FEHMANN et al. 1991). Furthermore, it has been shown that GIP and carbachol as well as glucose itself prime the B cells (ZAWALICH et al. 1989a,b; FEHMANN et al. 1990a). So far, this action has not been reproduced in man and, therefore, it is not established whether priming actions represent physiological mechanisms in the regulation of islet hormone release. Employing patch clamp techniques, it was possible to explain at least in part the sensitizing effect of GLP-1 on glucose-induced insulin secretion (HOLZ et al. 1993). This study shows that only a subgroup of pancreatic B cells are sensitive to glucose and that pretreatment of cells with GLP-1 increases the number of glucose-competent B cells. Therefore, it was proposed that GLP-1 induces a "glucose competence" at the endocrine pancreas.

This concept (HOLZ and HABENER 1992) envisages that the timing of the secretion of GLP-1 is such that circulating levels of the hormone rise concomitantly with the postprandial increase in the concentration of blood glucose. By binding to specific B-cell receptors and stimulating cAMP production, GLP-1 synergizes with glucose to induce insulin secretion. To achieve the GLP-1 effect, glucose is required, which is suggested to reflect the fact that the B-cell adenylate cyclase is also a calmodulin-regulated enzyme, the activity of which is stimulated by the rise in intracellular calcium that accompanies glucose-induced depolarization (HOLZ and HABENER 1992). A key point within this hypothesis is inhibition of an ATP-sensitive potassium channel by GLP-1-induced phosphorylation which should eventually result in membrane depolarization and entry of calcium into the cell.

Several experimental and clinical studies demonstrated that gut hormones and nervous stimuli act in concert at the endocrine pancreas. The combination of GLP-1 and GIP gives an additive effect in vitro (FEHMANN et al. 1989; SUZUKI et al. 1990). Carbachol induced a clear priming effect on GLP-1-stimulated insulin release from the perfused pancreas (FEHMANN et al. 1990a). In man, infusion of exogenous GIP and GLP-1 stimulates insulin secretion in an additive manner (NAUCK et al. 1993a). Furthermore, the insulinotropic effect of GLP-1 is potently inhibited by the arterial infusion of somatostatin into the isolated pancreas (GÖKE et al. 1989c).

Treatment of different insulinoma cell lines with GLP-1 increases the cellular levels of proinsulin mRNA (DRUCKER et al. 1987; FEHMANN and HABENER 1992; GÖKE et al. 1993c; FEHMANN et al. 1994c). Transfection experiments showed that this is a direct stimulatory action on proinsulin gene transcription rather than mRNA stabilization (FEHMANN et al. 1994c). In addition to these findings, it was shown that GLP-1 also stimulates pro/insulin biosynthesis, measured by the determination of the incorporation of ^{3}H-leucine into pro/insulin (FEHMANN and HABENER 1992). Therefore, GLP-1 is not only a powerful insulin-secretagogue but also contributes to the restoration of the intracellular insulin pool. This is most likely facilitated by means of an increased intracellular cAMP level, and is supported by the finding that insulinoma βTC1 cells transfected with a p[−1.1\] Glu-CAT proglucagon gene promoter construct showed an increased transcription upon GLP-1 stimulation (GHERZI et al. 1995). Mutational analysis of the proinsulin promoter showed that forskolin increases proinsulin promoter activity via a specific cAMP-responsive element present in this gene. In DNA/protein-binding assays at least two different nuclear proteins from B cells can be detected that bind to this specific sequence (PHILIPPE and MISSOTTEN 1991). Furthermore, the recently cloned cAMP-response element binding protein CREB 327/341 binds in vitro to this promoter element and is indeed expressed in insulin-producing cells (PHILIPPE and MISSOTTEN 1991; HABENER 1990). Therefore, it is possible that CREB 327/341 might act as at least one of several "third messengers" after binding of GLP-1 to its receptor at B cells.

Galanin, a recently discovered neuropeptide with strong inhibitory but species-dependent actions on insulin secretion, is able to inhibit GLP-1-stimulated cAMP production and, in addition, GLP-1-induced proinsulin gene expression (FEHMANN and HABENER 1990). Furthermore, somatostatin and epinephrine are able to reduce cellular levels of proinsulin mRNA (ZHANG et al. 1991). These findings support a model in which proinsulin gene expression is regulated by several stimulatory (e.g., glucose, GLP-1, GIP) and inhibitory (e.g., galanin, somatostatin, epinephrine, amylin) mediators.

III. Lung

Receptors for GLP-1 were detected in rat lung membranes (GÖKE et al. 1993d; RICHTER et al. 1990a, 1991; KANSE et al. 1988). The apparent molecular weight of this receptor is 55kDa (RICHTER et al. 1991). The isolation of a rat lung GLP-1 receptor cDNA revealed that the receptors in both B cells and lung possess identical sequences (LANKAT-BUTTGEREIT et al. 1994a). Therefore, it is likely that the GLP-1 receptor undergoes different post-transcriptional processing in lung and pancreas. The receptor in rat lung is located on mucus glands in the trachea and on vascular smooth muscle of the pulmonary arteries (RICHTER et al. 1993). GLP-1 stimulated mucus secretion from isolated rat trachea, and in isolated rings of pulmonary arteries it relaxed the pre-constricted vessels (RICHTER et al. 1993).

IV. Stomach

Specific binding sites for GLP-1 coupled to the adenylate cyclase system were shown on isolated rat gastric glands and the human gastric cancer cell line HGT-1 (UTTENTHAL and BLAZQUEZ 1990; HANSEN et al. 1988; SCHMIDTLER et al. 1991, 1994). Binding of GLP-1 to rat parietal cells resulted in an increase in cAMP formation and [14C]aminopyrine accumulation as a measure of H^+ production (SCHMIDTLER et al. 1991). Thus, GLP-1 exerts a histamine-like effect on parietal cells. Surprisingly, GLP-1 had no effect on basal and pentagastrin-stimulated acid secretion in rats but inhibited potently the pentagastrin-stimulated acid secretion in man (SCHJOLDAGER et al. 1989; O'HALLORAN et al. 1990). Furthermore, it has recently been shown that GLP-1 is a potent inhibitor of the cephalic phase of acid secretion in man (WETTERGREN et al. 1994). Thus, inhibitory effects might overshadow its stimulatory actions. This effect may be explained by a stimulatory action of GLP-1 on somatostatin release and an inhibitory action on gastrin release, which was shown in the isolated perfused rat stomach (EISSELE et al. 1990) and was also demonstrated with the GLP-1 receptor antagonist exendin (9-39) amide (EISSELE et al. 1994). However, GLP-1 had no effect on somatostatin (or gastrin release) from the antrum or from nonantral stomach in pigs (ORSKOV et al. 1988). These contradictory results may be due to species differences. Similarly, another study in man failed to confirm an inhibitory GLP-1 action at near-physiological concentrations of the peptide (NAUCK et al. 1992). However, a very recent study has suggested that the distal small intestine may participate in the late postprandial inhibitory regulation of gastric secretory function in humans, with GLP-1 possibly being an intermediary factor (LAYER et al. 1995). Thus, although the in vitro action of GLP-1 on parietal cells seems to be quite well understood, clearly more work is necessary to fully establish the in vivo effect of GLP-1 on gastric acid secretion. There is, in addition, evidence that GLP-1 inhibits gastric motility (WETTERGREN et al. 1993; WILLMS et al. 1996). Subcutaneous injection of pharmacological GLP-1 amounts prolonged the lag period of gastric emptying (SCHIRRA et al. 1994, 1996). Total emptying time (1% gastric retention) and post-lag emptying rate remained unaffected. In the same experiments, GLP-1 injection initially reduced but then transiently stimulated pancreatic enzyme delivery into the duodenum. This latter finding is very likely explained with the prolonged gastric lag period and a consecutively delayed maximal nutrient emptying into the duodenum (SCHIRRA et al. 1994).

V. Brain

The presence of cell bodies immunoreactive for GLP-1 was shown only in the nucleus tractus of solitarius and the ventral and dorsal part of the medullary reticular nucleus (SALAZAR and VAILLANT 1990; JIN et al. 1988; HAN et al. 1986). Nerve fibers immunoreactive for GLP-1 have been found throughout the brain, with the highest concentrations in midline hypothalamic structures (SALAZAR and VAILLANT 1990; JIN et al. 1988; HAN et al. 1986). Quantification of GLP-1 by radioimmunoassay revealed large concentrations in the hypothalamus and the brain stem (LUI et al. 1990; YOSHIMOTO et al. 1989). These data indicate that GLP-1 may play a role as a central neurotransmitter in the brain, probably involved in autonomic and neuroendocrine regulations. Confirming this assumption, it was demonstrated that GLP-1 induces a selective release of glutamine and glutamic acid in the basal ganglia (MORA et al. 1992). Since glutamine is synthesized almost exclusively in astrocytes, this suggests a stimulatory action of GLP-1 on astrocytes and/or neurons of the rat basal ganglia. Previous studies investigating brain GLP-1 receptors were performed with crude membrane preparations of large brain regions or gave only indirect evidence for the presence of GLP-1 receptors by the ability of GLP-1 to stimulate cAMP production in certain brain regions (HOOSEIN and GURD 1984; SHIMIZU et al. 1987). However, these findings probably need to be readdressed since in this study biologically inactive, nontruncated GLP-1 (1-36)amide was employed. Still, the existence of a receptor for nontruncated GLP-1 (1-37) is hypothetically possible. In other studies, GLP-1 receptors were localized on rat brain sections using quantitative autoradiography. Binding of radiolabeled GLP-1 was high in the mammillary nuclei, the lateral septum, the subfornical organ, the arcuate nuclei, the thalamus, the hypothalamus, the interpenduncular nucleus, the posterodorsal tegmental nucleus, the area postrema, the inferior olive, the nucleus of the solitary tract, the posterior lobe of the pituitary and the olfactory bulb (UTTENTHAL et al. 1992; GÖKE et al. 1995c, in press). Biochemical characterization by cross-linking studies revealed an M_r of the GLP-1-binding protein complex of 63000 (GÖKE et al. 1996), which is identical to that of receptors on RINm5F cells (GÖKE et al. 1989b). On the other hand, CALVO et al. (1995) reported an M_r of 56000 for the brain GLP-1 receptor. This difference may be due to variations in the PAGE technique and the presence or absence of reducing agents. In the hypothalamus and brain stem high- and low-affinity GLP-1 receptors are present (CALVO et al. 1995; GÖKE et al. 1995c). In contrast, only a single class of binding site was found on RINm5F cells (GÖKE and CONLON 1988), rat lung membranes (RICHTER et al. 1990a) and gastric glands (UTTENTHAL and BLÁZQUEZ 1990). So far, it is unclear whether these discrepancies are due to experimental conditions or environmentally dependent variations of the receptor conformation. However, the recent cloning of the human brain GLP-1 receptor revealed no differences in the amino acid sequence of the GLP-1 receptor in pancreas, heart and brain (WEI and MOJSOV 1994).

A number of functional implications emerge from the observed distribution of GLP-1-binding sites. For example, the presence of high densities of binding sites in the hypothalamus, the limbic system and in areas without blood-brain barrier such as the area postrema and the subfornical organ suggest GLP-1 influences the regulation of vital homeostatic functions such as body temperature, blood pressure, heart rate, blood osmolarity, and food and water intake. Confirming this assumption, intracerebroventricular (i.c.v.) injection of GLP-1 almost totally abolished food intake in free-feeding rats (TURTON et al. 1996; TANG-CHRISTENSEN, LARSEN, GÖKE, FINK-JENSEN, JESSUP, MØLLER, SHEIKH, unpublished data). In rats kept on a food restriction schedule, i.c.v. administration of GLP-1 dose-dependently inhibited food and water intake. Furthermore, angiotensin-II-induced drinking behavior in rats was blocked by i.c.v. GLP-1, and in rats kept on a water restriction schedule water intake was suppressed by exogenous GLP-1. Peripheral administration of GLP-1 inhibited both angiotensin-II-induced drinking and water intake in rats kept on a food restriction schedule. These findings correlate with the presence of GLP-1 receptors in the lateral hypothalamic area, the arcuate and ventromedial nuclei and the subfornical organ. Furthermore, the i.c.v. application of GLP-1 induced an increase in urine output and natriuresis (TANG-CHRISTENSEN, LARSEN, GÖKE, FINK-JENSEN, JESSUP, MØLLER, SHEIKH, unpublished data), suggesting that GLP-1 could influence magnocellular vasopressinergic neurons in the hypothalamo-neurohypophysial tract. Recently, it was reported that intravenous administration of GLP-1 increases blood pressure and heart rate in rats (BARRAGÁN et al. 1994). These effects may be due to binding of GLP-1 to the area postrema or the nucleus of the solitary tract. GLP-1 inhibits pancreatic secretion in humans in response to ingestion of a meal (WETTERGREN et al. 1993). Since GLP-1 has no effect on the exocrine secretion of the isolated rat pancreas (WEBER et al. 1991) and does not affect the secretory response to vagus stimulation in isolated pig pancreas with intact innervation (HOLST et al. 1993), the inhibitory action of GLP-1 is most likely centrally or neurally mediated. Therefore, GLP-1 may not only act as a regulator of autonomic and neuroendocrine functions but may also play a role in the putative "gut-brain axis."

VI. Adipose Tissue

In explanted rat adipose tissue, GLP-1 may stimulate fatty acid synthesis (OBEN et al. 1991). Furthermore, GLP-1 was reported to stimulate the lipolysis in isolated rat adipocytes (RUIZ-GRANDE et al. 1992). Recently, receptors for GLP-1 were described on solubilized membranes of rat epididymal adipose tissue (VALVERDE et al. 1993). Scatchard plot analysis of the corresponding binding data revealed the presence of high- and low-affinity binding sites (K_d 0.6 and 20 nM, respectively), which is in contrast to data obtained with other tissues (insulinoma cells, lung, stomach), where only a single class of high-affinity binding sites were identified. This difference may be due to a negative cooperativity between GLP-1 receptors on adipocytes. In any case, because of the importance of these findings these data need confirmation. Data from the same laboratory suggested a higher binding of GLP-1 to adipose tissue in diabetics than in healthy controls (VILLANUEVA-PENACARRILLO et al. 1994b).

VII. Skeletal Muscle

A previous study has claimed the existence of a potent stimulatory effect of GLP-1 on glycogen synthesis, glucose oxidation and lactate formation in rat skeletal muscle (VILLANUEVA-PENACARRILLO et al. 1994a). These data await confirmation. In a very recent study aiming to reproduce these effects, Fürnsinn and coworkers could not find any effect of GLP-1 on rat skeletal muscle glucose metabolism (FÜRNSINN et al. 1995). On the other hand, another recent study reported GLP-1 binding to rat skeletal muscle (DELGADO et al. 1995). Surprisingly, binding was not followed by an increase in intracellular cAMP, which led to the hypothesis of an alternative type of GLP-1 receptor different to the conventional species.

VIII. Others

1. Exocrine Pancreas

GLP-1 has no direct influence on exocrine rat pancreatic secretion in vitro (WEBER et al. 1991). Thus, it seems unlikely that GLP-1 acts as a direct inhibitor of exocrine pancreatic secretion, thereby mediating an "ileal brake" effect on the exocrine pancreas (LAYER and HOLST 1993).

2. Liver

GLP-1 does not possess specific receptors on hepatocytes and rat liver membranes (BLACKMORE et al. 1991) and, furthermore, GLP-1 does not directly influence hepatocyte function like cAMP formation, glycogen phosphorylase A activity and glucose release in cultured rat hepatocytes (MURAYAMA et al. 1990; GHIGLIONE et al. 1985; SHIMIZU et al. 1986). These data are in contrast to recent reports, which suggest that GLP-1 stimulates the formation of glycogen from glucose in isolated rat hepatocytes (VALVERDE et al. 1994) and that the GLP-1 receptor mRNA is present in liver of neonatal mice (2–5 weeks old) (CAMPOS et al. 1994).

D. Signal Transduction of the GLP-1 Receptor

I. cAMP Pathway

There is solid evidence that the GLP-1 receptor expressed on pancreatic B cells is functionally coupled to the adenylate cyclase system (Fig. 3). GLP-1 enhances intracellular cAMP in pancreatic islets cells and in several

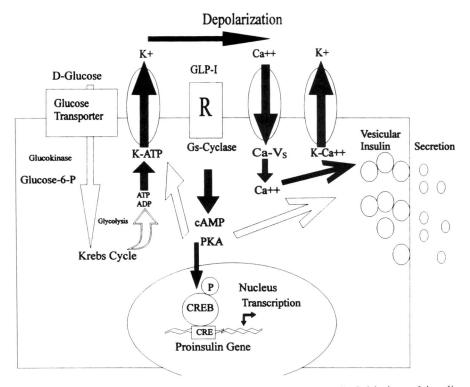

Fig. 3. Interplay of signal transduction systems in the B cell. Initiation of insulin secretion by glucose involves closure of K-ATP channels and an influx of Ca^{2+} via L-type Ca^{2+} channels. GLP-1 potentiatiates the glucose-induced insulin secretion after activating protein kinase A (*PKA*). This is a result of an activation of the adenylate cyclase (Gs-cyclase), which is linked to a stimulatory G protein (Gs). cAMP increases. PKA stimulates proinsulin gene transcription by phosphorylation of the transcription factor CREB (cAMP-responsive element), which binds to a cAMP-responsive element (*CRE*) in the proinsulin gene promoter. By a so far unresolved mechanism PKA activation also supports closure of the K-ATP channel. This leads to membrane depolarization and influx of calcium into the cell via a slow voltage-dependent calcium channel (Ca-Vs). Elevation of intracellular calcium is a prerequisite for insulin secretion. PKA may also stimulate other effector proteins with an impact on exocytosis

insulinoma cell lines (GÖKE and CONLON 1988; GÖKE et al. 1989a; DRUCKER et al. 1987; FEHMANN et al. 1994c). Stable guanidine nucleotide analogs reduced the GLP-1 receptor binding to RINm5F cells (GÖKE et al. 1989a). This finding suggests that the GLP-1 receptor is functionally coupled to the adenylate cyclase system by a stimulatory G-protein. Indeed, in response to GLP-1, a direct stimulation of adenylate cyclase and protein kinase A (PKA) was shown (FEHMANN et al. 1994c). Studies with insulinoma cells have demonstrated that activation of the cAMP pathway can induce phosphorylation as well as dephosphorylation. GLP-1 stimulation induced an increased phosphorylation of two proteins at molecular weights of 18.5kDa at pIs of 4.9, 5.1 and 5.3, respectively, and one with 23kDa, pI 5.5. One protein (23kDa, pI 5.5) was dephosphorylated (GÖKE et al. 1996; STEFFEN and GÖKE 1994). GLP-1 did not influence B-cell glucose utilization or inositol phosphate generation in [³H]inositol-prelabeled islets and HIT-T15 cells (ZAWALICH et al. 1993).

In addition, the receptor for GLP-1 undergoes a rapid and reversible homologous receptor desensitization in vitro, which is typical for G-protein-coupled receptors (FEHMANN and HABENER 1991b). Amylin and the structurally related hormone calcitonin-gene-related peptide inhibited GLP-1 but not forskolin-stimulated cAMP increase in RINm5F cells, whereas the molecular mechanism of this inhibition remained unclear (GÖKE et al. 1993a).

II. Calcium

Similar to the induction of insulin secretion by nutrient secretagogues, the stimulatory action of GLP-1 on insulin secretion also appears to be largely dependent on calcium influx across the B-cell plasma membrane. The calcium channel blocker nitrendipine ($5\,\mu M$) nearly abolished stimulation of insulin secretion by GLP-1 in perfused rat islets (ZAWALICH et al. 1993). Since insulin secretion in general is closely linked with the cytosolic calcium concentration (PRENTKI et al. 1987; GILON et al. 1993), the question arises as to whether GLP-1 acts by raising cytosolic calcium concentrations.

In fact, elevation of B-cell cytosolic calcium by GLP-1 has been described in several studies, using for example isolated single rat B cells (YADA et al. 1993), rat B-cell suspensions (FRIDOLF and AHRÉN 1993) or isolated single B cells from *ob/ob* mice (CULLINAN et al. 1994). In all these studies the GLP-1-induced calcium elevation was prevented completely (YADA et al. 1993; FRIDOLF and AHRÉN 1993) or almost completely (CULLINAN et al. 1994) by antagonists of L-type voltage-dependent calcium channels. This demonstrates that opening of L-type calcium channels is required for cytosolic calcium elevation by GLP-1 in B cells.

B-cell calcium elevation by GLP-1 was found to depend on the ambient glucose concentration. At a glucose concentration that does not elevate cytosolic calcium or induce insulin secretion ($3\,mM$), GLP-1 also did not raise calcium levels (YADA et al. 1993; FRIDOLF and AHRÉN 1993; CULLINAN et al. 1994). In contrast, at glucose concentrations that induce insulin secretion and cytosolic calcium increases, GLP-1 mediates typical additional, superimposed calcium elevations. Thus, the mode of GLP-1 action on B-cell cytosolic calcium corresponds to its facilitating effect on glucose-induced insulin secretion. Figure 4 shows a typical example of GLP-1-mediated cytosolic calcium elevations at stimulatory glucose concentrations, recorded with digital imaging fluorescence microscopy.

The glucose dependency of the calcium elevation mediated by GLP-1 is further emphasized by its action in insulinoma cells. In the various insulinoma cell lines that serve as models for B cells, a GLP-1-mediated calcium elevation

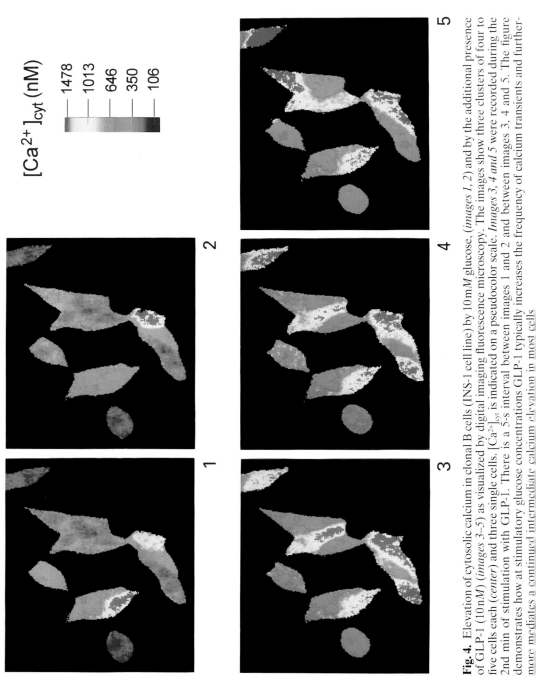

Fig. 4. Elevation of cytosolic calcium in clonal B cells (INS-1 cell line) by 10 mM glucose, (*images 1, 2*) and by the additional presence of GLP-1 (10nM) (*images 3–5*) as visualized by digital imaging fluorescence microscopy. The images show three clusters of four to five cells each (*center*) and three single cells. $[Ca^{2+}]_{cyt}$ is indicated on a pseudocolor scale. *Images 3, 4 and 5* were recorded during the 2nd min of stimulation with GLP-1. There is a 5-s interval between images 1 and 2 and between images 3, 4 and 5. The figure demonstrates how at stimulatory glucose concentrations GLP-1 typically increases the frequency of calcium transients and furthermore mediates a continued intermediate calcium elevation in most cells

is correlated with preservation of glucose sensitivity. In INS-1 rat insulinoma cells, which respond to glucose in the physiological concentration range with depolarization and cytosolic calcium elevation (ASFARI et al. 1992), GLP-1 has similar effects on cytosolic calcium as in B cells (BODE and GÖKE 1995) (Fig. 4). HIT-T15 hamster insulinoma cells display only a partially preserved response to glucose and, correspondingly, show less prominent calcium elevations by GLP-1 (LU et al. 1993). Finally, in the nearly glucose-insensitive rat insulinoma cell line RINm5F (PRAZ et al. 1983), no cytosolic calcium elevation by GLP-1 was found at all (GÖKE et al. 1989a).

The intracellular events that after binding of GLP-1 to its receptor finally lead to opening of L-type calcium channels are currently under debate. GLP-1 depolarizes rat B cells in the presence of stimulatory glucose concentrations, and this action was found to be associated with increased closure of ATP-sensitive potassium channels (HOLZ et al. 1993). The GLP-1-mediated depolarization could be based on closure-promoting phosphorylation of such K_{ATP} channels by protein kinase A. The recently determined nucleotide sequence of the sulfonylurea receptor (AGUILAR-BRYAN et al. 1995), which is closely associated with or identical to the K_{ATP} channel of B cells, suggests the existence of several protein kinase A phosphorylation sites in the protein. However, direct evidence for mediation of depolarization by GLP-1 via such a mechanism is as yet lacking. A study with mouse B cells failed to demonstrate an effect of GLP-1 on K_{ATP} channels (BRITSCH et al. 1995), although GLP-1 was also found to promote depolarization. The authors suggested that this effect was due to a slight slowing of calcium channel inactivation by GLP-1, leading to increased calcium influx. Another potential mechanism for a GLP-1-mediated depolarization can be inferred from the action of the related pituitary adenylate cyclase-activating polypeptide (PACAP) on B cells. PACAP, which like GLP-1 elevates cAMP, appeared to activate a nonselective cation channel, causing depolarization by sodium influx (LEECH et al. 1995). Interestingly, in one study the GLP-1-induced calcium elevation in B cells depended on the presence of extracelluar sodium (FRIDOLF and AHRÉN 1993). This could be explained by induction of the calcium rise by a depolarization based on sodium influx via nonselective cation channels. However, involvement of such channels in the calcium-raising action of GLP-1 remains to be proven.

In summary, the GLP-1-mediated calcium influx in B cells could originate from a depolarization by decreased activity of K_{ATP} channels or opening of nonselective cation channels, from directly increased activity of L-type calcium channels, or from a combination of these mechanisms (Fig. 5). Further studies will need to address whether the described discrepancies are due to different experimental conditions and species, and which mechanisms are relevant in vivo.

Agents that elevate cAMP as well as membrane-permeable cAMP analogs mimicked the calcium-elevating action of GLP-1 in B cells (YADA et al. 1993; FRIDOLF and AHRÉN 1993; CULLINAN et al. 1994). On the other hand, cAMP-elevating agents did not affect cytosolic calcium in the continued pres-

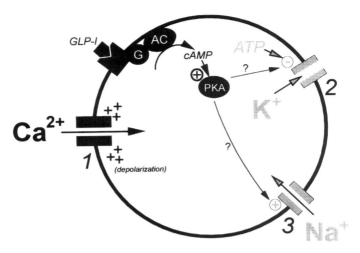

Fig. 5. Mechanisms of GLP-1-mediated calcium entry into B cells: GLP-1 stimulation induces membrane depolarization and thereby *1* opening of an L-type voltage-dependent calcium channel. This effect may be mediated by actions of protein kinase A on *2* the ATP-sensitive potassium channel or on *3* a nonselective cation channel

ence of a maximally effective GLP-1 concentration (YADA et al. 1993). These observations confirm mediation of GLP-1 signal transduction by cAMP also for the GLP-1-induced calcium rise. Several descriptive studies have already reported detailed effects of cAMP on B-cell cytosolic calcium, such as facilitation of slow calcium oscillations (GRAPENGIESSER et al. 1992), transition from an oscillatory state to a sustained elevation as observed in mouse (GRAPENGIESSER et al. 1991) or human B cells (HELLMAN et al. 1994), and induction of fast oscillations (GRAPENGIESSER et al. 1991). These effects will therefore likely be shared by GLP-1. The mentioned reports show that elevation of cAMP can entail very pronounced and physiologically relevant alterations of cytosolic calcium. This was less obvious for GLP-1 in the more basic and mechanistic studies on calcium elevation by this peptide (YADA et al. 1993; FRIDOLF and AHRÉN 1993; CULLINAN et al. 1994). The importance of cytosolic calcium in GLP-1 action is thus further emphasized by the reported effects of cAMP.

In contrast to mediation of calcium influx via voltage-dependent calcium channels, calcium release from intracellular stores does not appear to be a primary effect of GLP-1. In one study the rat GLP-1 receptor expressed in COS-7 cells was found to activate both adenylate cyclase and phospholipase C, with concomitant intracellular calcium release (WHEELER et al. 1993). This, however, is different from the behavior of the receptor in its natural environment, where no evidence exists for primary activation of phospholipase C and intracellular calcium release (YADA et al. 1993; GÖKE et al. 1989a). The findings with the reconstituted receptor (WHEELER et al. 1993) may result from the unnatural background represented by COS cells. Secondary calcium release from intracellular stores, subsequent to calcium influx via voltage-dependent calcium channels, is discussed as part of the effects of nutrient secretagogues (BERGGREN and LARSSON 1994). Secondary calcium release in this way could also be an effect of GLP-1. To clarify this, probably more knowledge about the relation of depolarization-induced calcium influx and release of stored calcium in B cells will be required.

While a rise in $[Ca^{2+}]_i$ is necessary for GLP-1 to potentiate glucose-induced insulin release, the effects of GLP-1 on insulin gene expression are mediated mainly through the cAMP-CRE pathway (FEHMANN and HABENER 1992).

E. Pathophysiological Relevance?

Increased levels of the incretin GIP were found in some obese human subjects as well as in a subgroup of patients with diabetes mellitus type II. Nevertheless, most studies emphasize that GIP levels are not elevated in diabetics (CREUTZFELDT 1979; CREUTZFELDT and EBERT 1985; CREUTZFELDT and NAUCK 1992; PEDERSON 1994). It was proposed that this phenomenon might at least in part be caused by an insensitivity of the K cells to insulin, resulting in an impaired feedback regulation between insulin and GIP release. Similar findings were made in studies with obese mice. Interestingly, these animals are characterized by a hyperplasia of the K cells (BROWN 1982; PEDERSON 1994).

GLP-1 plasma levels were also found slightly elevated in patients with diabetes mellitus type II (ORSKOV et al. 1991). Furthermore, the intestinal GLP-1 content was increased in streptozotocin-treated rats, but 3 weeks after the induction of diabetes no differences in proglucagon mRNA levels in the intestine were detected (BRUBAKER et al. 1989; KREYMANN et al. 1988a). This lack of effect was also seen in other diabetes models (BERGHÖFER et al. 1994). It is tempting to speculate that elevated levels of GLP-1 desensitize/downregulate its own receptor on B cells. However, such phenomena have only been described in vitro, and high concentrations of ligand are needed to induce such an effect (FEHMANN and HABENER 1991b). Recent genetic studies were unable to determine that inherited defects in the GLP-1 receptor gene are a major risk factor for type II diabetes (TANIZAWA et al. 1994; ZHANG et al. 1994).

In patients with dumping syndrome, high postprandial plasma levels of GLP-1 have been reported (MIHOLIC et al. 1991). This "oversecretion" of GLP-1 was suggested to be responsible for the hyperinsulinemia followed by hypoglycemia that causes the symptoms. Similarly, in patients who had undergone esophageal resection and replacement by the stomach, high GLP-1 plasma levels in the first 30 min after a meal were associated with low serum glucose when hypoglycemic episodes occurred (MIHOLIC et al. 1993). However, another recent communication disputed that GLP-1 is responsible for the reactive hypoglycemia after partial gastrectomy (ANDREASEN et al. 1994). Although plasma GLP-1 concentrations increased to much higher levels in the patients suffering from dumping symptoms, it was found by calculating the insulinogenic index that neither insulin nor C-peptide responses showed any correlation whatsoever with GLP-1 responses (ANDREASEN et al. 1994). In any case, keeping in mind the glucose dependency of the GLP-1 action on insulin secretion, it seems difficult to explain how GLP-1 would induce symptomatic hyperinsulinemia. However, high loads of GLP-1 were reported to enhance insulin levels under basal glucose levels in vitro (GÖKE et al. 1993b) and in vivo (KREYMANN et al. 1987). Another aspect in this context is the putative production and release of GLP-1 by gastroenteropancreatic endocrine tumors. In fact, immunohistochemical staining of 88 tumors revealed GLP-1 immunoreactivity in 17 neoplasias (19%), 7 out of 33 nonfunctioning tumors, 4 out of 20 gastrinomas, 4 out of 13 insulinomas, 1 out of 3 VIPomas, and 1 adrenocorticotropic hormone (ACTH)-producing tumor (EISSELE et al. 1994). Of the 17 GLP-1 immunoreactive tumors, 15 were primarily located in the pancreas. However, in only 2 out of 22 serum samples available were elevated GLP-1 levels found. Although it is possible that GLP-1 is secreted by a few tumors, no increased rates of hypoglycemic episodes were observed in the studied noninsulinoma patients. Another interesting observation was that patients with GLP-1-immunoreactive tumors were characterized by a significantly lower rate of distant metastases and a higher rate of curative resections (EISSELE et al. 1994). This indicates that the probability of producing GLP-1 is very likely correlated to the differentiation of the endocrine tumor.

As reviewed in Chap. 18 in this volume, GLP-1 is presently receiving a great deal of attention as a new tool for the treatment of diabetics (AMIEL 1994; NAUCK et al. 1993a,c,d; NATHAN et al. 1992; GUTNIAK et al. 1992, 1993; ELAHI et al. 1994). GLP-1 obviously has potent effects on endocrine pancreas and stomach (Fig. 2). Furthermore, additional effects may be important for its physiological/clinical significance. In recent studies, D'Alessio and coworkers demonstrated that GLP-1 has direct effects on tissues involved in glucose disposal since it enhanced glucose tolerance in healthy volunteers by significantly increasing glucose effectiveness (D'ALESSIO et al. 1994, 1995). Although so far it is only speculation, future work has to define whether GLP-1 is capable of inducing "insulin-sensitizing" effects in peripheral tissues.

References

Abello J, Ye F, Bosshard A, Bernard C, Cuber JC, Chayvialle JA (1994) Stimulation of glucagon-like peptide-I secretion by muscarinic agonist in a murine intestinal endocrine cell line. Endocrinology 134:2011–2017

Abrahamsen N, Nishimura E (1995) Regulation of glucagon and glucagon-like peptide-1 receptor messenger ribonucleic acid expression in cultured rat pancreatic islets by glucose, cyclic adenosine 3',5'-monophosphate, and glucocorticoids. Endocrinology 136:1572–1578

Adelhorst K, Hedegaard BB, Knudsen LB, Kirk O (1994) Structure-activity studies of glucagon-like peptide-I. J Biol Chem 269:6275–6278

Aguilar-Bryan L, Nichols CG, Wechsler SW, Clement JP IV, Boyd AE III, González G, Herrera-Sosa H, Nguy K, Bryan J, Nelson DA (1995) Cloning of the β cell high-affinity sulfonylurea receptor: a regulator of insulin secretion. Science 268:423–426

Amiel SA (1994) Glucagon-like peptide: a therapeutic glimmer. Lancet 343:4–5
Andreasen JJ, Orskov C, Holst JJ (1994) Secretion of glucagon-like peptide-I and reactive hypoglycemia after partial gastrectomy. Digestion 55:221–228
Asfari M, Janjic D, Meda P, Li G, Halban PA, Wollheim CB (1992) Establishment of 2-mercaptoethanol-dependent differentiated insulin-secreting cell lines. Endocrinology 130:167–178
Barragán JM, Rodríguez RE, Blázquez E (1994) Changes in arterial blood pressure and heart rate induced by glucagon-like peptide-1(7-36)amide in rats. Am J Physiol 266:E459–E466
Bell GI, Santerre RF, Mullenbach GT (1983) Hamster preproglucagon contains the sequence of glucagon and two related peptides. Nature 302:716–718
Benovic JL, Bouvier M, Caron MG, Lefkowitz RJ (1988) Regulation of adenylyl cyclase coupled β-adrenergic receptors. Annu Rev Cell Biol 4:405–428
Berggren PO, Larsson O (1994) Ca^{2+} and pancreatic B-cell function. Biochem Soc Trans 22:12–18
Berghöfer P, Schneider K, Linn T, Fehmann HC, Göke B (1994) Gewebespezifische Expression von Inkretinhormonen im Darm von NOD Mäusen. Diabetes Stoffwechsel 3:154
Blackmore P, Mojsov S, Exton JH, Habener JF (1991) Absence of insulinotropic glucagon-like peptide-I(7-37) receptors on isolated rat liver hepatocytes. FEBS Lett 283:7–10
Bode HP, Göke B (1995) Action of glucagon-like peptide (7-36)amide (GLP-1) on cytosolic calcium in insulin-secreting cells. Gut (abstract 4th UEGW)
Böttcher G, Sjölund K, Ekblad E, Hakanson R, Schwartz TW, Sundler F (1984) Coexistence of peptide YY and glicentin immunoreactivity in the endocrine cells of the gut. Regul Pept 8:261–266
Bradbury AF, Smyth DG (1991) Peptide amidation. Trends Biochem Sci 16:112–115
Britsch S, Krippeit-Drews P, Lang F, Gregor M, Drews G (1995) Glucagon-like peptide-1 modulates Ca^{2+} current in intact mouse pancreatic B-cells. Biochem Biophys Res Commun 207:33–39
Brown JC (1982) Gastric inhibitory polypeptide. Springer, Berlin Heidelberg New York
Brubaker PL (1991) Regulation of intestinal proglucagon-derived peptide secretion by intestinal regulatory peptides. Endocrinology 128:3175–3182
Brubaker PL, So DCY, Drucker DJ (1989) Tissue-specific differences in the levels of proglucagon-derived peptides in streptozotocin-induced diabetes. Endocrinology 124:3003–3009
Calvo JC, Yusta B, Mora F, Blázquez E (1995) Structural characterization by affinity cross-linking of glucagon-like peptide-1(7-36) amide receptors in rat brain. J Neurochem 64:299–306
Campos RV, Lee YC, Drucker DJ (1994) Divergent tissue-specific and developmental expression of receptors for glucagon and glucagon-like peptide-I in the mouse. Endocrinology 134:2156–2164
Clark SA, Burnham BL, Chick WL (1990) Modulation of glucose-induced insulin secretion from a rat clonal cell line. Endocrinology 127:2779–2788
Creutzfeldt W (1979) The incretin concept today. Diabetologia 16:75–85
Creutzfeldt W, Ebert R (1985) New developments in the incretin concept today. Diabetologia 28:565–573
Creutzfeldt W, Nauck MA (1992) Gut hormones and diabetes mellitus. Diabetes Metab Rev 8:149–177
Cullinan CA, Brady EJ, Saperstein R, Leibowitz MD (1994) Glucose-dependent alterations of intracellular free calcium by glucagon-like peptide-1(7–36amide) in individual ob/ob mouse β-cells. Cell Calcium 15:391–400
D'Alessio DA, Fujimoto WY, Ensinck JW (1989) Effects of glucagon-like peptide I(7-36) on release of insulin, glucagon, and somatostatin by rat pancreatic islet cell monolayer cultures. Diabetes 38:1534–1538
D'Alessio D, Thirlby R, Laschansky H, Zebrowski H, Ensinck J (1993) Response of tGLP-1 to nutrients in humans. Digestion 54:377–379

D'Alessio DA, Kahn SE, Lensuer CR, Ensinck IW (1994) Glucagon-like peptide 1 enhances glucose tolerance both by stimulation of insulin release and by increasing insulin-dependent glucose disposal. J Clin Invest 93:2263–2266

D'Alessio DA, Prigeon RL, Ensinck JW (1995) Enteral enhancement of glucose disposition by both insulin-dependent and insulin-independent processes. Diabetes 44:1433–1437

Deacon FD, Johnsen AH, Holst JJ (1995) Degradation of glucagon-like peptide-1 by human plasma in vitro yields an N-terminally truncated peptide that is a major endogenous metabolite in vivo. J Clin Endocrinol Metab 80:952–957

Delgado E, Luque MA, Alcantara A, Trapote MA, Clemente F, Galera C, Valverde I, Villanueva-Penacarillo ML (1995) Glucagon-like peptide-1 binding to rat skeletal muscle. Peptides 16:225–229

Dillon JS, Tanizawa Y, Wheeler MB, Leng XH, Ligon BB, Rabin DU, Warren HY, Permutt MA, Boyd AE III (1993) Cloning and functional expression of the human glucagon-like peptide-1 (GLP-1) receptor Endocrinology 133:1907–1910

Drucker DJ, Brubaker PL (1989) Proglucagon gene expresson is regulated by a cyclic AMP-dependent pathway in rat intestine. Proc Natl Acad Sci USA 86:3953–3957

Drucker DJ, Philippe J, Mojsov S, Chick WL, Habener JF (1987) Glucagon-like peptide I stimulates insulin gene expression and increases cyclic AMP levels in a rat islet cell line. Proc Natl Acad Sci USA 84:3434–3438

Ebert R, Creutzfeldt W (1982) Influence of gastric inhibitory polypeptide antiserum on glucose-induced insulin secretion in rats. Endocrinology 111:1601–1606

Ebert R, Unger H, Creutzfeldt W (1983) Preservation of incretin activity after removal of gastric inhibitory polypeptide (GIP) from rat gut extracts by immunoadsorption. Diabetologia 24:449–454

Efrat S, Teitelman G, Anwor M, Ruggiero D, Hanahan D (1988) Glucagon gene regulatory regions directs oncoprotein expression to neuron and pancreatic a-cells. Neuron 1:605–613

Eissele R, Koop H, Arnold R (1990) Effect of glucagon-like peptide-1 on gastric somatostatin and gastrin secretion in the rat. Scand J Gastroenterol 25:449–454

Eissele R, Göke R, Willemer S, Harthus HP, Vermeer H, Arnold R, Göke B (1992) Glucagon-like peptide 1(7-36)amide cells in the gastrointestinal tract and pancreas of rat, pig and man. Eur J Clin Invest 22:283–291

Eissele R, Bothe-Sandforth E, Göke B, Eng J, Arnold R, Koop H (1994a) Rat gastric somatostatin and gastrin release: interactions of exendin-4 and truncated glucagon-like peptide I(GLP-1). Life Sci 55:629–634

Eissele R, Göke R, Weichhardt U, Fehmann HC, Arnold R, Göke B (1994b) Glucagon-like peptide 1 immunoreactivity in gastroentero-pancreatic endocrine tumors: a light- and electron-microscopic study. Cell Tissue Res 276:571–580

Elahi D, McAloon-Dyke M, Fukagawa NK, Meneilly GS, Sclater AL, Minaker KL, Habener JF, Andersen DK (1994) The insulinotropic actions of glucose-dependent insulinotropic polypeptide (GIP) and glucagon-like peptide-1(7-37) in normal and diabetic subjects. Regul Pept 51:63–74

Elliot RM, Morgan LM, Tredger JA, Deacon S, Wright J, Marks V (1993) Glucagon-like peptide-1(7-36)amide and glucose-dependent insulinotropic polypeptide in response to nutrient ingestion in man: acute post-prandial and 24-h secretion patterns. J Endocrinol 138:159–166

Eng J, Kleinman WA, Singh L, Singh S, Raufman JP (1992) Isolation and characterization of exendin-4, an exendin-3 analogue, from Heloderma suspectum venom. J Biol Chem 267:7402–7405

Ensinck JW, D'Alessio DA (1992) The entero-insular-axis revisited – a novel role for an incretin. N Engl J Med 326:1352–1353

Fehmann HC, Habener JF (1990) Galanin inhibits proinsulin gene expression stimulated by the insulinotropic hormone glucagon-like peptide-I(7-37) in mouse insulinoma βTC-1 cells. Endocrinology 130:2890–2896

Fehmann HC, Habener JF (1991a) Functional receptors for the insulinotropic hormone glucagon-like peptide-I(7-37) on a somatostatin secreting cell line. FEBS Lett 279:335–340
Fehmann HC, Habener JF (1991b) Homologous desensitization of the insulinotropic glucagon-like peptide-I(7-37) receptor on insulinoma (HIT-T15) cells. Endocrinology 128:2880–2888
Fehmann HC, Habener JF (1992) Insulinotropic hormone glucagon-like peptide-I(7-37) stimulation of proinsulin gene expression and proinsulin biosynthesis in insulinoma βTC-1 cells. Endocrinology 130:159–166
Fehmann HC, Göke B, Göke R, Trautmann ME, Arnold R (1989) Synergistic stimulatory effect of glucagon-like peptide-1(7-36)amide and glucose-dependent insulin-releasing polypeptide on the endocrine rat pancreas. FEBS Lett 252:109–112
Fehmann HC, Göke R, Göke B, Arnold R (1990a) Carbachol priming increases glucose- and glucagon-like peptide-1(7-36)amide-, but not arginine-induced insulin secretion from the isolated perfused rat pancreas. Z Gastroenterol 28:348–352
Fehmann HC, Weber V, Göke R, Göke B, Arnold R (1990b) Cosecretion of amylin and insulin from isolated perfused rat pancreas. FEBS Lett 262:279–291
Fehmann HC, Göke R, Göke B, Bächle R, Wagner B, Arnold R (1991) Priming effect of glucagon-like peptide-1(7-36)amide, glucose-dependent insulinotropic polypeptide and cholecystokinin-8 at the isolated perfused rat pancreas. Biochim Biophys Acta 1091:356–363
Fehmann HC, Göke R, Göke B (1992) At the cutting edge: glucagon-like peptide-1(7-37)/(7-36)amide is a new incretin hormone. Mol Cell Endocrinol 85:C39–C44
Fehmann HC, Jiang J, Schweinfurth J, Dörsch K, Wheeler MB, Boyd AE III, Göke B (1994a) Ligand-specificity of the rat GLP-1 receptor recombinantly expressed in Chinese hamster ovary (CHO-) cells. Z Gastroenterol 32:203–207
Fehmann HC, Jiang J, Schweinfurth J, Wheeler MB, Boyd AE III, Göke B (1994b) Stable expression of the rat GLP-1 receptor in CHO-cells: activation and binding characteristics utilizing GLP1-(7-36)amide, oxyntomodulin, exendin-4 and exendin-(9–39). Peptides 15:453–454
Fehmann HC, Strowski M, Göke B (1994c) Interaction of glucagon-like peptide-I(7-37) and somatostatin-14 on signal transduction and proinsulin gene expression in βTC-1 cells. Metabolism 43:787–792
Fehmann HC, Strowski M, Lankat-Buttgereit B, Göke B (1994d) Molecular and functional characterization of insulin receptors present on hamster glucagonoma cells. Digestion 55:214–220
Fehmann HC, Hering B, Brandhorst D, Brandhorst T, Bretzel RG, Federlin K, Göke B (1995a) The effects of glucagon-like peptide-I (GLP-1) on hormone secretion from isolated human pancreatic islets. Pancreas 11:196–200
Fehmann HC, Strowski M, Göke B (1995b) Functional characterization of somatostatin receptors expressed on hamster glucagonoma cells. Am J Physiol 268:E40–E47
Fehmann HC, Göke R, Göke B (1995c) Cell and molecular biology of the incretin hormones glucagon-like peptide-I and glucose-dependent insulin releasing polypeptide. Endocr Rev 16:390–410
Fehmann HC, Jiang J, Pitt D, Schweinfurth J, Göke R, Göke B (in press) Ligand-induced regulation of glucagon-like peptide-1 receptor function and expression in insulin-secreting β-cells. Pancreas
Fieseler P, Bridenbaugh S, Nustede R, Martell J, Orskov C, Holst JJ, Nauck M (1995) Physiological augmentation of amino acid-induced insulin secretion by GIP and GLP-I but not by CCK-8. Am J Physiol 268:E949–E955
Fridolf T, Ahrén B (1993) Effects of glucagon like peptide-1(7-36)amide on the cytoplasmic Ca^{2+}-concentration in rat islet cells. Mol Cell Endocrinol 96:85–90
Fukase N, Igarashi M, Takahashi H, Manaka H, Yamatani K, Daimon M, Tominanga M, Sasaki H (1992a) Hypersecretion of truncated glucagon-like peptide I and gastric inhibitory polypeptide in obese patients. Diab Med 10:44–49

Fukase N, Takahashi H, Igarashi M, Yamatani K, Daimon M, Sugiyama K, Tominanga M, Sasaki H (1992b) Differences in glucagon-like peptide I and GIP responses following sucrose ingestion. Diab Res Clin Pract 15:187–195
Fuller PJ, Beveridge DJ, Taylor RG (1993) Ileal proglucagon expression in the rat: characterization in intestinal adaptation using in situ hybridization. Gastroenterology 104:459–466
Fürnsinn C, Ebner K, Waldhäusl (1995) Failure of GLP-1(7-36)amide to affect glycogenesis in rat skeletal muscle. Diabetologia 38:864–867
Gallwitz B, Schmidt WE, Conlon JM, Creutzfeldt W (1990) Glucagon-like peptide-1 (7-36)amide: characterization of the domain responsible for binding to its receptor on rat insulinoma RINm5F cells. J Mol Endocrinol 5:33–39
Gallwitz B, Witt M, Fölsch UR, Creutzfeldt W, Schmidt WE (1993) Binding specificity and signal transduction of receptors for glucagon-like peptide-I(7-36)amide and gastric inhibitory polypeptide. J Mol Endocrinol 10:259–268
Gallwitz B, Witt M, Paetzold G, Morys-Wortmann C, Zimmermann B, Eckart K, Fölsch UR, Schmidt WE (1994) Structure/activity characterization of glucagon-like peptide-I. Eur J Biochem 225:1151–1156
Gefel D, Hendrick GK, Mojsov S, Habener JF, Weir GC (1990) Glucagon-like peptide-I analogs: effects on insulin secretion and adenosine 3',5'-monophosphate formation. Endocrinology 126:2164–2168
Gherzi R, Fehmann HC, Volz A, Panassi M, Göke B (1995) The glucagon gene is transcribed in β-like pancreatic cells. Exp Cell Res 218:460–468
Ghiglione M, Blazquez E, Uttenthal LO, de Diego JG, Alvarez E, George SK, Bloom SR (1985) Glucagon-like peptide-1 does not have a role in hepatic carbohydrate metabolism. Diabetologia 28:920–921
Gilon P, Shepherd RM, Henquin JC (1993) Oscillations of secretion driven by oscillations of cytoplasmic Ca^{2+} as evidenced in single pancreatic islets. J Biol Chem 268:22265–22268
Göke B (ed) (1993) Abstracts of the international symposium on glucagon-like peptide-1. Digestion 54:337–397
Göke R, Colon JM (1988) Receptors for glucagon-like peptide-1(7–36)amide on rat insulinoma-derived cells. J Endocrinol 116:357–362
Göke R, Trautmann ME, Haus E, Richter G, Fehmann HC, Arnold R, Göke B (1989a) Signal transmission after GLP-1(7-36)amide binding in RINm5F cells. Am J Physiol 257:G397–G401
Göke R, Cole T, Conlon JM (1989b) Characterization of the receptor for glucagon-like peptide-1(7-36)amide on plasma membranes from rat insulinoma-derived cells by covalent cross-linking. J Mol Endocrinol 2:93–98
Göke R, Fehmann HC, Richter G, Trautmann M, Göke B (1989c) Interaction of glucagon-like peptide-1(7-36)amide and somatostatin-14 in RINm5F cells and in the perfused rat pancreas. Pancreas 4:668–673
Göke R, Richter G, Göke B, Trautmann M, Arnold R (1989d) Internalization of glucagon-like peptide-1(7-36)amide in rat insulinoma cells. Res Exp Med 189:257–264
Göke R, Fehmann HC, Göke B (1991) Glucagon-like peptide 1 (7-36) amide is a new incretin/enterogastrone candidate. Eur J Clin Invest 21:135–144
Göke R, Oltmer B, Sheikh S, Göke B (1992) Solubilization of active GLP-1(7-36)amide receptors from RINm5F plasma membranes. FEBS Lett 300:232–236
Göke R, McGregor GP, Göke B (1993a) Amylin alters the biological action of the incretin hormone GLP-1(7-36)amide. Life Sci 53:1367–1372
Göke R, Wagner B, Fehmann HC, Göke B (1993b) Glucose-dependency of the insulin stimulatory effect of glucagon-like peptide-1(7-36)amide on the rat pancreas. Res Exp Med 193:97–103
Göke R, Fehmann HC, Linn T, Schmidt H, Krause M, Eng J, Göke B (1993c) Exendin-4 is a high potency agonist and truncated exendin-(9-39)-amide an antagonist at the glucagon-like peptide 1-(7-36)amide receptor on insulin secreting cells. J Biol Chem 268:19650–19655

Göke R, Kolligs F, Richter G, Lankat-Buttgereit B, Göke B (1993d) Solubilization of active receptors for glucagon-like peptide-1(7-36)amide from rat lung membranes. Am J Physiol 264:L146–L152

Göke R, Just R, Lankat-Buttgereit B, Göke B (1994) Glycosylation of the GLP-1 receptor is a prerequisite for regular receptor function. Peptides 15:675–681

Göke R, Larsen PJ, Mikkelsen JD, Sheikh SP (1995a) Identification of specific binding sites for GLP-I on the posterior lobe of the rat pituitary. Neuroendocrinology 62:130–134

Göke R, Larsen PJ, Mikkelsen JD, Sheikh SP (1995b) Distribution of GLP-1 binding sites in the rat brain. Evidence that exendin-4 is a ligand of brain GLP-1 binding sites. Eur J Neuroscience 7:2294–2300

Göke B, Fuder H, Wiedhorst G, Theiβ U, Stridde E, Littke T, Kleist P, Arnold R, Lücker PW (1995c) Voglibose (AO-128) is an efficient $\propto$-glucosidase inhibitor and mobilizes the endogenous GLP-I reserve. Digestion 56:493–501

Göke B, Steffen H, Göke R (1996) The signal transduction of the glucagon-like peptide I receptor: fishing beyond the protein kinase level. Acta Physiol Scand (in press)

Grapengiesser E, Gylfe E, Hellman B (1991) Cyclic AMP as a determinant for glucose induction of fast Ca^{2+} oscillations in isolated pancreatic β-cells. J Biol Chem 266:12207–12210

Grapengiesser E, Gylfe E, Hellman B (1992) Glucose sensing of individual pancreatic β-cells involves transitions between steady-state and oscillatory cytoplasmic Ca^{2+}. Cell Calcium 13:219–226

Graziano MP, Hey PJ, Borkowski D, Chicci GC, Strader CD (1993) Cloning and functional expression of a human glucagon-like peptide-1 receptor. Biochem Biophys Res Commun 196:141–146

Gros L, Demiprence E, Bataille D, Kervran A (1992) Characterization of high affinity receptors for glucagon-like peptide-1(7-36)amide on a somatostatin-secreting cell line. Biomed Res 13 [Suppl 2]:143–150

Gros L, Thorens B, Bataille D, Kervran A (1993) Glucagon-like peptide-1-(7-36)amide, oxyntomodulin, and glucagon interact with a common receptor in a somatostatin-secreting cell line. Endocrinology 133:631–638

Gutniak M, Orskov C, Holst JJ, Ahren B, Efendic S (1992) Antidiabetogenic effect of glucagon-like peptide-1(7-36)amide in normal subjects and patients with diabetes mellitus. N Engl J Med 326:1316–1322

Gutniak MK, Linde B, Efendic S (1993) Subcutaneous injection of glucagon-like insulinotropic peptide reduces postprandial glycemia in non-insulin-dependent diabetes. Digestion 54:389–390

Habener JF (1990) Cyclic AMP response element binding proteins: a cornucopia of transcription factors. Mol Endocrinol 4:1087–1094

Han VKM, Hynes MA, Jin C, Towle AC, Lauder JM, Lund PK (1986) Cellular localization of proglucagon/glucagon-like peptide I messenger RNAs in rat brain. J Neurosci Res 16:97–107

Hansen AB, Gespach CP, Rosselin GE, Holst JJ (1988) Effect of truncated glucagon-like peptide 1 on cAMP in rat gastric glands and HGTβ1 human gastric cancer cells. FEBS Lett 236:119–122

Hayakawa T, Kondo T, Okumura N, Nagai K, Shibata T, Kitagawa M (1989) Enteroglucagon release in disaccharide malabsorption induced by intestinal alpha-glucosidase. Am J Gastroenterol 84:523–526

Heinrich G, Gros P, Habener JF (1984a) Glucagon gene sequence: four of six exons encode separate functional domains of rat preproglucagon. J Biol Chem 259:14082–14087

Heinrich G, Gros P, Lund PK, Bentley RC, Habener JF (1984b) Preproglucagon messenger ribonucleic acid: nucleotide and encoded amino acid sequences of the rat pancreatic complementary deoxyribonucleic acid. Endocrinology 115:2176–2181

Hellman B, Gylfe E, Bergsten P, Grapengiesser E, Lund PE, Berts A, Tengholm A, Pipeleers DG, Ling Z (1994) Glucose induces oscillatory Ca^{2+} signalling and insulin release in human pancreatic beta cells. Diabetologia 37 [Suppl 2]:S11–S20

Heller RS, Aponte GW (1995) Intra-islet regulation of hormone secretion by glucagon-like peptide-1 (7-36) amide. Am J Physiol 269:G852–G860

Herrmann C, Göke R, Richter G, Fehmann HC, Arnold R, Göke B (1995) Glucagon-like peptide-1 and glucose-dependent insulin releasing polypeptide secretion in response to nutrients. Digestion 56:117–126

Herrmann-Rinke C, Vöge A, Hess M, Göke B (1995) Regulation of glucagon-like peptide-1 (GLP-1) secretion from rat ileum by neurotransmitters and peptides. J Endocrinology 147:25–31

Hirota M, Hashimoto M, Hiratsuka M, Ohbishi C, Yoshimoto S, Yano M, Mizuno A, Shima K (1990) Alterations of plasma glucagon-like peptide I behaviour in noninsulin-dependent diabetes. Diab Res Clin Pract 9:179–185

Holst JJ, Orskov C (1994) Glucagon and other proglucagon-derived peptides. In: Walsh JH, Dockray GJ (eds) Gut peptides: biochemistry and physiology. Raven, New York, pp 305–340

Holst JJ, Orskov C, Vagn Nielsen O, Schwartz TW (1987) Truncated glucagon-like peptide I, an insulin-releasing hormone from the distal gut. FEBS Lett 211:169–174

Holst JJ, Rasmussen TN, Harling H, Schmidt P (1993) Effect of intestinal inhibitory peptides on vagally induced secretion from isolated perfused porcine pancreas. Pancreas 8:80–87

Holst JJ, Bersani M, Johnson AH, Kofod H, Hartmann B, Orskov C (1994) Proglucagon processing in porcine and human pancreas. J Biol Chem 269:18827–18833

Holz GG, Habener JF (1992) Signal transduction crosstalk in the endocrine system: pancreatic β-cells and the glucose competence concept. TIBS 17:388–393

Holz GG IV, Kuhtreiber WM, Habener JF (1993) Pancreatic beta-cells are rendered glucose-competent by the insulinotropic hormone glucagon-like peptide-I(7-37). Nature 361:362–365

Hörsch D, Fink T, Göke B, Arnold R, Büchler M, Weihe E (1994) Distribution and chemical phenotypes of neuroendocrine cells in the human anal canal. Regul Pept 54:527–542

Hoosein NM, Gurd RS (1985) Human glucagon-like peptide 1 and 2 activate rat brain a denylate cyclase. FEBS Lett 178:83–86

Hupe-Sodmann K, McGregor GP, Bridenbaugh R, Göke R, Göke B, Thole H, Zimmermann B, Voigt KH (1995) Characterisation of the processing by human neutral endopeptidase 24.11 of GLP-I(7-36)amide and comparison of the substrate specificity of the enzyme for other glucagon-like peptides. Regul Pept 58:149–156

Hvidberg A, Nielsen MT, Hilsted J, Orskov C, Holst JJ (1994) Effect of glucagon-like peptide-I (proglucagon 78-107amide) on hepatic glucose production in healthy man. Metabolism 43:104–108

Inoue K, Hisatomi A, Umeda F, Nawata H (1991) Effects of glucagon-like peptide 1(7-36)amide and glucagon on amylin release from perfused rat pancreas. Horm Metab Res 23:407–409

Jiang J, Fehmann HC, Loth H, Göke R, Göke B (1994) Regulation of GLP-1 receptor mRNA levels and GLP-1 binding sites by forskolin, PMA and dexamethasone in RINm5F cells. Exp Clin Endocrinol 102 [Suppl 1]:117

Jin SLC, Han VKM, Simmons JG, Towle AC, Lauder JM, Lund PK (1988) Distribution of glucagon-like peptide I (GLP-1), glucagon, and glicentin in the rat brain: an immunocytochemical study. J Comp Neurol 271:519–532

Kanse SM, Kreymann B, Ghatei MA, Bloom SR (1988) Identification and characterization of glucagon-like peptide-1 7-36 amide-binding sites in the rat brain and lung. FEBS Lett 241:209–212

Katschinski M, Schirra J, Weidmann C, Schäfer T, Arnold R, Göke B (1995) Incretin release to oral and duodenal glucose loads in man. Digestion 56:296

Kolligs F, Fehmann HC, Göke R, Göke B (1995) Reduction of the incretin effect in rats by the glucagon-like peptide 1 receptor antagonist exendin (9-39) amide. Diabetes 44:16–19

Kreymann B, Ghatei MA, Williams G, Bloom SR (1987) Glucagon-like peptide-1 7-36: a physiological incretin in man. Lancet II:1300–1303

Kreymann B, Yiangou Y, Kanse S, Williams G, Ghatei MA, Bloom SR (1988a) Isolation and characterization of GLP-1 7-36 amide from rat intestine: elevated levels in diabetic rats. FEBS Lett 242:167–170

Kreymann B, Yiangou Y, Kanse S, Williams G, Ghatei MA, Bloom SR (1988b) Isolation and characterization of GLP-1 (7-36) amide from rat intestine. FEBS Lett 242:167–170

Lankat-Buttgereit B, Göke R, Fehmann HC, Richter G, Göke B (1994a) Molecular cloning of a cDNA encoding the GLP-1 receptors expressed in rat lung. Exp Clin Endocrinol 102:341–347

Lankat-Buttgereit B, Göke R, Stöckmann F, Fehmann HC, Göke B (1994b) Detection of the human GLP-1(7-36)amide receptor on insulinoma-derived cell membranes. Digestion 55:29–33

Lauritsen KB, Holst JJ, Moody AJ (1981) Depression of insulin release by anti-GIP serum after oral glucose tolerance in rats. Scand J Gastroenterol 16:417–421

Lauritsen KB, Moody AJ, Christensen KC, Jensen SL (1980) Gastric inhibitory polypeptide (GIP) and insulin release ater small-bowel resection in man. Scand J Gastroenterol 15:833–840

Layer P, Holst JJ (1993) GLP-1: a humoral mediator of the ileal brake in humans? Digestion 54:385–386

Layer P, Holst JJ, Grandt D, Goebell H (1995) Ilead release of glucagon-like peptide-1 (GLP-1). Association with inhibition of gastric acid secretion in humans. Dig Dis Sci 40:1074–1082

Leech CA, Holz GG, Habener JF (1995) Pituitary adenylate cyclase-activating polypeptide induces the voltage-independent activation of inward membrane currents and elevation of intracellular calcium in HIT-T15 insulinoma cells. Endocrinology 136:1530–1536

Levin SR, Pehlevanian MZ, Lavee AE, Adachi RI (1979) Secretion of an insulinotropic factor from isolated, perfused rat intestine. Am J Physiol 236:E710–E720

Lok S, Kuijper JL, Jelinek LJ, Kramer JM, Whitmore TE, Sprecher CA, Mathewes S, Grant FJ, Biggs SH, Rosenberg GB, Sheppard PO, O'Hara PJ, Foster DC, Kindsvogel W (1994) The human glucagon receptor encoding gene: structure, cDNA sequence and chromosomal localization. Gene 140:203–209

Lu M, Wheeler MB, Leng XH, Boyd AE III (1993) The role the free cytosolic calcium level in β-cell signal transduction by gastric inhibitory polypeptide and glucagon-like peptide I(7-37). Endocrinology 132:94–100

Lui EY, Asa SL, Drucker DJ, Lee YC, Brubaker PL (1990) Glucagon and related peptides in fetal rat hypothalamus in vivo and in vitro. Endocrinology 126:10–117

Matsumara T, Itoh H, Watanabe N, Oda Y, Taneka M, Namba M, Kono N, Matsuyama T, Komatsu R, Matsuzawai Y (1992) Glucagonlike peptide-1(7-36)amide suppresses glucagon secretion and decreases cyclic AMP concentration in cultured InR1-G9 cells. Biochem Biophys Res Commun 186:503–508

McIntyre N, Holdsworth CD, Turner DS (1964) New interpretation of oral glucose tolerance. Lancet II:20–21

Miholic J, Orskov C, Holst JJ, Kotzerke J, Meyer HJ (1991) Emptying of the gastric substitute, glucagon-like peptide-1 (GLP-1), and reactive hypoglycemia after total gastrectomy. Dig Dis Sci 36:1361–1370

Miholic I, Orskov C, Holst JJ, Kotzerke I, Pichelmayer R (1993) Postprandial release of glucagon-like peptide 1, pancreatic glucagon, and insulin after esophageal resection. Digestion 54:73–78

Mojsov S, Weir GC, Habener JF (1987) Insulinotropin: glucagon-like peptide I(7-37) co-encoded in the glucagon gene is a potent stimulator of insulin release in the perfused rat pancreas. J Clin Invest 79:616–619

Mojsov S, Kopczynski MG, Habener JF (1990) Both amidated and nonamidated forms of glucagon-like peptide I are synthesized in the rat intestine and the pancreas. J Biol Chem 265:8001–8008

Moore B, Edie ES, Abram JH (1906) On the treatment of diabetes mellitus by acid extract of duodenal mucous membrane. Biochem J 1:28–38

Mora F, Exposito I, Sanz B, Blazquez E (1992) Selective release of glutamine and glutamic acid produced by perfusion of GLP-1(7-36)amide in the basal ganglia of the conscious rat. Brain Res Bull 29:359–361

Murayama Y, Kawai K, Suzuki S, Ohashi S, Yamashita K (1990) Glucagon-like peptide-1(7-37) does not stimulate either glycogenolysis or ketogenesis. Endocrinol Jpn 37:293–297

Nathan DM, Schreiber E, Fogel H, Mojsov S, Habener JF (1992) Preliminary studies of insulinotropic actions of glucagon-like peptide-I(7-37) administered to diabetic and non-diabetic human subjects. Diabetes Care 15:270–277

Nauck MA, Bartels E, Orskov C, Ebert R, Creutzfeldt W (1992) Lack of effect of synthetic human gastric inhibitory polypeptide and glucagon-like peptide 1(7-36 amide) infused at near-physiological concentrations on pentagastrin-stimulated gastric acid secretion in normal human subjects. Digestion 52:214–221

Nauck MA, Bartels E, Orskov C, Ebert R, Creutzfeldt W (1993a) Additive insulinotropic effects of exogenous synthetic human gastric inhibitory polypeptide and glucagon-like peptide-1-(7-36)amide infused at near-physiological insulinotropic hormone and glucose concentrations. J Clin Endocrinol Metab 76:912–917

Nauck MA, Heimesaat MM, Orskov C, Holst JJ, Ebert R, Creutzfeldt W (1993b) Preserved incretin activity of glucagon-like peptide 1(GLP-1)(7-36)amide but not of synthetic human gastric inhibitory polypeptide (GIP) in patients with type 2 diabetes mellitus. J Clin Invest 30:301–307

Nauck MA, Kleine N, Orskov C, Holst JJ, Willms B, Creutzfeldt W (1993c) Normalization of fasting hyperglycemia by exogenous glucagon-like peptide 1(7-36 amide) in type 2 (non-insulin-dependent) diabetic patients. Diabetologia 36:741–744

Nauck MA, Büsing M, Orskov C, Siegel EG, Talartschik J, Baartz A, Baartz T, Hopt UT, Becker HD, Creutzfeldt W (1993d) Preserved incretin effect in type 1 diabetic patients with end-stage nephropathy treated with combined heterotopic pancreas and kidney transplantation. Acta Diabetol 30:39–45

Oben J, Morgan L, Fletcher J, Marks V (1991) Effect of the entero-pancreatic hormones, gastric inhibitory polypeptide and glucagon-like polypeptide-1(7-36)amide, on fatty acid synthesis in explants of rat adipose tissue. J Endocrinol 130:267–272

O'Halloran DJ, Nikou GC, Kreymann B, Ghatei MA, Bloom SR (1990) Glucagon-like peptide-1(7-36)-NH2: a physiological inhibitor of gastric acid secretion in man. J Endocrinol 126:169–173

Orskov C, Nielsen JH (1988) Truncated glucagon-like peptide-1(proglucagon 78-107 amide), an intestinal insulin-releasing peptide, has specific receptors on rat insulinoma cells (RIN 5AH). FEBS Lett 229:175–178

Orskov C, Holst JJ, Poulsen SS, Kirkegaard P (1987) Pancreatic and intestinal processing of proglucagon in man. Diabetologia 30:874–881

Orskov C, Holst JJ, Nielsen OV (1988) Effect of truncated glucagon-like peptide-1[proglucagon-(78-107)amide] on endocrine secretion from pig pancreas, antrum, and nonantral stomach. Endocrinology 123:209–213

Orskov C, Bersani M, Johnson AH, Hojrup P, Holst JJ (1989) Complete sequences of glucagon-like peptide-1 from human and pig small intestine. J Biol Chem 264:12826–12829

Orskov C, Jeppesen J, Madsbad S, Holst JJ (1991) Proglucagon products in plasma of noninsulin-dependent diabetics and nondiabetic controls in the fasting state and after oral glucose and intravenous arginine. J Clin Invest 87:415–423

Orskov C, Wettergen A, Holst JJ (1993) Biological effects and metabolic rates of glucagonlike peptide-1 7-36 amide and glucagonlike peptide-1 7-37 in healthy subjects are indistinguishable. Diabetes 42:658–61

Orskov C, Rabenhoj L, Wettergren A, Kofod H, Holst JJ (1994) Tissue and plasma concentrations of amidated and glycine-extended glucagon-like peptide I in humans. Diabetes 43:535–539

Patzelt C, Weber B (1986) Early O-glycosidic glycosylation of proglucagon in pancreatic islets: an unusual type of prohormonal modification. EMBO J 5:2103–2108

Pederson RA (1994) Gastric inhibitory polypeptide. In: Walsh JH, Dockray GJ (eds) Gut peptides: biochemistry and physiology. Raven, New York, pp 217–259

Perley MJ, Kipnis DM (1967) Plasma insulin responses to oral and intravenous glucose: studies in normal and diabetic subjects. J Clin Invest 46:1954–1962

Philippe J, Missotten M (1991) Functional characterization of a cAMP responsive element of the rat insulin I gene. J Biol Chem 265:1465–1469

Praz GA, Halban PA, Wollheim CB, Blondel B, Strauss JA, Reynold AE (1983) Regulation of immunoreactive insulin release from a rat cell line (RINm5F). Biochem J 210:345–352

Prentki M, Matschinsky FM (1987) Ca^{2+}, cAMP, and phospholipid-derived messengers in coupling mechanisms of insulin secretion. Physiol Rev 67:1185–1248

Richter G, Göke R, Göke B, Arnold R (1990a) Characterization of receptors for glucagon-like peptide-1(7-36)amide on rat lung membranes. FEBS Lett 267:78–80

Richter G, Göke R, Göke B, Arnold R (1990b) Dexamethasone pretreatment of rat insulinoma cells decreases binding of glucagon-like peptide-I(7-36)amide. J Endocrinol 126:445–450

Richter G, Göke R, Göke B, Schmidt H, Arnold R (1991) Characterization of glucagon-like peptide-1(7-36)amide receptors of rat lung membranes by covalent cross-linking. FEBS Lett 280:247–250

Richter G, Feddersen O, Wagner U, Barth P, Göke R, Göke B (1993) GLP-1 stimulates secretion of macromolecules and relaxes pulmonary artery. Am J Physiol 265:L374–L381

Richter G, Schilling T, Göke B (1995) Challenge of the entero-insular axis in healthy volunteers by liquid diets modified for the enteral nutrition of patients with diabetes mellitus. Aktuel Ernahr Stoffw 20:122–126

Roberge JN, Brubaker PL (1993) Regulation of intestinal proglucagon-dervied peptide secretion by glucose-dependent insulinotropic peptide in a novel enteroendocrine loop. Endocrinology 133:233–240

Rountree DB, Ulshen MH, Selub S, Fuller CR, Bloom SR, Ghatei MA, Lund PK (1992) Nutrient-independent increases in proglucagon and ornithine decarboxylase messenger RNAs after jejunoileal resection. Gastroenterology 103:462–468

Ruiz-Grande C, Alarcon C, Merida E, Valverde I (1992) Lipolytic action of glucagon-like peptides in isolated rat adipocytes. Peptides 13:13–16

Salazar I, Vaillant C (1990) Glucagon-like immunoreactivity in hypothalamic neurons of the rat. Cell Tissue Res 261:355–358

Samols E, Bonner-Weir S, Weir GC (1986) Intra-islet insulin-glucagon-somatostatin relationships. clin Endocrinol Metab 15:33–58

Schirra J, Katschinski M, Kuwert P, Wank U, Arnold R (1994) Differential effects of subcutaneous GLP-1 on gastric emptying, insulin release and exocrine pancreatic secretion in man. Digestion 56:317–318

Schirra J, Katschinski M, Weidmann C, Schäfer T, Wank U, Arnold R, Göke B (1996) Gastric emptying and release of incretin hormones after glucose ingestion in humans. J Clin Invest 97:92–103

Schjoldager BTG, Mortensen PE, Christiansen J, Orskov C, Holst JJ (1989) GLP-1 (glucagon-like peptide 1) and truncated GLP-1, fragments of human proglucagon, inhibit gastric acid secretion in humans. Dig Dis Sci 34:703–708

Schmid R, Schusdziarra V, Aulehner R, Weigert N, Classen M (1990) Comparison of GLP-1(7-36amide) and GIP on release of somatostatin-like immunoreactivity and insulin from the isolated rat pancreas. Z Gastroenterol 28:280–284

Schmidtler J, Schepp W, Janczewska I, Weigert N, Fülinger C, Schusdziarra V, Classen M (1991) GLP-1-(7-36)amide, -(1-37), and -(1-36)admide: potent cAMP-dependent stimuli of rat parietal cell function. Am J Physiol 260:G940–G950

Schmidtler J, Dehne K, Allescher HD, Schusdziarra V, Classen M, Holst JJ, Polack A, Schepp W (1994) Rat parietal cell receptors for GLP-1-(7-36) amide: northern blot, cross-linking, and radioligand binding. Am J Physiol 267:G423–G432

Shima K, Hirota M, Ohboshi C (1988) Effect of glucagon-like peptide-I on insulin secretion. Regul Pept 22:245–250

Shimizu I, Hirota M, Ohboshi C, Shima K (1986) Effect of glucagon-like peptide-1 and -2 on glycogenolysis in clutured rat hepatocytes. Biomed Res 7:431–436

Shimizu I, Hirota H, Ohboshi C, Shima K (1987) Identification and localization of glucagon-like peptide-1 and its receptor in rat brain. Endocrinology 121:1076–1082

Steffen H, Göke B (1994) Regulation der Proteinphosphorylierung durch Glucagon-like Peptid-1 und Forskolin in Ratteninsulinomzellen. Diabetes Stoffwechsel 3:148

Stoffel M, Espinosa R III, Le Beau MM, Bell GI (1993) Human glucagon-like peptide-1 receptor gene. Localization to chromosome band 6p21 by fluorescence in situ hybridization and linkage of a highly polymorphic simple tandem repeat DNA polymorphism to other markers on chromosome 6. Diabetes 42:1215–1218

Suzuki S, Kawai K, Ohashi S, Mukai H, Yamashita K (1989) Comparison of the effects of various C-terminal and N-terminal fragment peptides of glucagon-like peptide-1 on insulin and glucagon release from the isolated perfused rat pancreas. Endocrinology. 125:3109–3113

Suzuki S, Kawai K, Ohashi S, Mukai H, Murayama Y, Yamashita K (1990) Reduced insulinotropic effect of glucagonlike peptide I-(7-36)-amide and gastric inhibitory polypeptide isolated perfused diabetic rat pancreas. Diabetes 39:1320–1325

Tanizawa Y, Riggs AC, Elbein SC, Whelan A, Donis-Keller H, Permutt MA (1994) Human glucagon-like peptide 1 receptor gene in NIDDM. Identification and use of simple sequence repeat polymorphisms in genetic analysis. Diabetes 43:752–757

Taylor RG, Beveridge DJ, Fuller (1992) Expression of ileal glucagon and peptide tyrosine-tyrosine genes. Response to inhibition of polyamine synthesis in the presence of massive small-bowel resection. Biochem J 286:737–741

Thorens B (1992) Expression cloning of the pancreatic β cell receptor for the glucoincretin hormone glucagon-like peptide 1. Proc Natl Acad Sci USA 89:8641–8645

Thorens B, Porret A, Bühler L, Deng SP, Morel P, Widman C (1993) Cloning and functional expression of the human islet GLP-1 receptor: demonstration that exendin-4 is an agonist and exendin-9(9-39) an antagonist of the receptor. Diabetes 42:1678–1682

Thornton K, Gorenstein DG (1994) Structure of glucagon-like peptide-I(7–36)amide in a dodecylphosphocholine micelle as determined by 2D NMR. Biochemistry, 33:3532–3539

Turton MD, O'Shea D, Guinn J, Beak SA, Edwards CMB, Meeran K, Choi SJ, Taylor GM, Heath MM, Lambert PD, Wilding JPH, Smith DM, Ghater MA, Herbert Y, Bloom SR (1996) A role for glucagon-like peptide-1 in the central regulation of feeding. Nature 379:69–72

Unger RH, Eisentraut AM (1969) Entero-insular-axis. Arch Intern Med 123:261–266

Unger RH, Ohneda A, Valverde I, Eisentraut AM, Exton J (1968) Characterization of the responses of circulating glucagon-like immunoreactivity to intraduodenal and intravenous administration of glucose. J Clin Invest 47:48–65

Uttenthal LO, Blazquez E (1990) Characterization of high-affinity receptors for truncated glucagon-like peptide-1 in rat gastric glands. FEBS Lett 262:139–141

Uttenthal LO, Ghiglione M, George SK, Bishop AE, Polak JM, Bloom SR (1985) Molecular forms of glucagon-like peptide-1 in human pancreas and glucagonomas. J Clin Endocrinol Metab 61:472–479

Uttenthal LO, Toledano A, Blazquez E (1992) Autoradiographic localization of receptors for glucagon-like peptide-1(7-36)amide in rat brain. Neuropeptides 21:143–146

Valverde I, Merida E, Delgado E, Trapotes MA, Villanueva-Penacarillo ML (1993) Presence and characterization of glucagon-like peptide-1(7-36)amide receptors in solubilized membranes of rat adipose tissue. Endocrinology 132:75–79

Valverde I, Morales M, Clemente F, Lopez-Delgado MI, Delgado E, Perea A, Villanueva-Penacarrillo ML (1994) Glucagon-like peptide I: a potent glycogernic hormone. FEBS Lett 349:313–316 van Eyll B, Lankat-Buttgereit B, Bode HP, Göke R, Göke B (1994) Signal transduction of the GLP-1 receptor cloned from a human insulinoma. FEBS Lett 348:7–13

Vertongen P, Ciccarelli E, Woussen-Colle MC, De Neef P, Robberecht P, Cauvin A (1994) A pituitary adenylate cyclase-activating polypeptide receptor of types I and II glucagon-like peptide-I receptors are expressed in the rat medullary carcinoma of the thyroid cell line 6/23. Endocrinology 135:1537–1542

Villanueva-Penacarrillo ML, Alcantara AL, Clemente F, Delgado E, Valverde I (1994a) Potent glycogenic effect of GLP-1 (7-36) amide in rat skeletal muscle. Diabetologia 37:1163–1166

Villanueva-Penacarrillo ML, Merida E, Delgado E, Molina LM, Arrieta F, Rovira A, Valverde I (1994b) Increased glucagon-like peptide I (7-36) amide binding in adipose tissue from non-insulin-dependent and insulin-dependent diabetic patients. Diab Nutr Metab 7:143–148

Watanabe Y, Kawai K, Ohashi S, Yokota C, Suzuki S, Yamashita K (1994) Structure-activity relationships of glucagon-like peptide-I (7-36) amide: insulinotropic activities in perfused rat pancreases, and receptor binding and cyclic AMP production in RINm5F cells. J Endocrinol 140:45–52

Weber V, Fehmann HC, Göke R, Göke B (1991) Effect of proglucagon-derived peptides on amylase release from rat pancreatic acini. Int J Pancreatol 325–330

Wei Y, Mojsov S (1994) Tissue-specific expression of the human receptor for glucagon-like peptide-I: brain, heart and pancreatic forms have the same deduced amino acid sequences. FEBS Lett 358:219–224

Weir GC, Mojsov S, Hendrick GK, Habener JF (1989) Glucagon-like peptide I(7–37) actions on endocrine pancreas. Diabetes 38:338–342

Wettergren A, Schjoldager B, Mortensen PE, Myhre J, Christiansen J, Holst JJ (1993) Truncated GLP-1 (proglucagon 78-107 amide) inhibits gastric and pancreatic functions in man. Dig Dis Sci 38:665–673

Wettergren A, Petersen H, Orskov C, Christiansen J, Sheikh SP, Holst JJ (1994) Glucagon-like peptide-I 7-36 amide and peptide YY from L cell of the ileal mucosa are potent inhibitors of vagally induced gastric acid secretion in man. Scand J Gastroenterol 29:501–505

Wheeler MB, Lu M, Dillon JS, Leng XH, Chen C, Boyd AE III (1993) Functional expression of the rat glucagon-like peptide-I receptor, evidence for coupling to both adenylyl cyclase and phospholipase C. Endocrinology 133:57–62

Willms B, Werner J, Holst JJ, Orskov C, Creutzfeldt W, Nauck MA (1996) Gastric emptying, glucose responses, and insulin secretion after a liquid test meal: effects of exogenous glucagon-like peptide-1 (GLP-1) (7-36) amide in type 2 (non-insulin dependent) diabetic patients. J Clin Endocrinol Metab 81:327–332

Yada T, Itoh K, Nakata M (1993) Glucagon-like peptide-1-(7-36)amide and rise in cyclic adenosine 3'5'-monophosphate increase cytosolic free Ca^{2+} in rat pancreatic β-cells by enhancing Ca^{2+} channel activity. Endocrinology 133:1685–1692

Yoshimoto S, Hirota M, Ohboshi C, Shima K (1989) Identification of glucagon-like peptide-1(7-36)amide in the brain. Ann Clin Biochem 26:169–171

Zawalich WS, Zawalich KC, Rasmussen H (1989a) Cholinergic agonists prime the beta cell to glucose stimulation. Endocrinology 125:2400–2406

Zawalich WS, Zawalich KC, Rasmussen H (1989b) Interactions between cholinergic agonists and enteric factors in the regulation of insulin secretion from isolated perifused rat islets. Acta Endocrinol (Copenh) 120:702–707

Zawalich WS, Zawalich KC, Rasmussen H (1993) Influence of glucagon-like peptide-1 on β-cell responsiveness. Regul Pept 44:277–283

Zhang HJ, Redmon JB, Andresen JM, Robertson RP (1991) Somatostatin and epinephrine decrease insulin messenger ribonucleic acid in HIT cells through a pertussis toxin-sensitive mechanism. Endocrinology 129:2409–2414

Zhang Y, Cook ITE, Hattersley AT, Firth R, Saker PI, Warren-Perry M, Stoffel M, Turner RC (1994) Non-linkage of the glucagon-like peptide 1 receptor gene with maturity onset diabetes of the young. Diabetologia 37:721–724

CHAPTER 18
Potential of GLP-1 in Diabetes Management

J.J. HOLST, M.A. NAUCK, C.F. DEACON, and C. ØRSKOV

A. Introduction

As reviewed in Chap. 3, this volume, glucagon-like peptide-1 (GLP-1) is the designation given to the sequence in proglucagon corresponding to residues Nos. 72–108 (BELL et al. 1983a, 1983b). In proglucagon, this sequence is flanked by pairs of basic amino acid residues, i.e., typical processing cleavage sites, and shows an almost 50% homology with glucagon itself; hence its designation. Early immunochemical studies indicated that the proteolytic processing of proglucagon would, indeed, lead to the formation of this peptide, but mainly in the intestinal mucosa, whereas in the pancreas this sequence was contained in the "major proglucagon fragment" as predicted by PATZELT (MOJSOV et al. 1986; ØRSKOV et al. 1986; PATZELT and SCHILTZ 1984). Synthetic replicas of this sequence, which were soon made available after the structure of proglucagon had been deduced (BELL et al. 1983b), were reported to exhibit weak insulinotropic activity (SCHMIDT et al. 1985), but most investigators found the peptide inactive (ØRSKOV 1992). Physiological and pathophysiological studies had clearly shown that the distal intestinal mucosa, in which the glucagon gene is being expressed in the so-called L cells (MOJSOV et al. 1986; NOVAK et al. 1987), contained an insulinotropic hormone distinct from gastric inhibitory polypeptide (GIP) (EBERT 1990; LAURITSEN et al. 1980; MOODY et al. 1970; NOVAK et al. 1987). Therefore, a search for alternative products of intestinal expression of the glucagon gene was carried out, and it turned out that intestinal extracts contained a peptide with GLP-1-like immunoreactivity which was potently insulinotropic (HOLST et al. 1986, 1987). Upon structural analysis, this peptide was found to correspond to a truncated form of GLP-1 (HOLST et al. 1986, 1987). Further chemical analysis showed that the structure corresponded to proglucagon 78-107 amide (ØRSKOV et al. 1989a). A synthetic replica of this peptide turned out to be the most potent insulinotropic peptide hitherto isolated (ØRSKOV 1992). The glycine-extended peptide, proglucagon 78-108, a probable processing intermediate, was equally insulinotropic (MOJSOV et al. 1987). This discovery, that alternative processing of the glucagon precursor gave rise to novel biologically active products, prompted further investigations of proglucagon processing in humans, and today a complete picture of proglucagon processing in human gut and pancreas can be assembled (BUHL et al. 1988; HOLST et al. 1994a; ØRSKOV et al.

1989a,b, 1987, 1994). This was possible by combining, on one hand, gel permeation chromatography of unfractionated extracts using processing-independent radioimmunoassays for the various regions of proglucagon and, on the other hand, isolation and chemical structural analysis of each of the products identified. Using this approach the possibility that some products might escape detection was minimized.

Thus, in humans (Fig. 1) the pancreatic processing of proglucagon (henceforth designated PG) leads to the formation of (a) glicentin-related pancreatic peptide corresponding to PG 1-30, originally isolated from porcine pancreas (Moody et al. 1981; Thim and Moody 1982); so far this peptide has not been associated with any known biological activity; (b) glucagon itself, occupying positions 33-61 in PG; (c) a hexapeptide (sometimes designated intervening peptide-1) corresponding to PG 64-69 (Holst et al. 1994a; Yanaihara et al. 1985; Kadowaki et al. 1985); and (d) the major proglucagon fragment (MPGF) corresponding to PG 72-158 (Holst et al. 1994a). A small fraction of MPGF may be further processed to PG 72-107, i.e., full-length GLP-1 (Holst et al. 1994a). In humans almost all of this peptide occurs in its amidated from, less than 5% remaining glycine-extended (= PG 72-108) (Ørskov et al. 1994). In agreement with the formation of small amounts of GLP-1, human pancre-

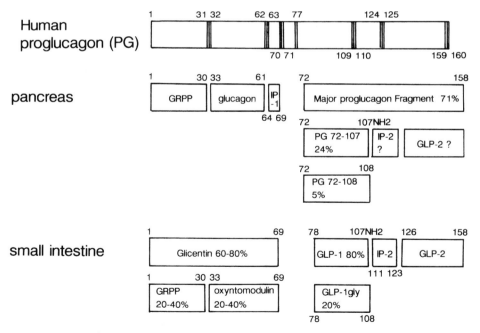

Fig. 1. Schematic representation of the post-translation processing of proglucagon as it occurs in the human pancreas and intestinal mucosa. *Numbers* indicate positions of amino acid residues in proglucagon. *IP*, intervening peptide. See text for further explanation. [From Ørskov (1994) with permission]

atic extracts also contain small amounts of a peptide corresponding to GLP-2, the other peptide with homology to glucagon, co-encoded in the glucagon gene (BELL et al. 1983a, 1983b; HOLST et al. 1994a).

In the human small intestine, proglucagon processing leads to the formation of (human) glicentin, corresponding to PG 1-69 (BELL et al. 1983b; THIM and MOODY 1981). A small fraction of this peptide may be processed further to GRPP (PG 1-30) and oxyntomodulin (PG 33-69) (BALDISSERA and HOLST 1984; HOLST 1983a). Exactly how much oxyntomodulin is formed in humans is difficult to determine; a spontaneous proteolytic cleavage at the pair of basic amino acids at positions 31 and 32 in proglucagon seems to occur rapidly unless the tissue is frozen immediately. Previous estimates of the amounts of oxyntomodulin formed in the human L cells may, therefore, have been exaggerated (BALDISSERA and HOLST 1984). At any rate, the quantities of oxyntomodulin circulating in humans are barely, if at all, detectable (HOLST 1983b; HOLST et al. 1983). As opposed to the pancreatic formation of mainly MPGF, the intestinal processing leads to the formation of each of the GLPs 1 and 2 (MOJSOV et al. 1986; ØRSKOV et al. 1986, 1987, 1994). Most of the GLP-1 corresponds to proglucagon 78-107 amide, but about one-fourth carries a C-terminal glycine extension (ØRSKOV et al. 1994). The structure of human GLP-2 corresponds to proglucagon 126-158 (BUHL et al. 1988). The intervening sequences corresponding to PG 72-76 and 111-123 are also found, and at least PG 111-123, which in humans is also amidated, is secreted (BUHL et al. 1988; L. THIM personal communication). Amongst all these products, interest focuses upon the truncated GLP-1 (proglucagon 78-107 amide) because of its documented biological activity. Glicentin (porcine) has been reported to inhibit gastric secretion in rats (KIRKEGAARD et al. 1982), but is unlikely to exhibit other glucagon-like activities because of its extensions at both termini. Oxyntomodulin shares a number of activities with glucagon [because it corresponds to glucagon plus a C-terminal extension (BALDISSERA et al. 1988)], and may exert specific action, at least in rats (see Chap. 19, this volume), but is, as already mentioned, present in very low concentrations in the circulation. So far, nothing is known about the biological actions of the intervening peptides or GLP-2. Unexpected biological activities of the "minor" products of proglucagon processing notwithstanding, it seems that the most important outcome of the differential processing of proglucagon is the formation of two different, biologically active peptides in the tissues, where the glucagon gene is expressed, viz., glucagon in the pancreas and GLP-1 in the gut, while the "other" active peptide is inactivated by bilateral extensions, glucagon as glicentin in the gut and GLP-1 as MPGF in the pancreas. The insulinotropic hormone GLP-1, therefore, is truly a gut hormone.

As already alluded to, the truncated GLP-1 corresponding to PG 78-107 amide (henceforth designated GLP-1; the peptide PG 72-107 amide should be designated "N-terminally extended GLP-1") is the most potent insulinotropic peptide known. In very careful recent studies by JIA et al. (1995) using isolated perfused rat pancreases exposed to gradients of glucose as well as peptide concentrations, the two peptides porcine GIP, the other important incretin hormone (see below), and GLP-1 (the sequence of the latter is identical in all mammals so far investigated) were equipotent, with a minimum effective concentration of 16 pmol/l, but in humans GLP-1 seems to be about five times more potent than (human) GIP (ELAHI et al. 1994; NAUCK et al. 1993b). In addition, three important features distinguish GLP-1 from GIP: (a) in addition to its insulinotropic activity GLP-1 also inhibits glucagon secretion (ØRSKOV et al. 1988); (b) it has pronounced effects on gastrointestinal secretion and motility (HOLST 1994; WETTERGREN et al. 1993); and (c) it is active in patients with diabetes mellitus (NAUCK et al. 1993b,c).

B. Actions of GLP-1 on Blood Glucose in Humans
(see also Chap. 17, this volume)

GLP-1 is insulinotropic in humans or, perhaps more correctly, it potentiates glucose-induced insulin secretion dose dependently (GEFEL et al. 1990; KREYMAN et al. 1987). This means that at decreasing blood glucose levels its effectiveness decreases, whereas at elevated glucose levels it is more efficient (GEFEL et al. 1990). Being released from the small intestine in response to ingestion of carbohydrates (KREYMAN et al. 1987; ØRSKOV et al. 1991), it is considered an incretin hormone (KREYMAN et al. 1987), and is likely to represent the "missing incretin" identified in earlier physiological studies (EBERT 1990; LAURITSEN et al. 1980; MOODY et al. 1970). Its effect on insulin secretion is additive to that of GIP, and a combined infusion of physiological amounts of the two peptides can fully account for the incretin effect (NAUCK et al. 1993a). In addition to being acutely insulinotropic, GLP-1 also enhances the rate of insulin gene expression (FEHMANN and HABENER 1992). An infusion of GLP-1 into normal human subjects in physiological amounts, i.e., at an infusion rate that increases the plasma concentrations to levels similar to those observed after meal ingestion, causes a decrease in blood glucose of about 1 mmol/l (HVIDBERG et al. 1994). A detailed analysis of the events underlying this decrease showed that insulin secretion was increased significantly during the infusion compared to a control experiment with saline infusion and that glucagon concentrations decreased (HVIDBERG et al. 1994). Thus, the molar ratio of the concentrations of insulin and glucagon in the plasma reaching the liver must have increased considerably, and as a result hepatic glucose production (estimated by a tracer technique) fell to about 70% of the preinfusion levels. Because glucose disposal remained unchanged, it could be calculated that glucose clearance increased during the GLP-1 infusion. This could be a result of the increased insulin secretion, but it also raises the possibility that GLP-1 might have effects on glucose turnover independent of its effects on the pancreatic glucoregulatory hormones. Several studies have indicated that this might be the case. Firstly, in studies by D'ALESSIO et al. on the effects of GLP-1 on glucose turnover in humans using the Bergman minimal model approach,

GLP-1 significantly increased S_g, the so-called glucose effectiveness, a measure of glucose-induced glucose disposal, whereas insulin sensitivity was unchanged (D'ALESSIO et al. 1994). In a series of studies, Valverde and coworkers (VALVERDE et al. 1993, 1994; VILLANUEVA-PENACARRILLO et al. 1994) have provided evidence that GLP-1 enhances glycogen formation in both skeletal muscle and in the liver, and other groups have reported GLP-1-induced increases in insulin-stimulated glucose transport in rat adipocytes (EGAN et al. 1994; MIKI et al. 1994). However, in a recent study in humans, GLP-1 had no influence on glucose disposal in response to an intravenous glucose tolerance test (IVGTT), when pancreatic hormone secretion was clamped at very low levels with somatostatin (TOFT-NIELSEN et al. 1995), and in another study a GLP-1 infusion affected neither hepatic glucose production, forearm metabolite balance nor glucose infusion rate in normal subjects during the conditions of a hyperinsulinemic euglycemic clamp as well as an endocrine clamp, i.e., a somatostatin infusion with careful i.v. substitutions of basal insulin, glucagon and growth hormone levels (ØRSKOV et al. 1995b). Thus, although these studies do not exclude that GLP-1 may influence insulin-dependent as well as insulin-independent glucose disposal in humans, they indicate that such effects are likely to be quantitatively minor.

Because of the glucose dependence of its effects, it is impossible to produce hypoglycemia by GLP-1 administration, regardless of dose (QUALMANN et al. 1995). This is clearly reflected by the glucagon and insulin responses during continued infusions of GLP-1 (HVIDBERG et al. 1994; ØRSKOV et al. 1993). As blood glucose decreases, insulin concentrations decrease and glucagon concentrations tend to increase in spite of ongoing or increased rates of GLP-1 infusion. Adrenergic counterregulation does not seem to contribute to the prevention of hypoglycemia, although GLP-1 infusion does increase plasma epinephrine concentrations (NIELSEN et al. 1994). Thus, in healthy volunteers the glucose as well as the hormone and glucose turnover responses to a GLP-1 infusion were virtually identical with and without extensive β-adrenergic blockade (NIELSEN et al. 1994). Nevertheless, several investigations have indicated that inappropriately increased secretion of GLP-1 might be responsible for postprandial hypoglycemia as observed, e.g., after gastrectomy (MIHOLIC et al. 1991, 1993). In these patients, meal-induced GLP-1 responses may be extremely exaggerated (up to 300 pmol/l) and have been shown to correlate to hypoglycemic events. However, a direct demonstration that GLP-1 is responsible has not been provided.

C. Gastrointestinal Effects of GLP-1 in Humans

In addition to its effects on the endocrine pancreas, GLP-1 powerfully influences gastrointestinal motility and secretion (WETTERGREN et al. 1993) and is, therefore, believed to act as (one of) the "ileal brake" hormone(s) (LAYER and HOLST 1993). Thus, in physiological amounts, GLP-1 inhibits pentagastrin- as well as meal-induced acid secretion and inhibits gastric-emptying rate (O'HALLORAN et al. 1990; SCHJOLDAGER et al. 1989; WETTERGREN et al. 1993). Its inhibitory effect on acid secretion is lost after vagotomy (ØRSKOV et al. 1995a), whereas sham-feeding-induced secretion is almost abolished by GLP-1 (WETTERGREN et al. 1994). These findings suggest that GLP-1 interacts with the autonomic nervous system, and in this connection it is of interest that a human brain GLP-1 receptor identical to the B-cell receptor has recently been cloned (WEI and MOJSOV 1995). Based on these findings and on autoradiographic studies, it has been suggested that GLP-1 secreted from the intestine interacts with receptors associated with neurons of the circumventricular organ around the third ventricle of the brain and influences autonomic functions in this way (ØRSKOV et al. 1996), Thus, GLP-1 may influence blood glucose not only by virtue of its effects on the endocrine pancreas but also because of its gastrointestinal actions.

D. GLP-1 and Diabetes

I. Secretion

Due to its profound effects on glucose metabolism, GLP-1 has attracted considerable interest with respect to its possible role in the pathogenesis of and possible effects in diabetes. Because of experience with GIP, which at best has weak effects on glucose metabolism in diabetics and does not seem to play a significant role in the pathogenesis (KRARUP 1988), expectations were not great. With respect to secretion of GLP-1 in diabetes patients no clear picture has yet emerged. Elevated levels were reported in patients with type II diabetes in the basal state as well as after oral glucose (HIROTA et al. 1990; ØRSKOV et al. 1991). However, in these studies, a processing-independent assay was used, which could not distinguish between fully processed GLP-1 (PG 78-107 amide) and other moieties containing the GLP-1 sequence, e.g., major proglucagon fragment from the pancreas. Since glucagon secretion (from the pancreas) is typically inappropriately elevated in diabetes, the increased GLP-1 levels measured with such assays may reflect a hypersecretion of pancreatic GLP-1-containing peptides. In a more recent study using an assay specific for fully processed amidated GLP-1, a significantly decreased response to oral glucose was found in type II diabetics (HOLST et al. 1994b). Clearly, further studies are needed in this area.

II. Receptors

The recent cloning of the human GLP-1 receptor (see Chap. 17, this volume) allowed studies of the hypothesis that defects in the GLP-1 receptor gene could be involved in the pathogenesis of non-insulin-dependent diabetes (NIDDM). However, TANIZAWA et al. (1994) and ZHANG et al. (1994) concluded that inherited defects in the GLP-1 receptor are not major risk factors for NIDDM in African Americans and Caucasians and in patients with maturity onset diabetes of the young.

III. Effects

As to the effects of GLP-1 in diabetes, GUTNIAK et al. in 1992 demonstrated that an infusion of GLP-1 in type II diabetic patients connected to an artificial pancreas significantly lowered meal-induced blood glucose excursions and increased the insulinogenic index almost ten fold. The calculated isoglycemic meal-related requirement of insulin was reduced from 17.4 ± 2.8 to 2.0 ± 0.5 U by GLP-1. Similarly, in patients with insulin-dependent diabetes, the insulin requirement was reduced from 9.4 ± 1.5 to 4.7 ± 1.4 U. NATHAN et al. (1992) showed that an infusion of glycine-extended GLP-1 (GLP-1 7-37) abolished meal-induced glucose excursions and stimulated insulin secretion in patients with NIDDM. NAUCK et al. (1993b) compared the effects of infusion of GIP and GLP-1 in normal subjects and in patients with moderate NIDDM during a hyperglycemic clamp at approximately 8 mmol/l to allow comparisons. In normal subjects, GLP-1 was strongly insulinotropic and much more potent than GIP, while in the diabetic patients GIP had no significant effects on insulin secretion or the amount of glucose required to maintain the clamp. By contrast, GLP-1 stimulated insulin secretion almost as much as in the normal subjects, and required significantly increased glucose infusion rates. In addition, whereas neither the glucose infusion (required to maintain the clamp) nor the additional GIP infusion had any effect on plasma glucagon concentrations in the diabetics (as opposed to the normal subjects in whom glucose with as well as without hormone addition inhibited glucagon secretion), GLP-1 strongly inhibited glucagon secretion. This study clearly showed that the "glucose blindness" of the A and B cells of the pancreatic islets could be corrected or ameliorated by GLP-1 in patients with moderate NIDDM. It was of great interest when, at the same time, HOLZ et al. (1993) presented the concept that GLP-1 might convey "glucose competence" to the pancreatic B cells. They demonstrated that only a small percentage of isolated, single B cells responded adequately to glucose (with membrane potential depolarization and closure of K^+-channels), whereas after exposure to GLP-1 almost all cells would respond. A similar effect of GLP-1 was observed in recent studies of glucose-induced intracellular Ca^{2+} increases in various B-cell lines (GROMADA et al. 1995; MONTROSE-RAFIZADEH et al. 1994). Although it has by no means been demonstrated that the unresponsiveness of the single cells to glucose is related to the defect of the diabetic B cell, the remarkable effectiveness of GLP-1 in diabetics should inspire further studies of its signal-transduction pathways in the B cell of the islets of Langerhans.

At this stage it was of great interest to investigate the effects of GLP-1 administration in patients with severe NIDDM. In patients with secondary failure to sulfonylurea treatment, fasting blood glucose levels around 13 mmol/

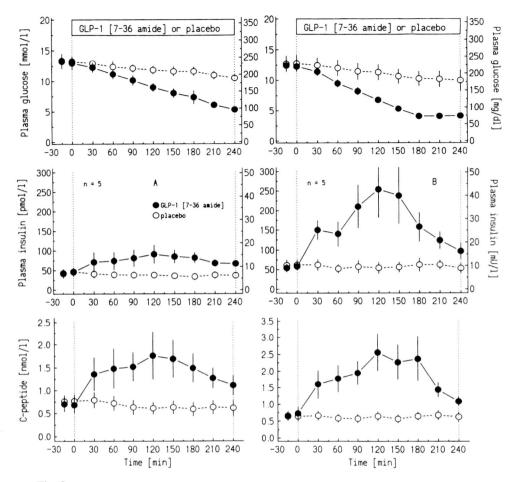

Fig. 2

l, and hemoglobin Alc levels above 11%, who were admitted to hospital for insulin therapy, a GLP-1 infusion of 1.2 pmol/kg per minute, which elevated their plasma GLP-1 concentrations to slightly supraphysiological values (around 100 pmol/l), completely normalized blood glucose concentrations in the course of less than 4h (NAUCK et al. 1993c) (see Fig. 2). All of the patients responded with an increased insulin secretion, some more than others. In patients with a relatively large insulin response, normoglycemia was reached after only 2h, but, in spite of ongoing GLP-1 infusion, blood glucose remained at normal fasting levels. At the same time, insulin secretion had decreased to preinfusion levels, illustrating the glucose dependency of the actions of GLP-1. Undoubtedly, another important contributory mechanism consisted of the pronounced inhibition of glucagon secretion observed in all patients. The importance of the glucagonostatic effects of GLP-1 was more clearly illus-

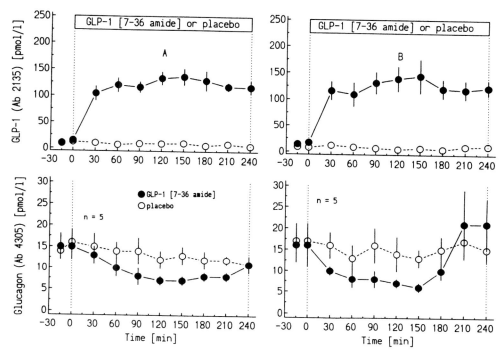

Fig. 2. Glucose, insulin, C-peptide, GLP-1 and glucagon concentrations in plasma in fasting patients with NIDDM during an infusion of GLP-1 (1.2 pmol/kg per minute). Based on data NAUCK et al. 1993c (with permission). The ten patients were divided into two groups of five according to their insulin responses. Those with the larger responses (*right panels*) reach euglycemia earlier, but in spite of ongoing GLP-1 infusion their glucose concentrations do not decrease further, illustrating the fact that in spite of potent glucose-lowering effects GLP-1 infusions do not produce hypoglycemia

trated in a subsequent study of the effects of a similar GLP-1 infusion in patients with IDDM (FOLEY 1992). In these patients, which could not respond with insulin secretion, the GLP-1 infusion, nevertheless, decreased blood glucose values by 50%. This occurred in parallel with a pronounced inhibition of glucagon secretion. It was concluded from these experiments that, in patients with NIDDM, GLP-1 effectively lowers blood glucose concentrations, most likely as a consequence of a stimulated insulin secretion and an inhibition of glucagon secretion, leading to a decreased hepatic glucose production. It has been suggested recently that it is the decreasing concentration of free fatty acids (FFAs) elicited by insulin rather than insulin itself that explains insulin's effect on the hepatic glucose production (FOLEY 1992). Whether this turns out to be the case or not, GLP-1 infusions also markedly lower plasma FFA concentrations in normal subjects as well as NIDDM patients (NAUCK et al. 1993c; ØRSKOV et al. 1993). In IDDM patients with no insulin secretion, however, GLP-1 administration does not change plasma FFA concentrations

(WILLMS et al. 1995), indicating that this effect of GLP-1 is mediated by its influence on the pancreatic glucoregulatory hormones. It will be of considerable interest to learn from future studies whether or not GLP-1 influences glucose metabolism by mechanisms independent of the pancreatic glucoregulatory hormones in patients with diabetes mellitus.

As mentioned above, GLP-1 administration effectively curtailed meal-induced blood glucose excursions in NIDDM patients (GUTNIAK et al. 1992; NATHAN et al. 1992). In further studies of NIDDM patients with secondary sulfonylurea failure, this effect of GLP-1 was examined in more detail (WILLMS et al. 1994). Whereas during control infusions with saline, blood glucose levels increased in response to a liquid meal, the GLP-1 infusion caused the blood glucose to fall to euglycemic levels, precisely as in fasting subjects. Surprisingly, however, insulin concentrations were similar or lower during the GLP-1 infusion than during the control study. Glucagon concentrations increased in the control study (the meal effect) and decreased during GLP-1 infusion. Measurements of the gastric-emptying rate provided the explanation: during the first 2h, emptying of the gastric contents could not be demonstrated at all, whereas in the control study close to 100% of the stomach contents were emptied during this period. This leaves little doubt that a large part of the acute GLP-1 effect on blood glucose during meal ingestion in diabetics is due to its inhibitory effects on gastric motility. The effect on gastric emptying is clearly dose dependent (NAUCK et al. 1995), and it is possible that doses that do not inhibit gastric emptying as much will still be sufficient to satisfactorily lower blood glucose in NIDDM patients. It will be important to address this issue in further studies.

E. GLP-1 Metabolism in Normal and Diabetic Subjects

Being a peptide, GLP-1 cannot be administered orally, at least not without substantial losses. Alternative routes of administration have, therefore, been studied. Subcutaneous administration of GLP-1 increases insulin secretion and inhibits glucagon secretion, but the rise in the plasma concentration of GLP-1 is short lived (basal levels being reached after 90–120min) as is the duration of action, and, with higher doses, side effects (nausea, vomiting) are common (RITZEL et al. 1995). In NIDDM patients a 25-nmol dose of GLP-1 injected s.c. completely abolished meal-induced hyperglycemia, but a lasting effect on blood glucose as noted during i.v. infusion was not observed (GUTNIAK et al. 1994). The short duration of action prompted studies of GLP-1 metabolism after subcutaneous as well as intravenous administration. Studies conducted in healthy volunteers indicated that both GLP-1 and glycine-extended GLP-1 are metabolized with a half-life of approximately 5min (ØRSKOV et al. 1993), and several studies have indicated that the renal clearance of GLP-1 may be important (RUIZ-GRANDE et al. 1993; ØRSKOV et al. 1992). However, studies by MENTLEIN et al. (1993) suggested that GLP-1 may also be a substrate for the ubiquitous enzyme dipeptidyl peptidase IV. By the actions of this enzyme, the peptide loses its two N-terminal amino acid residues, and studies by Grant et al. (1994) suggested that the peptide thereby turns into a GLP-1 receptor antagonist. This conversion also takes place in human plasma (DEACON et al. 1995a), with 62% intact GLP-1 and 38% metabolite being present in fasting plasma and 42% and 58%, respectively, being present after meal stimulation (DEACON et al. 1995a). A pronounced enzymatic conversion of exogenous GLP-1 is also observed in both normal subjects and diabetics (DEACON et al. 1995b). Thus, at plateau levels during intravenous infusion of GLP-1 only 20%–25% of total GLP-1 corresponded to the intact, biologically active peptide and after subcutaneous injections the area under the concentration curve of the intact peptide amounted to less than 10% of area of the curve for the metabolite. Thus, these studies indicate that precautions that might enhance the stability of the peptide would be helpful, particularly if subcutaneous administration of the peptide is considered. The biological significance of the enzymatically induced antagonist formation is presently unclear.

F. Conclusion and Outlook

In principle, GLP-1 has great potential in the treatment of the hyperglycemia of diabetic patients, predominantly patients with NIDDM, but possibly also patients with IDDM. In fact, GLP-1 is the only agent available that can completely normalize plasma glucose (by the mechanisms outlined above) administered in doses that are generally well tolerated (RITZEL et al. 1995). Among the drugs that act through a stimulation of insulin secretion (in concert with other mechanisms), the strict glucose dependency of this particular action is unique, and should limit the danger of producing hypoglycemia as a consequence of its administration.

The dual mode of action of GLP-1, i.e., the reduction of meal-related increments in plasma glucose due to inhibition of gastric emptying on one hand, and the normalization of fasting hyperglycemia via its actions on the secretion of the pancreatic glucoregulatory hormones on the other, probably has consequences for the dosing regimens. These should guarantee low-enough plasma levels during the daytime [perhaps around 50 pmol/l (NAUCK et al. 1995) in order not to interfere too much with gastric emptying] and probably somewhat higher levels (approximately 100 pmol/l) during the night. Along these lines of reasoning, a bedtime injection of a retarded-absorption GLP-1 preparation could be a reasonable and realistic therapeutic strategy for a peptide that can at present be administered only by a parenteral route. Other pharmacokinetic maneuvers to achieve appropriate 24 h plasma concentration profiles could be "microencapsulation" into self-dissolving materials that are slowly degraded in subcutaneous tissue, and transdermal/transepithelial administration. In the long run, however, the development of nonpeptide agonists for the GLP-1 receptor is probably the preferable strategy.

References

Baldissera FGA, Holst JJ (1984) Glucagon-related peptides in the human gastrointestinal mucosa. Diabetologia 26:223–228

Baldissera FGA, Holst JJ, Knuhtsen S, Hilsted L, Nielsen OV (1988) Oxyntomodulin (glicentin 33-69): pharmacokinetics; binding to liver cell membranes; effects on isolated perfused pig pancreas; secretion from isolated perfused lower small intestine of pigs. Regul Pept 21:151–166

Bell GI, Sanchez-Pescador R, Laybourn PJ, Najarian RC (1983a) Exon duplication and divergence in the human preproglucagon gene. Nature 304:368–371

Bell GI, Santerre RF, Mullenbach GT (1983b) Hamster preproglucagon contains the sequence of glucagon and two related peptides. Nature 302:716–718

Buhl T, Thim L, Kofod H, Ørskov C, Harling H, Holst JJ (1988) Naturally occurring products of proglucagon 111–160 in the porcine and human small intestine. J Biol Chem 263:8621–8624

D'Alessio DA, Kahn SE, Leusner C, Ensinck JW (1944) Glucagon-like peptide 1 enhances glucose tolerance both by stimulation of insulin release and by increasing insulin-independent glucose disposal. J Clin Invest 93:2263–2266

Deacon CF, Johnsen AH, Holst JJ (1995a) Degradation of glucagon-like peptide-1 by human plasma in vitro yields an N-terminally truncated peptide that is a major endogenous metabolite in vivo. J Clin Endocrinol Metab 80:952–957

Deacon CF, Nauck MA, Toft-Nielsen M, Pridal L, Willms B, Holst JJ (1995b) Both subcutaneous and intravenously administered glucagon-like peptide-1 are rapidly degraded from the amino terminus in type II diabetic patients and in healthy subjects. Diabetes (in press)

Ebert R (1990) Gut signals for islet hormone release. Eur J Clin Invest 20 (Suppl 1):S20–S26

Egan JE, Montrose-Rafizadeh C, Wang Y, Bernier M, Roth J (1994) Glucagon-like peptide-1 (7-36) amide (GLP-1) enhances insulin-stimulated glucose metabolism in 3T3-L1 adipocytes: one of several potential extrapancreatic sites of GLP-1 action. Endocrinology 135:2070–2075

Elahi D, McAloon-Dyke M, Fukagawa NK, Meneilly GS, Sclater AL, Minaker KL, Habener JF, Andersen D (1994) The insulinotropic actions of glucose-dependent insulinotropic polypeptide (GIP) and glucagon-like peptide-1 (7-37) in normal and diabetic subjects. Regul Pept 51:63–74

Fehmann H-C, Habener JF (1992) Insulinotropic hormone glucagon-like peptide-I (7–37) stimulation of proinsulin gene expression and proinsulin biosynthesis in insulinoma βTC-1 cells. Endocrinology 130:159–166

Foley JE (1992) Rationale and application of fatty oxidation inhibitors in treatment of diabetes mellitus. Diabetes Care 15:773–781

Gefel D, Hendrick GK, Mojsov S, Habener J, Weir GC (1990) Glucagon-like peptide-1 analogs: effects on insulin secretion and adenosine 3', 5'-monophosphate formation. Endocrinology 126:2164–2168

Grant D, Sieburg B, Sievert J et al. (1994) Is GLP-1 (9-36) amide an endogenous antagonist at GLP-1 receptors? Digestion 55:302

Gromada JL, Dissing S, Bokvist K, Renström E, Frøkjær-Jensen J, Wulff BS, Rorsman P (1995) Glucagon-like peptide 1 increases cytoplasmic calcium in insulin-secreting βTC3 cells by enhancement of intracellular calcium mobilization. Diabetes (in press)

Gutniak M, Holst JJ, Ørskov C, Ahren B, Efendic S (1992) Antidiabetogenic effect of glucagon-like peptide-1 (7-36) amide in normal subjects and patients with diabetes mellitus. N Engl J Med 236:1316–1322

Gutniak MK, Linde B, Holst JJ, Efendic S (1994) Subcutaneous injection of the incretin hormone glucagon-like peptide-1 reduces postprandial glycemia in non-insulin dependent diabetes. Diabetes Care 17:1039–1044

Hirota M, Hashimoto M, Hiratsuka M, Oboshi C, Yoshimoto S, Yano M, Mizuno A, Shima K (1990) Alterations of plasma immunoreactive glucagon-like peptide-1 behaviour in non-insulin-dependent diabetics. Diabetes Res Clin Pract 9:179–185

Holst JJ (1983a) Radioreceptor assays for glucagon. In: Lefèbvre PJ (ed) Glucagon. Springer Verlag, Berlin Heidelberg New York, pp 245–261 (Handbook of experimental pharmacology, vol 66/1)

Holst JJ (1983b) Molecular heterogeneity of glucagon in normal subjects and in patients with glucagon-producing tumours. Diabetologia 24:359–365

Holst JJ (1994) Glucagon-like peptide 1: a newly discovered gastrointestinal hormone. Gastroenterology 107:1848–1855

Holst JJ, Holst Pedersen J, Baldissera F, Stadil F (1983) Circulating glucagons after total pancreatectomy in man. Diabetologia 25:396–399

Holst JJ, Ørskov C, Schwartz TW, Buhl T, Baldissera FGA (1986) Proglucagon 78–107, a potent insulinotropic hormone from lower small intestine. Diabetologia 29:549A

Holst JJ, Ørskov AC, Schwartz TW, OV Nielsen (1987) Truncated glucagon-like peptide I, an insulin-releasing hormone from the distal gut. FEBS Lett 211:169–174

Holst JJ, Bersani M, Johnsen AH, Kofod H, Hartmann B, Ørskov C (1994a) Proglucagon processing in porcine and human pancreas. J Biol Chem 269:1827–1883

Holst JJ, Vaag A, Beck-Nielsen H (1994b) Decreased GLP-1 secretion in NIDDM – studies in identical twins discordant for NIDDM. Diabetes 43 (Suppl 1):65A

Holz GG, Kühltreiber WM, Habener JF (1993) Pancreatic beta-cells are rendered glucose competent by the insulinotropic hormone glucagon-like peptide-1 (7-37). Nature 361:362–365

Hvidberg A, Toft-Nielsen M, Hilsted J, Ørskov C, Holst JJ (1994) Effect of glucagon-like peptide-1 (proglucagon 78-107 amide) on hepatic glucose production in healthy man. Metabolism 43:104–108

Jia X, Brown JC, Pederson RA, McIntosh CHS (1995) The effects of glucose dependent insulinotropic polypeptide and glucagon-like peptide (7-36) on insulin secretion. Am J Physiol (in press)

Kadowaki M, Iguchi K, Yanaihara N (1985) Rat pancreas contains the proglucagon (64-69) fragment and arginine stimulates its release. FEBS Lett 187:307–310

Kirkegaard P, Loud FB, Moody AJ, Holst JJ, Christiansen J (1982) Gut GLI inhibits gastric acid secretion in the rat. Nature 297:156–157

Krarup T (1988) Immunoreactive gastric inhibitory polypeptide. Endocr Rev 9:122–134

Kreymann B, Ghatei MA, Williams G, Bloom SR (1987) Glucagon-like peptide-1 7-36: a physiological incretin in man. Lancet II:1300–1303

Lauritsen KB, Moody, AJ, Christensen KC, Jensen SL (1980) Gastric inhibitory polypeptide (GIP) and insulin release after small-bowel resection in man. Scand J Gastroenterol 15:833–840

Layer P, Holst JJ (1993) GLP-1: a humoral mediator of the ileal brake in humans? Digestion 54:385–386

Mentlein R, Gallwitz B, Schmidt W (1993) Dipeptidyl-peptidase IV hydrolyses gastric inhibitory polypeptide, glucagon-like peptide-1 (7-36) amide, peptide histidine methionine and is responsible for their degradation in serum. Eur J Biochem 214:829–835

Miholic J, Ørskov C, Holst JJ, Kotzer J, Meyer HJ (1991) Emptying of the gastric substitute, glucagon-like peptide-1 (GLP-1), and reactive hypoglycemia after total gastrectomy. Dig Dis Sci 36:1361–1370

Miholic J, Ørskov C, Holst JJ, Kotcerke J, Meyer HJ (1993) Postprandial release of glucagon-like peptide-1 (GLP-1), pancreatic glucagon, and insulin after esophageal resection. Digestion 54:73–78

Miki H, Nishimura T, Mineo I, Matsumura T, Namba M, Kuwajima M, Matsizawa Y (1994) Effects of glucagon-like peptide-1 (7-36) amide on glucose uptake in isolated rat adipocytes. Abstract vol, 15th congress of the International Diabetes Federation, Kobe, November 1994, p 125

Mojsov S, Heinrich G, Wilson IB, Ravazzola M, Orci L, Habener JF (1986) Preproglucagon gene expression in pancreas and intestine diversifies at the level of post-translational processing. J Biol Chem 261:11880–11889

Mojsov S, Weir GC, Habener JF (1987) Insulinotropin: glucagon-like peptide-I (7-37) co-encoded in the glucagon gene is a potent stimulator of insulin release in the perfused pancreas. J Clin Invest 79:616–619

Motrose-Rafizadeh C, Egan JM, Roth J (1994) Incretin hormones regulate glucose-dependent insulin secretion in RIN 1046-38 cells: mechanism of action. Endocrinology 135:589–594

Moody AJ, Markussen J, Schaich Fries A, Steestrup C, Sundby F, Malaisse W, Lalaisse-lagae F (1970) The insulin-releasing activities of extracts of pork intestine. Diabetologia 6:135–140

Moody AJ, Holst JJ, Thim L, Jensen SL (1981) Relationship of glicentin to proglucagon and glucagon in the porcine pancreas. Nature 289:514–516

Nathan DM, Schreiber E, Fogel H, Mojsov S, Habener JF (1992) Insulinotropic action of glucagonlike peptide-1-(7-37) in diabetic and nondiabetic subjects. Diabetes Care 15:270–276

Nauck MA, Bartels E, Ørskov C, Ebert R, Creutzfeldt W (1993a) Additive insulinotropic effects of exogenous synthetic human gastric inhibitory polypeptide and glucagon-like peptide-1 (7-36) amide infused at near-physiological insulinotropic and glucose concentrations. J Clin Endocrinol Metab 76:912–917

Nauck MA, Heimesaat MM, Ørskov C, Holst JJ, Ebert R, Creutzfeldt W (1993b) Preserved incretin activity of GLP-1 (7-36 amide) but not of synthetic human GIP in patients with type 2-diabetes mellitus. J Clin Invest 91:301–307

Nauck MA, Kleine N, Ørskov C, Holst JJ, Willms B, Creutzfeldt W (1993c) Normalization of fasting hyperglycemia by exogenous GLP-1 (7–36 amide) in type 2-diabetic patients. Diabetologia 36:741–744

Nauck M, Ettler R, Niedereichholz U, Ørskov C, Holst JJ, Schmiegel W (1995) Inhibition of gastric emptying by GLP-1 (7-36 amide) or (7-37): effects on postprandial glycemia and insulin secretion. Abstract to be presented at the UEGW meeting in Berlin, (in press)

Nielsen M, Hvidberg A, Hilsted J, Petersen HD, Holst JJ (1994) Effect of β-adrenergic blockade on the hypoglycemic effects of glucagon-like peptide-1 (GLP-1). Diabetologia 37 (Suppl 1):A119

Novak U, Wilks A, Buell G, McEwen S (1987) Identical mRNA for preproglucagon in pancreas and gut. Eur J Biochem 164:553–558

O'Halloran DJ, Nikou GC, Kreymann B, Ghatei MA, Bloom SR (1990) Glucagon-like peptide-1 (7-36)-NH2: a physiological inhibitor of gastric acid secretion in man. J Endocrinol 126:169–173

Ørskov C (1992) Glucagon-like peptide-1, a new hormone of the enteroinsular axis. Diabetologia 35:701–711

Ørskov C, Holst JJ, Knuhtsen S, Baldissera FGA, Poulsen SS, Nielsen OV (1986) Glucagon-like peptides GLP-1 and GLP-2, predicted products of the glucagon gene, are secreted separately from the pig small intestine, but not pancreas. Endocrinology 119:1467–1475

Ørskov C, Holst JJ, Poulsen SS, Kirkegaard P (1987) Pancreatic and intestinal processing of proglucagon in man. Diabetologia 30:874–881

Ørskov C, Holst JJ, Nielsen OV (1988) Effect of truncated glucagon-like peptide-1 (proglucagon 78-107 amide) on endocrine secretion from pig pancreas, antrum and stomach. Endocrinology 123:2009–2013

Ørskov C, Bersani M, Johnsen AH, Højrup P, Holst JJ (1989a) Complete sequences of glucagon-like peptide-1 (GLP-1) from human and pig small intestine. J Biol Chem 264:12826–12829

Ørskov C, Buhl T, Rabenhøj L, Kofod H, Holst JJ (1989b) Carboxypeptidase-B-like processing of the C-terminus of glucagon-like peptide-2 in pig and human small intestine. FEBS Lett 247:1932–106

Ørskov C, Jeppesen J, Madsbad S, Holst JJ (1991) Proglucagon products in plasma of non-insulin dependent diabetics and nondiabetic controls in the fasting state and following oral glucose and intravenous arginine. J Clin Invest 87:415–423

Ørskov C, Andreasen J, Holst JJ (1992) All products of proglucagon are elevated in plasma from uremic patients. J Clin Endocrinol Metab 74:379–384

Ørskov C, Wettergren A, Holst JJ (1993) The metabolic rate and the biological effects of GLP-1 7-36 amide and GLP-1 7-37 in healthy volunteers are identical. Diabetes 42:658–661

Ørskov C, Rabenhøj L, Kofod H, Wettergren A, Holst JJ (1994) Production and secretion of amidated and glycine-extended glucagon-like peptide-1 (GLP-1) in man. Diabetes 43:535–539

Ørskov C, Wettergren A, Poulsen SDS, Holst JJ (1995a) Is the effect of glucagon-like peptide-1 on gastric emptying centrally mediated? Diabetologia 38[Suppl 1]:A39 (Abstract)

Ørskov L, Holst JJ, Ørskov C, Møller N, Schmitz O (1995b) Acute GLP-1 administration does not affect insulin sensitivity in healthy man. Abstract to be presented at the EASD, Stockolm, September 1995. Diabetologia (in press)

Ørskov C, Poulser SS, Møller M, Holst JJ (1996) GLP-1 receptors in the subfornical organ and the area postrema are accessible to cirkulating glucagon-like peptide-1. Diabetes (in press)

Patzelt C, Schiltz E (1984) Conversion of proglucagon in pancreatic alpha cells: the major endpoints are glucagon and a single peptide, the major proglucagon fragment, that contains two glucagonlike sequences. proc Natl Acad Sci USA 81:5007–5011

Qualmann C, Nauck M, Holst JJ Ørskov C, Creutzfeldt W (1995) Insulinotropic actions of intravenous glucagon-like peptide-1 [7-36 amide] in the fasting state in healthy subjects. Acta Diabetol 32:13–16

Ritzel R, Ørskov C, Holst JJ, Nauck MA (1995) Pharmacokinetic, insulinotropic, and glucagonostatic properties of GLP-1 [7-36 amide] after subcutaneous injection in healthy volunteers. Dose-response relationships. Diabetologia 38:720–725

Ruiz-Grande C, Alarcon C, Alcantara A, Castilla C, Lopez Novoa JM, Villanueva-Penacarillo M, Valverde I (1993) Renal catabolism of truncated glucagon-like peptide 1. Horm Metab Res 25:612–616

Schjoldager BTG, Mortensen PE, Christiansen J, Ørskov C, Holst JJ (1989) GLP-1 (glucagon-like peptide-1) and truncated GLP-1, fragments of human proglucagon, inhibit gastric acid secretion in man. Dig Dis Sci 35:703–708

Schmidt WE, Siegel EG, Creutzfeldt W (1985) Glucagon-like peptide-1 but not glucagon-like peptide-2 stimulates insulin release from isolated rat pancreatic islets. Diabetologia 28:704–707

Tanizawa Y, Riggs AC, Elbein SC, Whelan A, Donis-Keller H, Permutt MA (1994) Human glucagon-like peptide-1 receptor gene in NIDDM. Identification and use of simple sequence repeat polymorphisms in genetic analysis. Diabetes 43:752–757

Thim L, Moody AJ (1981) The primary structure of glicentin (proglucagon). Regul Pept 2:139–151

Thim L, Moody AJ (1982) Purification and chemical characterization of a glicentin-related pancreatic peptide (proglucagon fragment) from porcine pancreas. Biochim Biophys Acta 703:134–141

Toft-Nielsen M, Madsbad S, Holst JJ (1995) The effect of GLP-1 on glucose elimination. Abstract to be presented at the EASD, Stockholm, September 1995. Diabetologia (in press)

Valverde I, Merida E, Delgado E, Trapote MA, Villanueva-Penacarillo ML (1993) Presence and characterization of glucagon-like peptide-1 (7-36) amide receptors in solubilized membranes of rat adipose tissue. Endocrinology 132:75–79

Valverde I, Morales, Clemente F, Lopez-Delgado MI, Delgado E, Perea A, Villanueva-Penacarrillo ML (1994) Glucagon-like peptide 1: a potent glycogenic hormone. FEBS Lett 349:313–316
Villanueva-Penacarrillo ML, Alcantara AI, Clemente F, Delgado E, Valverde I (1994) Potent glycogenic effect of GLP-1 (7-36) amide in rat skeletal muscle. Diabetologia 37:1163–1166
Wei Y, Mojsov S (1995) Tissue-Specific expression of the human receptor for glucagon-like peptide-I: brain heart and pancreatic forms have the same deduced amino acid sequences. FEBS Lett (in press)
Weir GC, Mojsov S, Hendrick GK, Habener JF (1989) Glucagon-like peptide I (7-37) actions on the endocrine pancreas. Diabetes 38:338–342
Wettergren A, Schjoldager B, Mortensen PE, Myhre J, Christiansen J, Holst JJ (1993) Truncated GLP-1 (proglucagon 72-107 amide) inhibits gastric and pancreatic functions in man. Dig Dis Sci 38:665–673
Wettergren A, Petersen H, Ørskov C, Christiansen J, Sheikh SP, Holst JJ (1994) Glucagon-like peptide-1 (GLP-1) 7-36 amide and peptide YY from the L-cell in the ileal mucosa are potent inhibitors of vagally induced gastric acid in man. Scand J Gastroenterol 29:501–505
Willms B, Werner J, Creutzfeldt W, Ørskov C, Holst JJ, Nauck M (1994) Inhibition of gastric emptying by glucagon-like peptide-1 (7-36 amide) in patients with type-2-diabetes mellitus. Diabetologia 37 [Suppl 1]:A118
Willms B, Kleine N, Creutzfeldt W, Ørskov C, Holst J, Nauck M (1995) Glucagon-like peptide 1 (7-36 amide) lowers blood glucose also in type-1-diabetic patients. Diabetologia 38[Suppl 1]:A40 (Abstract)
Yanaihara C, Matsumoto T, Kadowaki M, Iguchi K, Yanaihara N (1985) Rat pancreas contains the proglucagon (64-69) fragment and arginine stimulates its release. FEBS Lett 187:307–310
Zhang Y, Cook JTE, Hattersley AT, Firth R, Saker PJ, Warren-Perry M, Stoffel M, Turner RC (1994) Non-linkage of the glucagon-like peptide 1 receptor gene with maturity onset diabetes of the young. Diabetologia 37:721–724

CHAPTER 19
Oxyntomodulin and Its Related Peptides

D. BATAILLE

A. Introduction

Oxyntomodulin is a 37-amino-acid peptide which consists of glucagon (29 amino acids) C-terminally elongated by an octapeptide rich in basic amino acids; it corresponds to the (33-69) proglucagon fragment (Fig. 1). This glucagon-containing peptide was originally isolated from the porcine gut according to its glucagon-like properties, namely its ability to interact with glucagon receptors present in rat liver plasma membranes and, consequently, to activate adenylate cyclase linked to the glucagon receptor (BATAILLE et al. 1981a; BATAILLE et al. 1981b; BATAILLE et al. 1982a, 1982b). Together with the N-terminally elongated form glicentin (THIM and MOODY 1981), it forms the main part of "gut glucagon" (review in BATAILLE 1989), also called "gut GLI" (review in MOODY and THIM 1983) or "enteroglucagon" (review in BLOOM and POLAK 1978). The basis for these terms is the presence in both oxyntomodulin and glicentin of a common epitope (often called "central" epitope) present in all glucagon-containing peptides (including proglucagon itself) and the absence of the C-terminal glucagon epitope, which is masked by the octapeptide in oxyntomodulin and glicentin (Fig. 2). As developed in Sect. B.VII, the "gut GLI" or "enteroglucagon" terminologies, based on immunological characteristics, unrelated to the biological features of the C-terminally elongated peptides, should be replaced by the "oxyntomodulin-like immunoreactivity" (OLI) concept (LE QUELLEC et al. 1992, 1993).

B. Biological Characteristics of Oxyntomodulin

I. Receptors

Since oxyntomodulin (OXM) was purified according to the ability of chromatographic fractions to interact with glucagon receptors, the first study performed in our laboratory after having obtained the pure peptide was to compare the newly isolated 37-amino-acid peptide to glucagon in its affinity towards the glucagon receptors and its capacity to produce cyclic AMP, the glucagon second messenger (review in RODBELL 1983). The results indicated that OXM was five to ten times less potent than glucagon in these actions

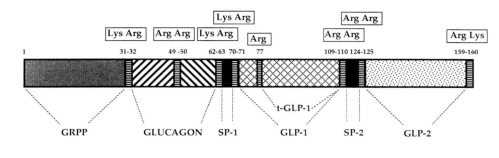

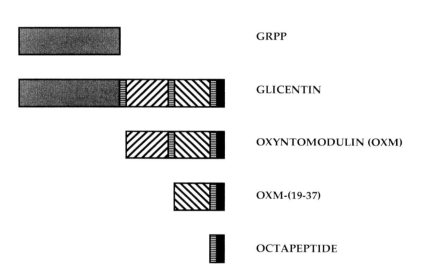

Fig. 1. General structure of mammalian proglucagon and the fragments discussed in this chapter. *GRPP*, glicentin-related pancreatic peptide; *SP-1*, *SP-2*, spacer peptides -1 and -2; *GLP-1*, *GLP-2*, glucagon-like peptides -1 and -2

(BATAILLE et al. 1981a, 1982b; BATAILLE 1989). This was true when most of the glucagon target tissues (liver, fat, heart, kidney) were tested. A single exception was the result of studies conducted with glands isolated from the gastric mucosa (BATAILLE et al. 1981a, 1982b; BATAILLE 1989). Indeed, in this preparation, OXM was 10–20 times more potent than glucagon, indicating that the C-terminal octapeptide modified significantly the biological "personality" of the elongated peptide, justifying the new acronym "oxyntomodulin." This term came from the observation that OXM displays an original biological profile, directed towards the acid-secreting oxyntic glands (from the Greek οχυντοσ = acid) of the gastric mucosa. Whereas the description of the

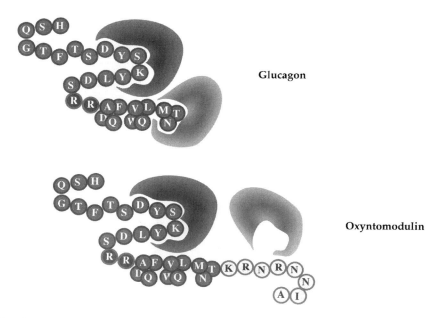

Fig. 2. Immunological basis for the "enteroglucagon" entity. C-terminally extended molecules (such as oxyntomodulin) are equally recognized by antibodies directed towards the "central" glucagon epitope but not by antibodies directed towards the C-terminal glucagon epitope, which is masked in C-terminally elongated molecules

"oxyntomodulin receptor" in the gastric mucosa turned out to be due to a misinterpretation (see Sect. B.V), these observations were at the origin of the search for a specific biological role of OXM and, more generally, of the peptides bearing the C-terminal octapeptide.

II. Acid Secretion

The "oxyntomodulin" specificity at the receptor level of the 37-amino-acid peptide induced a search for a specific effect on gastric acid secretion, a major biological response of the gastric mucosa to endocrine or paracrine regulators. This was of particular importance in view of the known inhibitory effect of glucagon on gastric acid secretion (CHRISTIANSEN et al. 1976). Comparing pure oxyntomodulin, either natural (BATAILLE et al. 1982a, 1982b) or synthetic (AUDOUSSET-PUECH et al. 1986), to glucagon indicated that the C-terminally elongated peptide was ten times more potent than glucagon in inhibiting pentagastrin-stimulated acid secretion in a model of perfused stomach from the anesthetized rat (DUBRASQUET et al. 1982). It was very tempting to relate the two observations and to conclude that the inhibitory effect of OXM was obtained via the "oxyntomodulin receptor." The fact, however, that an increase in gastric mucosal cyclic AMP is usually correlated with a stimulation rather than an inhibition of acid secretion (e.g., the stimulatory H_2 histamine-receptors are positively linked to adenylate cyclase; BLACK et al. 1972) led us to analyze further this question. Indeed, it is well known that the anesthetized animal is an excellent model for pharmacological approaches but that it suffers from several drawbacks, in particular a disappearance of most of the nervous system dependent-regulatory processes. To analyze the mechanisms that underlie the inhibitory effect of OXM, we started a series of experiments in the conscious rat equipped with a gastric fistula. The results (JARROUSSE et al. 1985, 1993; AUDOUSSET-PUECH et al. 1985; CARLES-BONNET et al. 1991) indicated that oxyntomodulin is a potent inhibitor of pentagastrin- or histamine-stimulated gastric acid secretion in the conscious rat. Measurements of the circulating levels after injection or perfusion of the peptide indicated that oxyntomodulin significantly inhibited gastric acid secretion (JARROUSSE et al. 1993) when the circulating doses of the peptide reached levels found under physiological conditions (KERVRAN et al. 1987). The potential physiological role of oxyntomodulin and its related peptides as a negative regulator of gastric secretion was thus demonstrated. Since similar data were obtained when oxyntomodulin was injected into human subjects (SCHJOLDAGER et al. 1988), this role is not restricted to laboratory animals. These observations opened up a series of studies in human physiopathology (see Sect. B.VII). It must also be noted that, from the intestinal L cells, both oxyntomodulin and glicentin are released under physiological conditions in a synchronous manner, suggesting that the 69-amino-acid and the 37-amino-acid peptides correspond to a single physiological entity. This assessment found strong experimental support from the observation that glicentin is also an inhibitor of gastric acid secretion at molar concentrations where glucagon is inactive (KIRKEGAARD et al. 1982).

III. Biologically Active Moiety of Oxyntomodulin

The original effect of oxyntomodulin, which contains the glucagon structure plus a C-terminal octapeptide, observed in gastric glands and in the anesthetized rat model argued for the search for the molecular moiety (or moieties) responsible for the oxyntomodulin specificity. Since the single structural difference between oxyntomodulin and glucagon was the octapeptide, we tested C-terminal fragments, which may carry this molecular specificity. Whereas the C-terminal hexapeptide coreleased with glucagon (YANAIHARA et al. 1985) had no effect, the octapeptide obtained by chemical synthesis (AUDOUSSET-PUECH et al. 1985) displayed inhibitory effects similar to that observed with the whole 37-amino-acid peptide (JARROUSSE et al. 1985; AUDOUSSET-PUECH et al. 1985; JARROUSSE et al. 1986; BATAILLE et al. 1988, 1989; CARLES-BONNET et al. 1991). The effectiveness of the octapeptide in inhibiting pentagastrin-stimulated gastric acid secretion was also observed in normal human subjects in a double-blind study (VEYRAC et al. 1989). The biological potency of the octapeptide observed in vivo was about two orders of magnitude lower than that of the whole molecule. On the other hand, comparing the effectiveness of the (19-37) oxyntomodulin fragment to that of intact oxyntomodulin indicated that the C-terminal nonadecapeptide was almost as potent as the whole hormone in inhibiting pentagastrin- or histamine-stimulated gastric acid secretion in the conscious rat (CARLES-BONNET et al. 1991; BATAILLE et al. 1989; JARROUSSE et al. 1993). Because of the slightly lower half-life of the (19-37) fragment observed in vivo than that of oxyntomodulin, it was necessary to compensate for this difference. The intrinsic potency of the fragment turned out to be identical to that of the whole hormone (JARROUSSE et al. 1993), clearly indicating that the N-terminal portion of oxyntomodulin is not significantly involved in its activity. Altogether, the data obtained were the basis for the concept that the biological activity of the peptides arising in intestinal L cells from the N-terminal portion of proglucagon is borne by their C-terminus. In other words, the "glucagon-like" character, both biological and immunological, of these peptides is anecdotal and these peptides should be classified according to the presence in their structure of the biologically active C-terminal moiety (see Sect. B.VII).

IV. In Vivo Mode of Action: Interactions with Other Peptides

The search for the mode of action of oxyntomodulin in vivo led to the concept that its inhibitory effect on gastric acid secretion was reinforced by interaction with other peptides. Thus, somatostatin, a known inhibitor of gastric secretion (ROBEIN et al. 1979), increased, at subthreshold doses, the inhibitory action of oxyntomodulin (Dubrasquet et al. 1986; BATAILLE 1989; BATAILLE et al. 1989). The result was a real potentiation, oxyntomodulin being 3.5 times more potent when injected together with doses of somatostatin inactive by themselves. Furthermore, oxyntomodulin and its related peptides were shown to release somatostatin locally from the gastric mucosa (BADO et al. 1989). The release by oxyntomodulin, at the vicinity of its target cells, of a peptide which potentiates its own effect was the basis for the "autopotentiation" concept (BADO et al. 1989; BATAILLE et al. 1989). Other interactions between intestinal peptides (such as PYY or neurotensin) on the same parameter were noted (BADO et al. 1993). This led us to conclude that the negative endocrine intestinal-gastric axis, a classical physiological concept known as "enterogastrone" (review in ANDERSSON 1967), was not borne by a single endocrine entity, but rather was the result of complex interactions between several peptides potentiating each other. This idea was further substantiated by the observations that, whereas oxyntomodulin was able, at physiological circulating concentrations, to inhibit gastric acid secretion stimulated by a single stimulant (pentagastrin or histamine), higher doses of the peptides were necessary to inhibit the secretion induced by a meal which involves stimulation by a set of stimulants (essentially gastrin, histamine and the acetylcholine-linked vagal tone) which potentiate each other (JARROUSSE et al. 1994a). The concept "a stimulus-an inhibitor/several stimuli-several inhibitors" was thus born (JARROUSSE et al. 1994a). In spite of this, oxyntomodulin and its related peptides maintain a leading role in the "enterogastrone" entity according to the data on the relationship between their circulating levels and human physiopathology (see Sect. B.VII).

V. In Vitro Mode of Action

The receptor, linked to adenylate cyclase observed in isolated gastric glands and initially considered as the "oxyntomodulin receptor," was the initial basis for the oxyntomodulin concept (BATAILLE et al. 1981a, 1982a,b). As indicated in the preceding sections, the doubt about a direct relationship between an inhibition of gastric secretion and an increase in gastric mucosal cyclic AMP was an early concern. The effectiveness of C-terminal fragments of oxyntomodulin without any modification of cellular cyclic AMP, particularly with the (19-37) fragment (JARROUSSE et al. 1993), was a direct proof for a dissociation between the two phenomena. Later studies indicated that the adenylate cyclase-linked receptor present in the gastric mucosa (DEPIGNY et al. 1984) and in the somatostatin-secreting RIN T3 cell line (TANI et al. 1991; GROS et al. 1992a) was a t-GLP-1-preferring receptor which, for still unexplained structural reasons, was more easily recognized by oxyntomodulin than by glucagon (GROS et al. 1992b, 1993). This t-GLP-1 receptor, the cDNA of which was cloned from another tissue (THORENS 1992), is also present in the parietal cell (GROS et al. 1995). Up to now, no physiological explanation for its presence in the parietal cell has been available, although some hypotheses have been raised concerning its possible role in controlling parietal cell metabolism, growth or differentiation (GROS et al. 1995). In the somatostatin-secreting D cell, the presence of this receptor is a likely explanation for the inhibitory effect of t-GLP-1 on gastric acid secretion (SCHJOLDAGER et al. 1989). It is unlikely, however, that, under physiological conditions, oxyntomodulin stimulates somatostatin secretion via this receptor, inasmuch as, in a conscious animal, the (19-37) fragment is as potent as oxyntomodulin itself (BADO et al. 1989), whereas it has no interference with the t-GLP-1 receptor. The authentic oxyntomodulin receptor, the presence of which is expected in both somatostatin-secreting gastric D cells and in the parietal cell or in its close vicinity, is still to be directly demonstrated in the gastric mucosa. A very convincing, although indirect, proof for its existence in the gut wall was recently obtained (see Sect. B.VIII).

VI. Pharmacology: Search for a "Minimal Oxyntomodulin"

The demonstration that the oxyntomodulin activity may be mimicked by shorter C-terminal peptides urged us to find the smallest fragment with oxyntomodulin activity. As already noted, the C-terminal octapeptide displayed similar effects, although at much higher concentrations (AUDOUSSET-PUECH et al. 1985; JARROUSSE et al. 1985, 1986; DUBRASQUET et al. 1986; VEYRAC et al. 1989; CARLES-BONNET et al. 1991). The rapid in vivo metabolism of the octapeptide may only partially explain this decreased activity. Indeed, it ap-

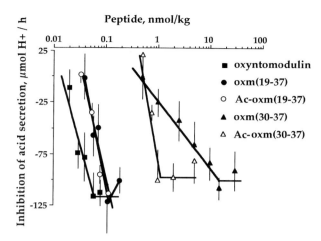

Fig. 3. Comparative effects of oxyntomodulin, and its 19-37 [oxm(19-37)] or 30-37 [oxm(30-37)] fragments or its N-acetylated counterparts [Ac-oxm(19-37) and Ac-oxm(30-37)] on gastric acid secretion in the conscious rat. (Data from JARROUSSE et al. 1993; CARLES-BONNET et al. 1991; CARLES-BONNET et al. 1992)

peared that a N-terminal elongation, even as short as an acetyl group (CARLES-BONNET et al. 1992), increases the potency (partially by reducing N-terminal degradation) and, more importantly, induces a recovery of the parallelism between the dose-response curve of the octapeptide and that of oxyntomodulin or of its (19-37) fragment, parallelism which was lost under certain conditions (CARLES-BONNET et al. 1992). Analyzing the molecular structure of the octapeptide in relation to its activity indicated that a β-turn is a prominent feature of the C-terminus of oxyntomodulin (AUMELAS et al. 1989). It was also shown that the octapeptide may be further simplified to a hexapeptide while keeping its activity and that the amino acid side chains, in particular that of the basic amino acids, are crucial (CARLES-BONNET et al. in preparation). Figure 3 summarizes the comparative data obtained with oxyntomodulin, and its (19-37) and (30-37) fragments, acetylated or not.

VII. Human Physiology and Pathophysiology: Oxyntomodulin-Like Immunoreactivity

Many studies were devoted to analysis of glucagon-related peptides of intestinal origin both in physiological and pathological states (see BLOOM and POLAK 1978; MOODY and THIM 1983). Most of these studies were performed using an immunological approach consisting of measuring the "total GLI" (of pancreatic and of intestinal origin) with a centrally directed antibody and the "pancreatic glucagon" using an antibody directed towards the C-terminus of glucagon. The difference between the two measurements was known as "gut GLI" or "enteroglucagon" (see Fig. 2). This approach was valid until we

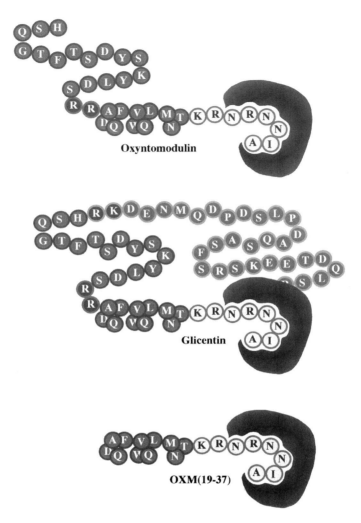

Fig. 4. Oxyntomodulin-like immunoreactivity (*OLI*) concept (Le Quellec et al. 1992): the molecules which have in common the C-terminal octapeptide (which bears the biological activity) and thus have the same biological specificity are measured in plasma by using an antibody directed specifically towards the octapeptide. This concept should replace that of "enteroglucagon," a heterogeneous, immunologically based concept (see text)

showed (see above sections) that the biologically active moiety of the peptides of intestinal origin was their C-terminal end and that peptides, such as oxyntomodulin (19-37), totally devoid of glucagon-like activity or immunoreactivity, not only were fully active but represent in fine the oxyntomodulin archetype. Furthermore, a C-terminally truncated form of glucagon, glucagon (1-21) displaying a unique biological specificity, is an "enteroglucagon" molecule according to its immunological features (Matsuyama et al. 1988). In view of these ambiguities, we defined a new endocrine entity which corresponds to the whole of peptides ending by the C-terminal octapeptide, which bears the biological specificity of the molecules [oxyntomodulin, oxyntomodulin (19-37), oxyntomodulin (30-37) and glicentin], which have their biological specificity directed towards the inhibition of gastric acid secretion. This entity may be measured in tissue extracts and in plasma, thanks to the development of a C-terminally directed oxyntomodulin antibody obtained by immunizing rabbits with a short synthetic immunogen consisting of the C-terminal octapeptide coupled through its N-terminus to a carrier protein (BLACHE et al. 1988); see Fig. 4. This new entity was called "oxyntomodulin-like immunoreactivity," "OLI" in short (LE QUELLEC et al. 1992). Analysis of the diurnal profile of OLI in plasma of hospitalized patients with no history of digestive disease indicated that plasma OLI faithfully followed natural oral nutrition, with a relatively rapid response to food intake (LE QUELLEC et al. 1992), well before nutrients reach the distally located intestinal L cells. This suggests the existence of a relay mechanism originating in the upper part of the GI tract, as suggested previously for proglucagon-derived peptides (OHNEDA 1987). The diurnal profile of OLI followed a rhythm similar to that of gastric acid secretion (MERKI et al. 1988) or that of plasma somatostatin (BURHOL et al. 1984). These data were the basis for the idea that, instead of representing an "end-of-digestion" signal stopping acid secretion and possibly other parameters in the gastrointestinal tract, the OLI peptides play an important role in an endocrine inhibitory control mechanism which avoids the consequence of an excess in the parameters induced by the stimulatory mechanisms within the gastrointestinal tract, in particular acid secretion. Following this idea, it was of particular interest to analyze the OLI profile in patients who have secretory disfunctions in the GI tract. The first obvious syndrome of interest was the duodenal ulcer. It is well known that only a proportion of duodenal ulcer patients have an increase in peak acid output (PAO), an index of the maximal capacity of the stomach to secrete HCl following gastrinergic stimulation (BARON 1973). Assay of OLI throughout a 24-h period in a series of patients suffering from duodenal ulcer (LE QUELLEC et al. 1993) allowed us to conclude that those who displayed an increased PAO had a decreased OLI profile, while those who had normal or sub-normal PAO parameters had normal or increased OLI profiles. A highly significant negative correlation between the PAO and the OLI profiles was noted from these studies (LE QUELLEC et al. 1993, see Fig. 5), suggesting that the total capacity of a human stomach to secrete HCl in response to physiological stimuli is under the control of the OLI peptides (LE QUELLEC et al. 1993). Studies on a larger scale of other patients with various PAO values as well as of normal subjects will allow this hypothesis to be confirmed.

VIII. Recent Developments

Still unpublished studies from our laboratory indicate that, in addition to the negative control of gastric acid secretion, oxyntomodulin and its related pep-

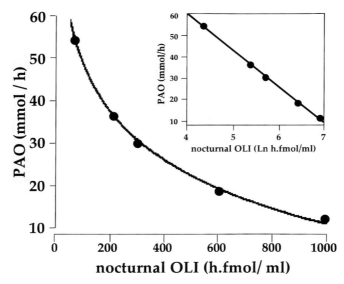

Fig. 5. Relationship between the maximal capacity of a human stomach to secrete HCl (peak acid output or *PAO*) and the quantity of circulating oxyntomodulin-like immunoreactivity (*OLI*) measured during the night (*nocturnal OLI*). *Inset* shows a semilogarithmic representation of the same results. (Data from LE QUELLEC et al. 1993)

tides may have other roles in the GI tract: The OLI peptides are likely to play a role, besides other known hormonal or local regulators, in the control of ionic transfer though the intestinal wall (JARROUSSE et al. 1994b). Another newly recognized target for the OLI peptides is the smooth muscles of the gut wall. The pharmacology observed for the regulation of contraction of these isolated smooth muscle cells by oxyntomodulin and its related peptides corresponds exactly to that expected for cells containing the authentic oxyntomodulin receptor (G. Rodier et al., in preparation). Further studies on this model will allow the direct study of the characteristics of this receptor as well as the second messengers and the intracellular pathways leading to the final biological responses.

C. Conclusions

The question about the existence of glucagon-like peptides of intestinal origin is a long-lasting story. Besides "true" glucagon, present at some places of the GI tract in several species (LEFÈBVRE and LUYCKX 1983), it has been known for a long time that peptides displaying some glucagon-like characteristics but different from true glucagon are present in the gut (UNGER et al. 1966; VALVERDE et al. 1970; BLOOM and POLAK 1978; MOODY and THIM 1983). Isolation of two forms of these molecules, glicentin (THIM and MOODY 1981) and oxyntomodulin (BATAILLE et al. 1981b, 1982a,b), clearly indicated that these peptides were glucagon-containing peptides with an additional C-terminal octapeptide. The comparison between oxyntomodulin and glucagon, which differ only in the octapeptide, was the basis for the discovery of the specificity of the glucagon-like peptides from the gut. It was only when testing the biological activities of the C-terminal fragments that have lost the glucagon-like character of the parent hormones that it clearly appeared that their biological specificity resides in their C-terminal half. The obvious conclusion of these findings was that these peptides should be measured in plasma using a radioimmunoassay recognizing specifically this portion of the molecule which carries its biological character. The application of this "oxyntomodulin-like immunoreactivity" (OLI) concept to humans opens up a new era in the clinical studies on gastrointestinal physiology and pathophysiology. If the present knowledge on how the OLI peptides act on its target tissues still shows wide gaps, the next few years will certainly provide us with a large part of the missing information thanks to the developments of models derived from the newly discovered regulatory processes in which these peptides are involved, in particular the regulation of intestinal absorption and/or secretion and gut motility.

Acknowledgements. I would like to thank Drs. Claire Jarrousse and Alain Le Quellec for helpful discussions during the preparation of the manuscript.

References

Andersson S (1967) Gastric and duodenal mechanisms inhibiting gastric secretion of acid. In: Code CF (ed) Alimentary canal II. American Physiology Society, Washington DC, pp 865–877 (Handbook of physiology, vol 48)

Audousset-Puech M-P, Jarrousse C, Dubrasquet M, Aumelas A, Castro B, Bataille D, Martinez J (1985) Synthesis of the C-terminal octapeptide of pig oxyntomodulin, Lys-Arg-Asn-Lys-Asn-Asn-Ile-Ala: a potent inhibitor of pentagastrin-induced acid secretion. J Med Chem 28:1529–1533

Audousset-Puech M-P, Dufour M, Kervran A, Jarrousse C, Castro B, Bataille D, Martinez J (1986) Solid-phase peptide synthesis of human (Nle-27)-oxyntomodulin. Preliminary evaluation of its biological activities. FEBS Lett 200:181–185

Aumelas A, Audousset-Puech M-P, Heitz A, Bataille D, Martinez J (1989) ^{1}H NMR conformational studies on the C-terminal octapeptide of oxyntomodulin, a β-turn locked by a salt bridge. Int J Pept Protein Res 34:268–276

Bado A, Bataille D, Accary JP, Lewin MJM, Dubrasquet M (1989) Luminal gastric somatostatin-like immunoreactivity in response to oxyntomodulin derivatives in cat. Biomed Res 9 [Suppl 3]:195–199

Bado A, Cloarec D, Moizo L, Laigneau J-P, Bataille D, Lewin MJM (1993) Neurotensin and oxyntomodulin$^{30-37}$ potentiate peptide YY regulation of gastric acid and somatostatin secretions. Am J Physiol 265 (Gastrointest Liver Physiol 28):G113–G117

Baron JH (1973) The clinical application of gastric secretion measurements. Clin Gastroenterol 2:293–314

Bataille D (1989) Gut glucagon. In: Schulz SG, Makhlouf G, Rauner BB (eds) Handbook of physiology – sect 6: The gastrointestinal system, vol II. Neural and endocrine biology neuroendocrinology of the gut, American Physiological Society, Bethesda, pp 455–474

Bataille D, Gespach C, Coudray AM, Rosselin G (1981a) "Enteroglucagon": a specific effect on gastric glands isolated from the rat fundus. Evidence for an "oxyntomodulin" action. Biosci Rep 1:151–155

Bataille D, Tatemoto K, Coudray AM, Rosselin G, Mutt V (1981b) "Enteroglucagon" bioactif (oxyntomodulin): evidence pour une extension C-terminale de la molécule de glucagon. CR Acad Sci Paris [III] 293:323–328

Bataille D, Coudray AM, Carlqvist M, Rosselin G, Mutt V (1982a) Isolation of glucagon-37 (bioactive enteroglucagon/oxyntomodulin) from porcine jejunoileum. Isolation of the peptide. FEBS Lett 146:73–78

Bataille D, Tatemoto K, Gespach C, Jörnvall H, Rosselin G, Mutt V (1982b) Isolation of glucagon-37 (bioactive enteroglucagon/oxyntomodulin) from porcine jejunoileum. Characterisation of the peptide. FEBS Lett 146:79–86

Bataille D, Blache P, Mercier F, Jarrousse C, Kervran A, Dufour M, Mangeat P, Dubrasquet M, Mallat A, Lotersztajn S, Pavoine C, Pecker F (1988) Glucagon and related peptides: molecular structure and biological specificity. Ann NY Acad Sci 527:168–185

Bataille D, Jarrousse C, Blache P, Kervran A, Dufour M, Mercier F, Le-Nguyen D, Martinez J, Bado A, Dubrasquet M, Mallat A, Pavoine C, Lotersztajn S, Pecker F (1989) Oxyntomodulin and glucagon: are the whole molecules and their C-terminal fragments different biological entities? Biomed Res 9 [Suppl 3]:169–179

Blache P, Kervran A, Martinez J, Bataille D (1988) Development of an oxyntomodulin/glicentin C-terminal radio-immunoassay using a "thiol-maleoyl" coupling method for preparing the immunogen. Anal Biochem 173:171–179

Black JW, Duncan WA, Durant CJ, Ganellin CR, Parsons EM (1972) Definition and antagonism of histamine H_2-receptors. Nature 236:385–390

Bloom SR, Polak JM (1978) Gut hormone overview. In: Bloom DR (ed) Gut hormones. Churchill-Livingston, Edinburgh, pp 3–18

Burhol PG, Jorde R, Jenssen TG, Lygren J, Florholmen J (1984) Diurnal profile of plasma somatostatin in man. Acta Physiol Scand 120:67–70

Carles-Bonnet C, Jarrousse C, Niel H, Martinez J, Bataille D (1991) Oxyntomodulin and its (19-37) and (30-37) fragments inhibit histamine-stimulated gastric acid secretion in the conscious rat. Eur J Pharmacol 203:245–252

Carles-Bonnet C, Jarrousse C, Niel H, Martinez J, Rolland M, Bataille D (1992) N-Acetyl oxyntomodulin (30-37): pharmacokinetics and activity on gastric acid secretion. Naunyn-Schmiedebergs Arch Pharmacol 345:57–63

Christiansen J, Holst JJ, Kalaja E (1976) Inhibition of gastric acid secretion in man by exogenous and endogenous pancreatic glucagon. Gastroenterology 70:688–692

Depigny C, Lupo B, Kervran A, Bataille D (1984) Mise en évidence d'un site récepteur spécifique du glucagon-37 (oxyntomoduline/entéroglucagon bioactif) dans les glandes oxyntiques de rat. C R Acad Sci III 299:677–680

Dubrasquet M, Bataille D, Gespach C (1982) Oxyntomodulin (glucagon-37 or bioactive enteroglucagon): a potent inhibitor of pentagastrin-stimulated acid secretion in rat. Biosci Rep 2:391–395

Dubrasquet J, Audousset-Puech M-P, Martinez J, Bataille D (1986) Somatostatin enhances the inhibitory effect of oxyntomodulin and its C-terminal octapeptide on acid secretion. Peptides 7 [Suppl 1]:257–259

Gros L, Demirpence E, Jarrousse C, Kervran A, Bataille D (1992a) Characterization of binding sites for oxyntomodulin on a somatostatin-secreting cell line (RIN T3). Endocrinology 130:1263–1270

Gros L, Demirpence E, Bataille D, Kervran A (1992b) Characterization of high affinity receptors for glucagon-like peptide-1 (7-36) amide on a somatostatin-secreting cell line (RIN T3). Biomed Res 13 [Suppl 2]:143–150

Gros L, Thorens B, Bataille D, Kervran A (1993) Glucagon-like peptide-1(7-36)amide, oxyntomodulin and glucagon interact with a common receptor in a somatostatin-secreting cell line. Endocrinology 133:631–638

Gros L, Hollande F, Thorens B, Kervran A, Bataille D (1995) Comparative effects of GLP-1-(7-36)amide, oxyntomodulin and glucagon on rabbit gastric parietal cell function. Eur J Pharmacol 288:319–327

Jarrousse C, Audousset-Puech M-P, Dubrasquet M, Niel H, Martinez J, Bataille D (1985) Oxyntomodulin (glucagon-37) and its C-terminal octapeptide inhibit gastric secretion. FEBS Lett 188:81–84

Jarrousse C, Niel H, Audousset-Puech M-P, Martinez J, Bataille D (1986) Oxyntomodulin and its C-terminal octapeptide inhibit liquid meal-stimulated acid secretion. Peptides 7 [Suppl 1]:253–256

Jarrousse C, Carles-Bonnet C, Niel H, Sabatier R, Audousset-Puech M-P, Blache P, Kervran A, Martinez J, Bataille D (1993) Inhibition of gastric acid secretion by oxyntomodulin and its [19-37] fragment in the conscious rat. Am J Physiol 264 (Gastrointest Liver Physiol 27):G816–G823

Jarrousse C, Carles-Bonnet C, Niel H, Bataille D (1994a) Activity of oxyntomodulin on gastric acid secretion induced by histamine or a meal in the rat. Peptides 15:1415–1420

Jarrousse C, Pansu D, Bataille D (1994b) Mise en évidence d'un effet régulateur de l'oxyntomoduline sur l'activité sécrétoire de l'intestin grêle chez le rat. Gastroenterol Clin Biol 18 (2bis):A101 (abstract)

Kervran A, Blache P, Bataille D (1987) Distribution of oxyntomodulin and glucagon in the gastro-intestinal tract and the plasma of the rat. Endocrinology 121:704–713

Kirkegaard P, Moody AJ, Holst JJ, Loud FB, Olsen PS, Christiansen J (1982) Glicentin inhibits gastric acid secretion in the rat. Nature 297:156–157

Lefèbvre PJ, Luyckx AS (1983) Extrapancreatic glucagon and its regulation. In: Lefèbvre PJ (ed) Glucagon II. Springer, Berlin Heidelberg New York, pp 205–216 (Handbook of experimental pharmacology, vol 66/11)

Le Quellec A, Kervran A, Blache P, Ciurana A, Bataille D (1992) Oxyntomodulin-like immunoreactivity: diurnal profile of a new potential enterogastrone. J Clin Endocrinol Metab 74:1405–1409

Le Quellec A, Kervran A, Blache P, Ciurana AJ, Bataille D (1993) Diurnal profile of oxyntomodulin-like immunoreactivity in duodenal ulcer patients. Scand J Gastroenterol 28:816–820

Matsuyama T, Itoh H, Watanabe N, Namba M, Komatsu R, Miyagawa J-I, Kono N, Tarui S (1988). Glucagon-(1-21)-peptide as an active enteroglucagon. Biomed Res 9 [Suppl 3]:137–142

Merki HS, Fimmel CJ, Walt RP, Harre K, Röhmel J, Witzel L (1988) Pattern of 24 hour intragastric acidity in active duodenal ulcer disease and in healthy controls. Gut 29:1583–1587

Moody AJ, Thim L (1983) Glucagon, glicentin and related peptides. In: Lefèbvre PJ (ed) Glucagon I. Springer, Berlin Heidelberg New York, pp 139–174 (Handbook of experimental pharmacology, vol 66/I)

Ohneda A (1987) Response of plasma glicentin to intraduodenal administration of glucose in piglets. Diab Res Clin Pract 3:97–102

Robein MJ, De La Mare MC, Dubrasquet M, Bonfils S (1979) Utilization of the perfused stomach in anesthetized rats to study the inhibitory effect of somatostatin in gastric secretion. Agents Actions Suppl 9:415–421

Rodbell M (1983) The actions of glucagon at its receptor: regulation of adenylate cyclase. In: Lefèbvre PJ (ed) Glucagon. Springer, Berlin Heidelberg New York, pp 263–290 (Handbook of experimental pharmacology, vol 66/I)

Schjoldager BTG, Baldissera FGA, Mortensen PE, Holst JJ, Christiansen J (1988) Oxyntomodulin: a potential hormone from the distal gut. Pharmacokinetics and effects on gastric acid secretion in man. Eur J Clin Invest 18:499–503

Schjoldager BTG, Mortensen PE, Christiansen J, Ørskov C, Holst JJ (1989) GLP-1 (glucagon-like peptide 1) and truncated GLP-1, fragments of human proglucagon, inhibit gastric acid secretion in humans. Dig Dis Sci 34:703–708

Tani T, Le Quellec A, Jarrousse C, Sladeczek F, Martinez J, Estival A, Pradayrol L, Bataille D (1991) Oxyntomodulin and related peptides control somatostatin secretion in RIN T3 cells. Biochim Biophys Acta 1095:249–254

Thim L, Moody AJ (1981) The primary structure of porcine glicentin (proglucagon). Regul Pept 2:139–151

Thorens B (1992) Expression cloning of the pancreatic b cell receptor for the gluco-incretin hormone glucagon-like peptide-1. Proc Natl Acad Sci USA 89:8641

Unger RH, Ketterer H, Dupré J, Eisentraut AM (1966) Distribution of immunoassayable glucagon in gastrointestinal tissues. Metabolism 15:865–867

Valverde I, Rigopoulo D, Marco J, Faloona GR, Unger RH (1970) Characterization of glucagon-like immunoreactivity. Diabetes 19:614–623

Veyrac M, Ribard D, Daures JP, Mion H, Le Quellec A, Martinez J, Bataille D, Michel H (1989) Inhibitory effect of the C-terminal octapeptide of oxyntomodulin on pentagastrin-stimulated gastric acid secretion in man. Scand J Gastroenterol 24:1238–1242

Yanaihara C, Matsumoto T, Kadowaki M, Iguchi K, Yanaihara N (1985) Rat pancreas contains the proglucagon (64-69) fragment and arginine stimulates its release. FEBS Lett 187:307–310

Subject Index

acetylcholine 150
acetylsalicylic acid 123
achalasia 182
adenylyl cyclase 75, 76 (fig.), 81, 96, 174, 283
  activation vs cAMP-phosphodiesterase inhibition in heart 79–83
adipocyte 116, 259
adrenaline see epinephrine
adrenergic counterregulation 315
β-adrenergic agonists 7
β-adrenergic blockade 75, 177
β-adrenergic receptor 177
adrenocorticotropic hormone (ACTH) 31
adrenocorticotropic hormone (ACTH)-producing tumor 298
adult respiratory distress syndrome 182
alanine 150
alcoholic hepatitis 183
alloxan 231
amino acid deduced sequence (rat) 56 (fig.)
36-amino acid peptide hormone 142
amrinone 176, 177–178
anaphylaxis 184, 205
angiography 202
anion exchange HPLC method 6
anorexia nervosa 218
antiglucagon antibodies 35
antisecretory peptide therapy 247–248
arginine 216
asthma 181
a stimulus-an inhibitor/several stimuli-several inhibitors 331
astrocytes 289
atrial fibrillation 184
atrial flutter 184
atropine 123
atropine methyl nitrate 232
autopotentiation 331

barium enema 199–200

barium sulfate suspension 205
biguanides 124
biliary colic (pain) 172 (table), 182–183
biliary spasm 183
biliary tract imaging 201
β-blocker overdose 172 (table)
blood pressure (rat) 290
brain 289–290
8-bromo-cyclic AMP 91
bronchospasm 181, 182
bronchospastic disease 181

calcitonin 76
  receptor 76
calcium 175, 203–297
  B cell cytosolic 293, 294 (fig.), 296
  B cell elevation by GLP-1 293–295
calcium-blocker overdose 172 (table)
Ca^{2+}-mobilising agonists 78–79
cAMP 77, 79, 80
cAMP-dependent protein kinase 79
cAMP-phosphodiesterase inhibition vs adenylyl cyclase activation in heart 79–83
carboxypeptidase 35
carboxypeptidase E (H) 39
carcinoid syndrome 184
  leukocyte interferon therapy 246
carcinoid tumor 205
cardiogenic shock 75, 175
cardiovascular insufficiency 174–178
catecholamines 175
CGI-PDE 82
  inhibition by glucagon 82
8-p-chlorophenyl-thio-cAMP 79
chlorozotocin 246
cholamidopropyl-dimethylammonio-propane sulfonate 43
cholecystitis, acute 182–183
cholecystokin-B-receptor blocker 142
cholera toxin 88, 89
cholinergic agonists 282

chronic obstructive pulmonary
    disease   182
chymotrypsin   4
cisplatin   246
computed tomography   201
conjunctival mucosa   166
coronary artery disease   175, 184
cortisol   150, 151, 154–155
CP96, 345   142
CP-99, 711   142
CRE-binding protein (CREB)   25, 287
C-terminal nonadecapeptide   35
C-terminal peptide   32
cyclic AMP   75
    analogues   77
    pools   96
    synthesis   75
cyclic AMP-response element   24, 25
cyclic GMP-inhibited
    phosphodiesterase   79
cyclic nucleotide phosphodiesterases   81
α-cyclodextrin   165
cystic fibrosis   218

dacarbazine   245–246
dDAVP   162
des-His-1 glucagon   140
des-His[Glu-9] glucagon   140
dexamethasone   284
diabetes mellitus   115–125, 133
    A cell   117, 118, 119
        dysfunction   120
    B cell   117
    circulating glucagon levels   117–118
    diabetogenic effects of
        glucagon   115–117
    glucagon dysfunction in   118–121
    hyperglucagonemia in   118–120
    insulin-dependent   115, 119, 121–122,
        133, 134
        deficient glucagon responses   155
        GLP-1 infusion   319–320
        glucagon test   216–217
        treatment   123
    ketoacidosis   171, 204–205
    maturity onset of young   317
    medical imaging   204–205
    non-insulin dependent   115, 119–120,
        122, 134, 137
        GLP-1 effect   268, 320
        glucagon test   217
        late-onset   233
        obesity associated   233
        pathogenesis   316
        postprandial glucagon   233
        treatment   123–125
    peptide treatment   138–139

postpancreatectomy   121
    type 1   117
    type 2   117
        GLP-1 plasma levels   297
diazepam   123
didecanoyl-L-α-phosphatidyl-
    choline   165
diltiazem   176
dipeptidyl peptidase IV   320–321
diphenylhydantoin   123
diverticulitis, acute   172 (table), 183
dobutamine   175
dopamine   175
dumping syndrome   1297

emergency medicine, glucagon
    role   171–184
endocrine cells proglucagon
    containing   41–42
endopeptidase   84, 174
endoproteolytic cleavage   31
endoscopy
    bile ducts   182
    intestines   182
enhancer-like elements   20
enteroclysis   198
enterogastrone   331, 332
enteroglucagon   327
entero-insular axis   275, 277
epinephrine (adrenaline)   150, 151, 154,
    175
    release   184
erythema multiforme   184, 205
Escherichia coli   1
esophageal impaction   172 (table)
esophageal resection   297
etoposide   246
exendin   261
exendin-4   279, 283

5-fluorouracil   246
forskolin   77, 284, 287
fructose-1,6-bisphosphatase   150
furin   39

galanin   287
gallstones   182
gastrectomy, partial   287
gastric inhibitory polypeptide   14, 16
gastrinoma   298
gastro-enteropathic endocrine
    tumors   298
genes encoding peptide hormones of
    glucagon superfamily   14–17
glicentin   32, 34 (fig.), 35, 37, 45, 313
    epitope   35
    L cell production   278

Subject Index

porcine 313
glicentin-KR 37, 41
glicentin-like immunoreactivity 35
glicentin-related pancreatic peptide 312
GLP-1-(7-36) amide 38
GLP-1-immunoreactive tumor 298
glucagon 79–83, 97 (fig.)
   actions receptor mediated 76–83
   administration *see* glucagon administration
   β-adrenergic blocker toxicity treatment 177
   adverse effects 184
   aggregation 3
   appetite control 223–234
   assay 9
   binding 68
   Ca^{2+} mobilization in hepatocytes 77–78
   Ca^{2+} release 77
   cardiac action 79–83, 97 (fig.), 174–178
   cardiac CGI-PDE inhibitor 82–83
   crystallization 2
   C-terminal region 14
   C-terminally truncated form 334
   deamination sites 6–8
   [D-Gln³] 84
   des-His-1 140
   des-His [Glu-9] glucagon 140
   diabetogenic effects 115–117
   drug target 133–135
   effects 135–136
   expression 1
   fasting control 134
   fermentation 1
   gastrointestinal effects 182–184
   gluconeogenesis regulation 13
   glycemic action 150–151
   glycogenolysis regulation 133
   glycogenolytic analogue 78
   glycogenolytic effect 9
   hepatic action 77–78, 97 (fig.)
   hepatic gluconeogenesis 75
   hepatic glycogenolysis activation 75
   hospital supply 178
   hypersensitivity reactions 205
   hypoglycemia correction 154
   hypoglycemia prevention 154
   immunoreactivity 18
   intravenous, effect on appetite 226–227 (table)
   isolation 2
   low concentrations 78
   metabolic actions 171
   mini-glucagon action in concert 94–96
   mode of action 75–98
      receptor-mediated 75
      secondary to processing into carboxy-terminal fragment 75
   mRNA 17–18
   myocardial carbohydrate uptake 176
   non-elongated 46
   peptide analogues 139–141
   pharmaceutical preparations 3
   pharmacological actions 9
   physiological actions 9
   portal vein concentration 83
   positive inotropic effect 89–91
   postprandial secretion 223–224
   potentiation of effect of Ca^{2+} mobilizing agonists 78–79
   processing 137–139
      by target cells 83–85
   production 1–3
   prohormone action 98
   pulsatility *see* pulsatility of glucagon
   purification 2–3
   receptor *see* glucagon receptor
   recombinant human *see* recombinant human glucagon
   recovery 1–2
   regulation of cyclic-GMP-inhibited phosphodiesterase 79
   release mechanism 119–120
   renal effects 180
   respiratory effects 181–182
   satiety *see* glucagon satiety
   secretion 137–139, 151–153
      glycemic thresholds 153
      regulatory mechanisms 151–152
   side effects 204–205
   stability 9–10
   superfamily, genes encoding peptide hormones 14–17
   synthesis 137–139
   test *see* glucagon test for evaluation of insulin secretion
   therapeutic agent in critical care 75
   tissue concentrations 83–84
   urologic effects 180
   vascular effects 178–180
glucagon administration 159–169
   eye drops 166–167
   food intake 225–229
      animal studies 225–229
      human studies 229
   intramuscular 160–161
   intranasal 162–166
   intrvenous 160
   rectal 167–168
   subcutaneous 161–162
glucagon antagonists 133–144

glucagon antagonism, food intake 229–231
glucagon antagonist targets 137–143
glucagon-containing peptides 32–37
glucagon gene 11–27
  expression 17–26
    A cell specific 19–23
    inhibition 137
    pancreatic 18
  expression regulation 23–26
    by insulin 23–24
    by second messenger CAMP 24–26
  tissue-specific 17–19
  5'-flanking sequences 19
  regulation at G2 20
  structure 12–14
  transcriptional start site 18
glucagon-glucose test 214
glucagon-like immunoreactive material 278–279
glucagon-like peptide-1 (GLP-1) 11, 13–14, 32, 34 (fig.), 57, 85, 123, 255–256
  action on blood glucose 314–315
  ATP-sensitive potassium channel inhibition 286
  autonomic nervous system interaction 316
  binding characteristics 261–262
  cross-talk between GLP-1 and glucose signaling pathways 264–266
  diabetes mellitus associated 298, 316–320
    effects 317–320
    receptors 316–317
    secretion 316
  gastro-intestinal effects 315–316
  glucagon secretion effect 285–286
  glucagon secretion inhibition 125
  "glucose competence" 317
  gut hormone 313
  "ileal brake" 315
  immunoreactivity 278–279
  incretin hormone 275–277
  insulin-sensitizing effect 298
  metabolism in diabetic subjects 320–321
  metabolism in normal subjects 320–321
  in non-insulin-dependent diabetes mellitus 268
  origin 277–280
  oversecretion 297
  pathophysiology 275–298
  physiology 275–298
  plasma levels 297
  post-translational modification 278
  post-translational product of proglucagon processing, post-secretory fate 277–280
  receptor see glucagon-like peptide-1 receptor
  secretagogue (rat) 282
  secretion 280–282
  similarity to glucagon 279
  tissue distribution 258–260
  truncated 313
  vs GIP in stimulation of insulin secretion 266
glucagon-like peptide-1(7-37)/(7-36) amide 278
glucagon-like peptide-1 receptor 57, 76, 255–269
  biological actions 282–291
    adipose tissue 290–291
    brain 289–290
    endocrine pancreas 285–287
    exocrine pancreas 291
    liver 291
    lung 287
    skeletal muscle 291
    stomach 288
  coupling to intracellular second messenger 262–263, 264 (fig.)
  desensitization 267
  gene 283, 297
  inherited defects 317
  internalization 267–268
  regulated expression 267
  regulation 267–268
  signal transduction 291–297
    calcium 293–297
    cAMP pathway 291–293
  structure 256–258
  tissue distribution 282–291
    adipose tisue 290–291
    brain 298–290
    endocrine pancreas 285–287
    exocrine pancreas 291
    liver 291
    lung 287
    skeletal muscle 291
    stomach 288
glucagon-like peptide-2 11, 13–14, 31, 54 (fig.)
glucagon meal test 214
glucagonoma 26, 239–250
  antisecretory peptide therapy 247–248
  biological therapy 246–247
  chemotherapy 244–246
    chlorozotocin 246
    cisplatin 246

Subject Index

dacarbazine 245–246
etoposide 246
5-fluorouracil 246
lomustine 246
streptozotocin 244–245
diagnosis 240–241
hepatic artery
 chemoembolization 243
^{131}I-labeled Tyr^3-octreotide scan 241
localization 240, 241
management 242–248
nuclear magnetic resonance
 imaging 240
peptide receptor scintigraphy 241
percutaneous transhepatic portal
 venous sampling 240–241
prognosis 248–249
radiation therapy 243–244
surgical treatment 242
symptomatic treatment 248
ultrasonography 240
vascular occlusion 243
glucagon polyclonal antibody 230
glucagon receptor 76–77
 adipose tissue 61–62
 amino acid sequence (rat) 55 (fig.),
  59 (fig.)
 brain 63
 cloning 53–60
 function analysis 67–69
 gene see glucagon receptor gene
 heart 63
 human 69
 inhibition 139–143
 intestinal tract 63
 kidney 62
 ontogenesis 65
 pancreatic islets 62–63
 pathological conditions 66
 structure analysis 67–69
 transcripts, tissue distribution 63–65
glucagon receptor gene 53–70
 expression regulation 65–67
 missense mutation 233
 organization 60–61
 structure 64–65
glucagon-receptor knockout mice 136–137
glucagon-related peptides 38–39
glucagon satiety 231–234
 hepatic vagal afferents 232–234
 pathophysiology 233
 site of action 231
 therapeutic potential 233–234
 transduction 231–232
glucagon-SDV40 large T antigen
 transgene 19

glucagon test for evaluation of insulin
 secretion 211–219
classical test 211–212
clinical applications 216–218
 anorexia nervosa 218
 chronic pancreatitis 218
 cystic fibrosis 218
 hypoglycemia 217–218
 insulin-dependent diabetes
  mellitus 216–217
 non-insulin-dependent diabetes
  mellitus 217
 obesity 218
 polycystic ovary syndrome 218
comparisons 214–216
 intravenous glucose tolerance
  test 215
 meal 215
 oral glucose tolerance test 214–215
 other tests 216
dose-response curve 212
glucagon-glucose test 214
glucagon-meal test 214
influence of prevailing glucose
 level 213–214
reproducibility 212–213
glucose 149
 autoregulation 155
 glucagon action 150–151
 systemic balance 149
glucose counterregulation 120–121,
 149–155
 defective 120–121
 glucagon role 153–155
  pathophysiology 155
  physiology 153–155
 humans 152 (fig.)
α-glucosidase inhibitors 134
GLUT 2 265–266
glycemic thresholds 153
glycerol 150
G protein 83
 αs/$\beta\gamma$ subunits 87–89
G protein-coupled transmembrane
 domain receptors 76
G protein-linked receptor 142, 143
growth hormone 150, 151, 154–155
growth hormone releasing-factor
 receptor 57, 76
growth hormone releasing-hormone
 receptor 76
Gs protein 88, 139
guanosine triphosphate-binding
 heterotrimeric G protein
 complex 53
gut-brain axis 290
gut glucagon 327

gut glucagon-like immunoreactivity 280

heart 75
    CGI-PDE inhibitors 82
    cyclic nucleotide phosphodiesterases
        action 81–82
    glucagon action 79–83
    rate (rat) 290
heart failure 75, 172 (table), 175
helodermin 283
helospectin I/II 283
hepatic artery 243
hepatic failure 183
hepatic β-fatty acid oxidation 232
hepatic steatosis 183
hepatic vagal afferents 232–233
hepatitis 183
    alcoholic 183
hepatocyte 115
    nuclear factor 3β1 20
    vagal innervation 233
hepatoma cell line 84
hexapeptide 34 (fig.)
high-performance size exclusion
        chromatography 8
histricomorph rodents 14
    insulin genes 14
humanized mice 143–144
hypercalcemia 175
hypergluconemia 118–120
    relative 118
hyperglycemia 173
    postprandial 256
    stress 172
hyperinsulinemia 153, 155
hypoglycemia 153, 171–174
    glucagon action 171–174
    glucagon test 217–218
    iatrogenic 155
hypoglycemic cognitive
        dysfunction 153, 155
hysterosalpingography 202–203

^{123}I-labelled Tyr^3-octreotide scan 241
impaired glucose tolerance 133
incretin 38, 275–277, 297
    genes encoding 280
indomethacin 123, 167
inositol phosphates 78
inisitol 1,4,5-triphosphate 76
InsP$_3$/Ca pathway 76
insulin 133–134, 153–154
    glucagon gene expression
        regulation 23–24, 66
    overdose 173
    pulsatile 105, 106, 119

secretion, GLP-1 vs GIP-2 66
insulin-degrading enzyme 83
insulinoma 184, 195, 298
insulinotropic hormone 311
insulin-response element 24
insulin sensitizers 134
intestinal resection patient 275
intestinal surgery 183, 275
intravenous glucose tolerance
        test 215
intussusception reduction 200–201
isobutylmethylxanthine 78
isoproterenol 76, 79, 89
isradipine 95 (fig.)

Karl Kischer titration 6
kinase A 54

L364718 142
lactate 150
large intestine, neuroendocrine
        carcinoma 25
latex 205
L cells 278
    gastric 18
    intestinal 35, 41–42, 44–45, 278, 311
leucocyte interferon therapy 246–247
limulus amebocyte lysate test 5
γ-lipotropic hormone 31
liver 75, 291
    disease 183
    glucagon action 77–79
    see also hepatic
LLC-P1 cells 76–77
lomustine 246
lower esophageal sphincter
        relaxation 182
lung, GLP-1 receptor in 287–288
luteinizing hormone releasing
        hormone 162
lysosomes 267

magnetic resonance imaging 202
major proglucagon fragment 38
malonyl-CoA 116
meal test 215
Meckel's diverticulum 203
medical imaging 195–205
    barium enema 199–200
    biliary tract 201
    contraindications 204–205
    duodenum 196–187
    enteroclysis 198
    esophagus 195–196
    intussusception reduction 200–201
    large bowel 199–201

Subject Index

peroral pneumocolon 199
retrograde ileography 198–199
side effects 204–205
small bowel 198–199
stomach 196–197
melanocyte-stimulating hormone 31
metalloproteases, classical 43
microadenoma 40
milrinone 82
mini-glucagon 34 (fig.), 42–44, 174–175
  accumulation in cell environment 89
  action on heart 89–94, 175
    cell contractility decrease 90
    cell death 90
    sarcolemmal Ca^{2+} pump as target 91–93
  action on liver 85–89
  Ca^{2+} accumulation in sarcoplasmic reticulum stores 93–94
  component of positive inotropic effect of glucagon 89–91
  glucagon action in concert 94–96
  liver Ca^{2+} pump inhibition, G protein subunit mediation 87–89
  liver plasma membrane Ca^{2+} pump effector 86–87
  liver plasma membrane Ca^{2+} pump inhibitor 85–86
miniglucagon-generating endopeptidase 43, 44
morphine 174
mother-hormone 44
myocardial dysfunction, Ca^{2+} channel blocker-induced 75

necrolytic migratory erythema 248
neuroglycopenia, severe 155
neurokinin-1 receptor blocker 142
neutral endopeptidase 279
nifedipine 176
non-occlusive acute mesenteric ischemia 178
norepinephrine 150, 151

obesity 218, 233
octapeptide, synthetic 84–85
octreotide (Sandostatine) 247–248
oligonucleotide primers 54
oligosaccharides, N-linked 54
opiate peptides 31
oral glucose tolerance test 214–215
ouabain 175, 176
oxyntomodulin 32, 33 (fig.), 37, 41, 44, 45, 327–337
  acid secretion 329–330
  biologically active moiety 330–331
  circulating amount 313
  C-terminus 333
  in vitro mode of action 332
  in vivo mode of action 331–332
  "minimal" 332–333
  production 275
  recent developments 335–336
  receptors 327–329, 332
  release into blood 35
oxyntomodulin-like immunoreactivity 327, 333–335
oxyntomodulin-like peptide 46

paired basic amino acid cleaving enzyme 39
pancreas
  A cells 25, 37, 41–42, 44–45, 117, 118, 119
    dysfunction 120
  artificial 118
  B cells 117, 263, 268
    receptors 267
  cellular expression of GLP-1, GIP and glucagon receptors in islet cells 260
  endocrine 285–287
  exocrine 291
  GLP-1 receptor distribution 285–287
  islet hormone genes 18
  islets of Langerhans 44, 65
  secretions 183
pancreatic polypeptide, plasma level oscillation 105, 106
pancreatitis
  acute 183–184
  chronic 218
paraben derivative 205
parathyroid hormone receptor 76
parathyroid hormone-related receptor 76
pentagastrin 203
peptide(s) 138–139
  with GLP-1-like immunoreactivity 311
peptide analogues 138
peptide histidine isoleucine 279
peptide hormones 31, 278
peptide map/mapping 4
peptide receptor radiotherapy 244
peptide receptor scintigraphy 241
peptidyl glycine α-amidating mono-oxygenase 39
pheochromocytoma 184, 205
phentolamine 205
phorbol 12-myristate 13-acetate 284
phosphodiesterase inhibition 174

phosphoinositide metabolism 75
phospholipase C 76, 78, 263, 267
pituitary adenylate cyclase activating polypeptide (factor) 14, 17, 295
  receptor 76
polyacrylamide gel electrophoresis 6
polycystic ovary syndrome 218
preproglucagon 12, 31–46
  chromosome location 31
procainamide 175, 176
proglucagon 11, 138, 278, 311
  160-amino acid 31
  cDNAs 18–19
  endoproteases role 39
  gene 31, 40
  ^{3}H-tryptophan-labelled 40–41
  major fragment 38, 311, 312
  mRNA 280
  pancreatic 278
  processing 311–313
    in pancreactic A cell 40
    prohormone convertases in 39–41
    tissue-specific 44–45
    tissue-specific posttranslational 32–39
  structure 33 (fig.), 34 (fig.)
  tree 45 (fig.)
proglucagon72-158 (MPGF) 278
proglucagon-derived peptides 25
  secondary, postsecretory 41–44
prohormone convertases 39–41
proinsulin 31
proopiomelanocortin 31
prostaglandin synthesis inhibitors 123
proteases, M16 family 43
protein
  activation by post-translational events 31
protein drug substances 3
protein kinase A 295
protein kinase C 268
pulmonary hypertension 182
pulsatility of glucagon 105–112
  administration in dogs 109–111
  delivery in vitro 107
  delivery in vivo 107–109
    glucagon-insulin combined 108–109, 110 (fig.), 111 (fig.)
    normal man 107–108
  diabetic patient 109
  oscillations in plasma levels 105–106
    animal studies 105–106
    human studies 106
  secretion in vitro 106–107

quinidine 175, 176

radiographic studies 184
rat insulin I gene 24
β-receptor-blocking agents 123
recombinant human glucagon 3–9
  chemical impurities 6–8
    aggregation 8
    foreign impurities 6
    related impurities 6–8
  identification 3–4
  microbiological impurities 4–5
  purity 4–8
  water impurities 5–6
renal calculi 180
renal colic, acute 172 (table)
retrograde ileography 198–199
reversed-phase-high-performance liquid chromatography 4, 5 (fig.), 6,9
ryanodine receptor 266

Saccharomyces cerevisiae 1
Sandostatine (octerotide) 247–248
sarcoplasmic reticulum store, Ca^{2+} accumulation 93–94
scintigraphy 203
second messenger cAMP, glucagon gene expression regulation 24–26
secretin 14, 17, 55, 57
  receptor 76
serine proteases 39
  subtilisin family, enzymes belonging to 85
shock 180–181
signal peptide, 19 amino acids 54
simian virus, T antigen 18
skeletal muscle 291
slow calcium channel blockers 175–176, 177
small molecular weight synthetic compounds 141–143
sodium glycholate 162–163
  nasal tolerance 164
somatostatin 123, 124, 138, 331
  analogues 123, 124
  gene (SMS-UE) 24
  plasma level oscillation 105–106
  receptor 138
SR120819A 142
Stevens-Johnson syndrome 184
stomach
  endocrine L cell 18
  GLP-1 effect 288
  receptor mRNA in 259
streptozotocin 244–245
stress hyperglycemia 172
styryl quinoxaline 142

Subject Index

subcutaneous heparin anticoagulation 248
sulfonylureas 124, 134
superior mesenteric artery occlusion 178, 179

terbutaline 216
thyrotropin-releasing hormone 162
tolbutamide 216
total parenteral nutrition 183
tricyclic antidepressants 75, 175
1-N^a-trinitrophenylhistidine, 12-homoarginine 78
1-N-trinitrophenylhistidine, 12-homoarginine-glucagon 140

ultrasonography 201–202
ureteral calculi 180

urography 204

vagus nerve 224
  dorsal motor nucleus 224
  hepatic afferents 232–233
vascular occlusion therapy 243
vasoactive intestinal peptide (VIP) 14–16, 55, 276, 283
  receptor 76
vasoactive intestinal peptidoma (VIPoma) 298
verapamil 123
  overdose 176

YY1 receptor blocker 142

Zollinger-Ellison syndrome 246

Springer-Verlag and the Environment

We at Springer-Verlag firmly believe that an international science publisher has a special obligation to the environment, and our corporate policies consistently reflect this conviction.

We also expect our business partners – paper mills, printers, packaging manufacturers, etc. – to commit themselves to using environmentally friendly materials and production processes.

The paper in this book is made from low- or no-chlorine pulp and is acid free, in conformance with international standards for paper permanency.

Printing: Saladruck, Berlin
Binding: Buchbinderei Lüderitz & Bauer, Berlin